The DEFENDER

ROGER
FRANKLIN

The DEFENDER

The Story of GENERAL DYNAMICS

1817

HARPER & ROW, PUBLISHERS, New York
Cambridge, Philadelphia, San Francisco, London
Mexico City, São Paulo, Singapore, Sydney

To my three favorite critics,
my parents, John and Billie,
and my wife, Joanne

 For information address Harper & Row, Publishers, Inc., 10 East 53rd Street, New York, N.Y. 10022. Published simultaneously in Canada by Fitzhenry & Whiteside Limited, Toronto.

FIRST EDITION

Designed by Ruth Bornschlegel

Library of Congress Cataloging-in-Publication Data

Franklin, Roger.
The defender.

Includes index.
1. General Dynamics corporation—History. 2. Munitions—United States—History. I. Title.
HD9743.U8G42 1986 338.7'6234'0973 85-45192
ISBN 0-06-015510-8

86 87 88 89 90 RRD 10 9 8 7 6 5 4 3 2 1

Contents

Acknowledgments

When I embarked on this book early in 1983, it was with no small degree of anxiety. To examine the U.S. defense establishment by a close look at one of the most resilient of its components was to take on a subject so broad that often I was tempted simply to abandon it. That the undertaking came to fruition at all is less a reflection of my talents than a tribute to those whose support, encouragement, and guidance were of more value than they will ever know. It is impossible to thank by name all who helped; given the Pentagon's disdain for those who speak out of turn, it would also be inappropriate. There are, however, many whose names I am glad to be able to mention.

First thanks, now and always, are due my wife, Joanne Carlini. Her willingness to live among the tools of my trade—the cluttered strata of clippings, notes, and source materials that seemed often at the point of engulfing our home—went beyond any accepted definition of tolerance. *The Defender* is as much hers as mine; it could not have been done without her. My agent, Andrew Wylie, also warrants special mention. It was at his suggestion that I began work and through his efforts I was able to continue. Over the past three years he has been a constructive critic, a sage adviser, and, above all, my dear friend.

At Harper & Row, I am indebted to Aaron Asher, my editor, for his faith and counsel, and to his assistant, Claire Reich, for her patience. Dot Gannon, together with the copy editor and production staff, command my respect for their attention to detail and perseverance with a manuscript several times more grubby than it should have been.

In Washington, I owe much to Dina Rasor and the people at the Project on Military Procurement, particularly Paul Hoven, who was always there to explain the sometimes subtle complexities of modern weaponry; similarly, Andrew Cockburn for his general assistance and specific recollections of the events that figure in Chapter 11. I profited immensely from the insights of A. Ernest Fitzgerald and Pierre Sprey, who both helped me to understand some of the more oblique forces that influence weapons during their conception and, later, determine their worth on the field of combat. Knut Royce and Clark R. Mollenhoff also figured as sources of invaluable guidance on specific points.

Much of the latter third of the book is based upon research and documentary evidence compiled by two people, Richard Kaufman, of Senator Proxmire's Joint Economic Committee, and Peter Stockton, of Rep. John D. Dingell's Energy and Commerce Committee. Their labors not only resulted in the recovery of many millions of dollars in incorrect expense charges but provided a treasure trove of General Dynamics' internal correspondence, all now on the public record, which made my task a good deal easier. Lisa Hovelson and her boss, Senator Charles E. Grassley, were equally kind in dealing with my queries and gracious in accommodating my requests for interviews.

All writers owe a tremendous debt to those on whose work they build and I am no exception. Without the books of Richard K. Morris, John Holland's biographer, and Gaddis Smith's study of the clandestine arms trade that existed between the U.S. and Britain during the early years of the Great War, my task would have been infinitely more difficult, if not impossible. Richard Austin Smith, the *Fortune* journalist, who wrote several long series on General Dynamics' woes during the early sixties, also provided a font of invaluable information. Finally, it would be ill-mannered not to mention the countless librarians and archivists who worked on my behalf, specifically the staff at the Federal Court archives in New Jersey who risked injury in manhandling the fifty-odd boxes of documents relating to the Convair 880–990 fiasco. I am no less thankful for the efforts of the librarians and curators at the Paterson Museum, Paterson, New Jersey, and those at the American Institute of Aeronautics and Astronomics. Finally, a word of praise for the staff at the New York Public Library, where so much of this book was written. As they worked under conditions that are seldom pleasant and often difficult, their cheerful assistance never varied.

—Roger Franklin

Introduction

It is only from the air, several thousand feet above somewhere like General Dynamics' Fort Worth plant, that the statistics begin to make some sort of sense. Trying to find an image that does justice to the concept of the world's most diverse arms company is a hopeless task when inspiration is limited to profit statements, employment rolls, and backlogged military contracts, each worth more than the annual budgets of most third world countries. Seen from above, however, the spectacle of just one part of the empire renders the abstract concrete: This is what the money buys.

Strung out alongside an airfield that it shares with the B-52s of the adjacent Carswell Air Force Base, General Dynamics' Fort Worth operation looms out of the Texas plain like some pharaonic tribute to the utility of aluminum siding. In a good year, when defense bills are fashionable on Capitol Hill and appropriations flow largely unrestrained, perhaps twenty thousand people work beneath that vast acreage of industrial roofing, many in a main assembly building that is itself more than one mile long. This is the home of the West's most popular fighter aircraft, the F-16, and as good a place as any to begin a short tour of a modern armaments empire that dwarfs the fallen kingdoms of the Krupps, Britain's Vickers, Alfred Nobel, and the others known to an earlier generation as the merchants of death.

Four hours or so to the west, two more of General Dynamics' most active fiefdoms figure prominently among the corporate residents of southern California. Here the emphasis is on missiles, menageries of them. There are Vipers and Stingers for use against tanks and aircraft, Standards for combat at sea, and Tomahawk cruise missiles for just about every strategic purpose one might care to mention. Like the F-16, the Tomahawk is another of General Dynamics' great success stories. Originally a vague, war-gamers' notion of a handy weapon to have in a crisis, the idea of a small, relatively slow, pilotless aircraft carrying a nuclear bomb has demonstrated a universal appeal. By the 1990s, when most of the research and development work will have drawn to a close, Tomahawks will be found beneath the wings of Air Force jets installed, with their Army handlers, at ground sites throughout Europe and, far out to sea, aboard the Navy's battleships and submarines. There was once a time when General Dynamics' Convair division in San Diego ranked with Boeing and Douglas as one of the leading manufacturers of commercial passenger planes. That chapter of its history came to an abrupt end in the early sixties when an attempt to break into the jetliner market collapsed in a shambles of disorganization and atrocious planning. Today, with the exception of some subcontracting work for McDonnell Douglas, Convair and its sister plant at Pomona on the southeast fringe of Los Angeles look solely to the Pentagon for their support.

At the end of another cross-country hop, this one touching down in Warren, Michigan, the latest addition to General Dynamics' family turns out M-1 tanks. This is another of the company's military monopolies, since Land Systems, as it was rechristened after its purchase from Chrysler in 1982, is the sole supplier of heavy battle armor to the U.S. Army. Some seven thousand tanks are on order here, each worth $3 million, and General Dynamics will build them all.

When Chrysler's enfeeblement obliged it to auction off the tank works and its only other profitable division, there were many who wondered if the policymakers at General Dynamics' Saint Louis headquarters had not taken temporary leave of their senses. The final price of $336.1 million in cash was at least $30 million above the most common estimates of the division's worth, while the M-1 itself was the object of intense criticism from those inside and outside Congress who depicted it as inferior in every way to the older and far cheaper M-60 tank, which it was due to replace. General Dynamics' affable chairman, David Lewis, shrugged off the doubters, explaining that the M-1 was a "target of opportunity" too good to pass up. With a team of imported managers applying the lessons learned on the F-16 production line in Fort Worth, the Warren operation would settle into a steady groove, permitting a start to research and development on the next generation of heavy battle armor. In this light, the premium purchase price amounted to little more than a registration fee for the privilege of doing business with the Army, the one branch of the armed forces General Dynamics had neglected in the past.

The journey continues to its final destination in New England, looping back to pass briefly over Chicago, the home of both General Dynamics' chief stockholder, the octogenarian millionaire Henry Crown, and its one notable civilian division, Material Services. One of the largest building-supply outfits in the Midwest, with interests in coal, sand, gravel, cement, and, until recently, asbestos, Material Services has rarely contributed more than a slight 10 percent to the group's total sales.

At Quincy, on the Fore River in Massachusetts, the company builds barges and supply ships for the Rapid Deployment Force, occasionally diverging from the service of the Navy to work on oceangoing tugs and liquefied natural gas tankers. Here too, though the money does not flow from the Pentagon, federal government maritime construction subsidies make these excursions into the civilian market possible.

Finally, there is Groton, Connecticut, and the Electric Boat Company, where the tour ends and the story proper begins. Twenty thousand people work amid the dockside clutter to maintain the one truly inviolate leg of the nuclear deterrent triad, the Trident ballistic missile submarine. There are smaller, 688-class attack submarines under construction here as well, but they seem decidedly puny in comparison with the immense Tridents. No other shipyard in the country has the capacity to build these behemoths, each larger than the Washington monument and all equipped to carry as many as 2,400 independently targetable nuclear warheads. The average price is around $2 billion and as of the mid-1980s, General Dynamics held high hopes of building perhaps two dozen over the next few years.

Electric Boat is both the nucleus around which the company grew and the one single possession that makes General Dynamics unique among the leading defense contractors. Almost without exception, the others are relative newcomers to the field, airframe companies for the most part, whose reliance in part or whole on the Pentagon dates back no further than World War II. By then, Electric Boat had been a seasoned veteran of the procurement system for more than forty years. It is this longevity that lies at the root of the company's appeal to students of the modern Pentagon. Much more than the sum of its parts—an astonishing array in themselves—the history of Electric Boat and its cold war progeny, General Dynamics, is the story of America's defense establishment from the days of Teddy Roosevelt's Big Stick to the trillion-dollar spending spree initiated by Ronald Reagan and his cabinet. Through all the fluctuations in policy and public attitude, the wars and isolationist fervor, the family line that begins with Electric Boat has boomed or foundered in step with the times. From a small, undercapitalized organization using every fair and devious means at its disposal to see submarines added to the surface fleet, it has grown to become a charter member of Eisenhower's fabled military-industrial complex. It is a saga punctuated by scandals, by inconceivably expensive weapons that fail to perform yet continue to be funded. It is also the story of procurement decisions made for all the wrong reasons, of lives needlessly jeopardized, and, most recently, of waste, fraud, and contract abuse centering on the same Electric Boat division where it all began so many years ago. Above all else, it remains a most improbable tale.

I ANCIENT HISTORY

1. Fenian Fantasies

It began with a great day for the Irish in the early fall of 1876, when the *Catalpa,* a battered New England whaler, tied up in New York Harbor at the end of a voyage that had taken it halfway around the world and back again. The ship's imperious and abstemious master, Captain George Anthony, could hardly have expected the reception awaiting him at dockside. He had brought the *Catalpa* into port with half a mind to charge the crew and his troublesome Irish passengers with mutiny on the high seas. The scenes of exultant tumult on the wharf dispelled that notion at a stroke. While he may have regarded his charges as drunken, bristling liabilities, it was evident from the frenzied welcome that much of America considered them heroes. A brass band and an assembly of Irish pipers competed with each other, while the crowd surged around a hastily erected podium graced by an assortment of city fathers and elected officials. Anthony made a tactful decision to swallow his anger and do what little he could to curb the acquisitive impulses of the spectators, who were already pouring up the gangplank intent on collecting pieces of the ship as mementos of the great occasion.

Captain Anthony's troubles had begun one year earlier, when he chartered his boat to a ruggedly handsome Irish adventurer, John Breslin. The Irishman was vague about his intentions, saying only that he needed a reliable captain to take him and a small party of fellow countrymen to the British penal colony of Fremantle, on the desolate west coast of Australia. Breslin offered a considerable sum, more than Anthony could have made searching the seas for whales, and he agreed to put the boat at the Irishman's disposal. It was some two months later and several days after the ship pointed its bow toward the Cape of Good Hope that Breslin explained his plan in detail. The *Catalpa* was to raid the prison settlement and snatch a group of Irish rebels who had been transported in chains from England over the previous decade. The prisoners were his revolutionary comrades in arms, Breslin, asserted, adding that he was determined to effect their liberation come what may. Captain Anthony had little choice but to agree.

Unlike most previous Irish-American attempts to wage an unofficial war against Britain, the *Catalpa*'s enterprise proved a conspicuous success. At the conclusion of an uneventful outward voyage, Breslin made contact with the rebels and led them back through the bush to the hidden longboat and the waiting whaler. It was then that things started to go wrong.

Along with the Irish prisoners, Captain Anthony found himself saddled with several native-born Australian criminals, who had decided that the uncertainties of a new life in America were infinitely more appealing than continued incarceration in their homeland. The party was scarcely on board before discovery was made of the captain's small store of medicinal rum. Half an hour later, while the newly liberated prisoners continued to celebrate their escape by combing the ship for further quantities of alcohol, Captain Anthony was confronted by a bristling British frigate, which he faced down only by running up the Stars and Stripes and informing the captain that any attempt to board his ship would result in an international incident. Anthony won the confrontation, but only at great cost to his future. The seas around Fremantle constituted some of the richest whaling grounds in the world; if Anthony ever chose to return, it was obvious he could expect a most unfriendly reception. The incident also set the tone of barely suppressed hostility and near-violence that was to dominate the homeward leg of the expedition. When Anthony tried to deviate from the most direct route to pursue a school of whales, Breslin and his colleagues threatened to lock him in the wheelhouse. The captain backed down and the *Catalpa* continued on its way, the chief topic of conversation being Anthony's inadequate supply of hard liquor.

The merrymakers assembled on the New York wharf knew nothing of the conflicts that had dominated life on the *Catalpa* for the previous three months. It was enough that Breslin had achieved what he set out to do. There had been precious few opportunities to celebrate British humiliation in the past. Twice over the course of the previous ten years, armed bands of Irish-Americans had set out to invade Canada, with the objective of capturing a major city or an Atlantic seaport where they could establish a government in exile, one they hoped would be blessed by official recognition from Washington. The first foray ended in a blood-soaked debacle when British and Canadian troops descended on the poorly trained and ill-equipped invaders, killing sixty in the course of a brief battle and capturing a further two hundred in the days that followed. The second attempt proved even more embarrassing. Acting on information supplied by the British Secret Service, President Grant warned that Washington would take a dim view of a second attempted invasion and backed up his words by ordering the principal leaders arrested when it became clear that his cautions were about to be ignored. While the rebels cursed the treachery of their adopted homeland, Grant's actions undoubtedly saved the Fenians from another crushing defeat, since one of their own strategists, a purported Frenchman who called himself Henri LeCarron, was really Thomas Beach, the son of a Colchester squire and the chief British agent in the United States.

After the two humiliations, Breslin's expedition to Australia came as a massive boost to the movement's morale. News of the success reached New York long before the *Catalpa* completed its ten-thousand-mile voyage, and by the time it arrived, an unprecedented outbreak of solidarity had united the mutually abusive factions that dominated the political life of the teeming Irish immigrant communities in the major cities of the Northeast. With the old feuds temporarily swept aside, there arose a wave of expectant enthusiasm for the next great blow against England. Money poured into a rejuvenated skirmishing fund so quickly that there was little reason for those in charge to dismiss even the most fantastic schemes. One group of emissaries was sent off to Saint Petersburg to negotiate a secret agreement with the czar. Another delegation left for Mexico City to pledge Irish-American support for the Mexican government should an ongoing dispute with Britain over the ownership of the Yucatán peninsula ever erupt into open war. A third squad, representing what was by far the most militant wing of the movement, went even farther to the south, scouting the small British colony of what is now Belize with a view to making yet another attempt to establish an Irish government in exile.

No proposal was too farfetched to be given serious consideration, even the one being advanced by John Phillip Holland, a former Christian Brother who asserted that he could build a submarine capable of sinking the largest British battleship. Breslin was to learn much more of the scheme during the evening of festivities that followed the *Catalpa*'s arrival.

Holland had been discreetly promoting his idea for more than a year and had succeeded in making a considerable impression but very little headway. A younger brother, Michael Holland, was responsible for the original introduction, having brought the inventor to the home of O'Donovan Rossa, arguably the most fiery of the Irish-American leaders. Rossa—his real name was Jeremiah O'Donovan—had spent more than seven years in British jails for his involvement with various revolutionary groups and was one of the most prominent exiles active in New York. Fond of punctuating his speeches with calls for a campaign of terrorism against Britain, he was known by the nickname Dynamite and distinguished by a long, goatlike beard and red-rimmed eyes that were said to blaze with equal fury when he railed against Britain or, almost as frequently, his fellow Irish-American leaders. Holland and Rossa were opposites in almost every respect, but the submarine brought them together. At the conclusion of their first meeting, Rossa dashed off a short letter introducing the quiet-spoken inventor to Jerome Collins, science editor of the *New York Herald* and a founder of the largest Irish organization, Clan-na-Gael.

Collins was more inclined to skepticism than Rossa, although he did agree to lay the proposal before the other members of Clan-na-Gael's executive committee. And there the idea languished until the *Catalpa*'s triumphant return provided Holland with the opportunity to bring it back to life.

At that evening's reception, Holland delivered what had become his standard pitch, this time for the benefit of Breslin and yet another influential Fenian, John Devoy. Holland was "cool and good tempered and talked to us as a school-master would to his children," Devoy wrote in his diary. "He was well informed of Irish affairs and was anti-English with clear and well defined ideas about the best way of fighting England." One gathers that if it had been Devoy's decision alone, the necessary funds would have been approved without further delay. But Breslin needed something more than a classroom lecture before he would agree to commit a sizable chunk of the skirmishing fund to such an apparently outrageous project. Holland countered by offering to arrange a demonstration. Breslin agreed, and the trio parted on the understanding that they would meet again when the inventor had constructed a working scale model.

It is easy to understand Breslin's skepticism. Dreamers had been trying to perfect a workable submarine since the days of the ancient Greeks, when Alexander is said to have had himself lowered beneath the Aegean in a primitive diving bell. Leonardo toyed with the idea, as did several other inventors, whose experiments met with varying degrees of official approval. Cornelius Drebbel, a Dutchman with a flair for showmanship, astounded the court of James I by stretching greased leather over a frame containing twelve nervous oarsmen and remaining beneath the Thames for periods of several minutes at a time. As an attention-grabbing novelty, Drebbel's boat was a resounding success; as a potential instrument of war it was totally useless.

It took an American, David Bushnell, to design the first submarine with anything resembling the ability to fight. An egg-shaped submersible propelled by a hand-cranked propeller, the *Turtle,* as he called his boat, was launched on the outgoing tide in New York Harbor early one night in 1776. No sailor would volunteer for the mission, so the job fell to an innocent infantryman, one Sergeant Ezra Lee, who was supposed to bring the unwieldy craft alongside one of the British ships then blockading the harbor, submerge by flooding a series of ballast tanks, and attach a keg of gunpowder to the keel of the intended victim. Lee was swept past his first target and was closing on a second ship when some British sailors noticed the windowed observation turret and raised an alarm. Deciding that discretion was the better part of valor, Lee cut loose his gunpowder charge and disappeared. The exhausted infantryman managed to bring his strange craft back to shore several hours later, no doubt unconvinced of the submarine's military usefulness.

All the early boats were fatally flawed by their inventors' inability to suggest a practical means of propulsion or a method of maintaining longitudinal and lateral stability. Bushnell's ballast tanks presented a partial solution to the latter quandary, until the development of a practical electric motor more than a century later. When the Confederate submarine *Hunley* destroyed both itself and the Union blockade ship *Housatonic* in 1862, it was driven to its target by a crank mechanism that differed from Bushnell's only in that it endangered the lives of not one crewman but nine.

For want of a viable alternative, Holland's earliest sketches also suggest the use of a hand crank. However, he rejected all the earlier nostrums of submarine handling by incorporating two horizontal diving surfaces, which he hoped would serve as underwater wings. Combined with fore and aft ballast tanks capable of taking on and releasing water on command, the hydroplanes became the basis of the code of fundamental principles to which submarine designers have adhered ever since. Drawing in part on his earlier studies of soaring birds, Holland reasoned that with the planes angled down, the pressure of running water would be enough to force a boat beneath the surface; angled up, they would cause the submarine to rise. In an emergency, with the propeller dead, the diving planes would lose their bite and the boat would float slowly to the surface.

Since inspiration is nothing without opportunity, Holland might well have been forced early on to abandon his dream as an impossible fantasy had it not been for the hidden blessing of chronic ill health. At the tender age of fourteen, he had been placed by his family in a special training academy for future Christian Brothers. There was no trace of deep-seated religious conviction in the decision; it was a matter of survival. The boy's father, his uncles, and most of his male relatives had fallen victim to the disease and general privation that racked Ireland during the famine-stricken 1840s. To a youth who was myopic, weak in the chest, and cursed with brittle, easily broken bones, the teaching order promised a measure of personal security, while the small stipend he received went part of the way toward the support of his widowed mother and several equally penniless aunts. Fortunately for the future of the company that would become General Dynamics, even the relatively comfortable life of a Christian Brother proved too much for the young man's delicate constitution. Of the fifteen years he was to spend within the order, almost half the time was spent in convalescent homes, where he passed bedridden days immersed in private speculation about heavier-than-air flying machines, astrolabes, and, of course, the "underwater wrecking boats" that would make him famous.

It was this last and most enduring fixation that best captured the streak of romantic nationalism underlying the inventor's mild public demeanor. As a Christian Brother, Holland was forbidden to take an active part in the tempestuous Irish politics of the day, but there can be no doubt about his revolutionary sympathies. Both his elder and his younger brothers were members of secret Fenian societies, and there is evidence the British authorities' growing interest in the family may have prompted at least one member of the clan to flee his homeland for America. Limited by his health and the constraints of his calling, John Holland appears to have found consolation in his dreams of a weapon capable of sinking the awesome battleships that were the imperial symbols of his country's oppressor. Many years later, he wrote that his growing fixation with submarines was the force that led him to seek dispensation from his vows and follow the other members of his family to the United States. Unveiling his theories in Ireland, Holland explained, would have been tantamount to handing the Admiralty a wonderful new weapon. "The idea of making John Bull any stronger or more domineering than we already found him," he wrote, "ran counter to the nature of a man whose sympathies were with my own country."

Late in 1872, while Holland was recovering from his third attack of pneumonia in as many years, his superiors approved his request to leave the order. Three months later, having checked into yet another nursing home to prepare for the ordeal ahead, he booked passage in steerage on a boat bound for Boston. The inventor's life in his new homeland was to begin inauspiciously: Weakened by the rigors of the voyage, he slipped on a stretch of icy pavement and broke a leg. Laid up in bed and overflowing with the enthusiastic innocence of a new arrival, Holland called for his folio of submarine notes and sketches, and compiled a lengthy report on his work for the information and consideration of the U.S. Navy. His first disappointment followed. Far from leaping at the chance to acquire his designs, the Navy dismissed Holland as a harmless crank and committed his correspondence to the Torpedo College, where it was filed away with the unsolicited brainstorms of other naval fantasists. Thus rebuffed, and disenchanted with Boston, Holland left for New York with the coming of spring. Eventually he would find a job as a teacher of music and mathematics at the Christian Brothers college in Paterson, New Jersey.

Nurtured by the unaccustomed freedom of his new homeland and encouraged by the atmosphere of hothouse conspiracy that dominated the politics of the Irish-American community, Holland continued to work on his submarine sketches. It had become something of an obsession, and the planned demonstration for Breslin and Devoy became the fulcrum on which his future life and career would turn.

The reunion took place late in the fall of 1877 and must rank as one of the most bizarre ethnic gatherings ever seen on the beach at Coney Island. Holland arrived from Paterson for the Saturday morning rendezvous, wearing his customary derby, squinting through bottle-bottom spectacles, and carrying a large leather case containing a thirty-inch clockwork model. He was met by Breslin, Devoy, and a retinue of less important Fenian officials, many no doubt anticipating a few laughs at the expense of the quiet teacher and his ambitious contraption. Derby hats bobbing along the sand, the party ambled off to a relatively secluded stretch of beach, where Breslin posted lookouts to guard against the possibility that British spies might be lurking on the boardwalk. The leather satchel was unlocked and a metal and wooden model produced for the guarded inspection of those not engaged in counterespionage. Holland explained the basic principles, pointing out the rudder, the diving planes, and the location of ballast tanks, engines, and crew quarters. Then Holland, flanked by Devoy and Breslin, their trousers rolled above their knees, marched into the shallows, set the controls, and activated the clockwork motor. For the next hour the trio played boats, while the others looked on from the shore.

Breslin was utterly delighted by the model's ability to dive, rise, and execute complex underwater maneuvers in response to Holland's adjustments to the steering surfaces. The meeting concluded with a historic promise: Devoy and Breslin would do everything in their power to obtain sufficient funds for the construction of a small, one-man submarine to serve as a working test bed for a later, full-sized version.

Their enthusiasm must have been contagious, because a few weeks later another senior Fenian, Dr. William Carroll, wrote to a friend: "We can do it and we mean to try." Confidence in the outcome was so high, Carroll predicted the chief problem would not concern technical points but the questionable ability of those in the know to keep their mouths shut. "No loose pavement or barroom palaver will do for that work," Carroll cautioned in the same letter. In the interests of secrecy, Breslin was given the distinctly un-Irish code name "Jacobs Senior" and the unfettered use of some six thousand dollars in cash.

Holland wasted no time lest his benefactors undergo a change of heart. A Paterson foundry was contracted to do the basic construction work, while George Brayton, the inventor of a primitive gasoline engine, was commissioned to provide a two-cylinder power plant to replace the hand crank originally envisioned.

The launching took place on the Passaic River, not far from Holland's home, on the afternoon of May 22, 1878. The inventor had chosen a site near his home and hoped to get the boat into the water without attracting too much attention. It was not to be. The moment the dray arrived at the riverbank, a large and increasingly boisterous crowd began to assemble on a nearby bridge. What they saw tended to confirm a suspicion, widely held around Paterson, that the retiring Irish teacher was more than slightly demented.

The boat itself resembled a coffin equipped with a domed and windowed observation turret that jutted from the lid about two thirds of the way back from the blunt, boxlike bow. It measured fourteen feet in length, weighed two tons, and stood roughly two feet from the keel to the base of the observation turret into which the pilot was obliged to jam his head and shoulders. The *Holland I* slid from the cart and, to the accompaniment of howls from the bridge, sank like a stone.

If the Fenians had deliberately set out to attract publicity, they could not have been more successful. The next day, a Paterson newspaper archly reported that Holland's project was an obvious success, since the wrecking boat went immediately to the bottom "without even the aid of its captain."

Eight days later, Holland tried again. The leaking hull had been plugged and the boat ballasted with pig iron. The chief disappointment was the Brayton engine, which consistently refused to start and forced Holland to make some last-minute, stopgap repairs. By running a steam hose from the boiler of an accompanying service launch and connecting it to the engine's inlet valves, Holland was able to turn the useless power plant into a serviceable steam engine. The little submarine was to remain tethered to its mother ship, but at least there would be power sufficient to demonstrate the soundness of Holland's general concepts.

After a few brief preliminaries Holland clambered into the cramped pilot's space and the two boats chugged slowly upstream, rounding a bend and leaving the gawkers behind. Holland flooded both tanks, depressed the diving planes, and vanished on a maiden voyage of less than sixty seconds. The sub went down to about twelve feet below the surface, leveled off while executing a shallow turn, and reemerged perhaps thirty yards upstream. By the end of the day he had remained under water for several hours, including one dive of sixty-two minutes that so alarmed Devoy, Breslin, and the other officials that they were about to start dragging the river when he reappeared.

Any remaining doubts the backers may have harbored disappeared during the course of the afternoon. When the demonstration concluded, Holland found himself praised as an Irish hero and, more important, assured that he could count on a much larger sum to finance construction of a battle-ready, full-sized submarine to be used against England at some future date.

The boat's active life was short and its end inglorious. Holland conducted several additional tests before bowing to Breslin's demand that the boat be destroyed in the interests of secrecy. Like most of Clan-na-Gael's attempts at subterfuge, it proved less than successful. After being stripped of its fittings and scuttled in the Passaic, the hulk was dragged by currents to the bridge overlooking the original launching site, where it proved an irresistible target for locals who tried unsuccessfully to salvage it for scrap. There it was to remain for almost fifty years, until a group of students finally hauled it to the surface at the request of the Paterson Museum, where it remains today.

If Holland was pleased by the tests, the management of Breslin's Jacobs and Company was positively ecstatic. Realizing the second and much larger boat would be many times more expensive than the first, they nevertheless committed themselves to the project with scarcely a thought for the final cost. The trustees' eagerness to part with what eventually turned out to be more than sixty thousand dollars testifies to their mood of unrestrained and optimistic enthusiasm, since the question of how, and when, the submarine would go into action was still the subject of bitter debate. O'Donovan Rossa's so-called Dynamitards advocated open war: Any British ship, merchant or naval, might be sunk without warning at any time or place. Devoy and the more cautious leaders insisted that the submarine be perfected and placed in mothballs, ready and waiting for the moment when it could do the greatest damage. A future war involving Britain and Russia would provide just such an opportunity, as might any escalation in the tension between London and Washington over the question of English influence in South and Central America.

Holland's design for his second boat did not reflect any obvious sympathy for either view; it was built to be practical, and in this respect it was a complete success. At nineteen feet and some twenty tons, it was small enough to be transported inside a sealed boxcar or a deck cargo on a medium-sized coastal trader. As its main armament Holland designed a compressed-air cannon, which was to prove capable of lofting a hundred-pound charge of dynamite more than three hundred feet through the air. The boat was to be safe at a depth of up to sixty feet, able to remain submerged as long as twenty-four hours, and powerful enough to make seven knots on the surface and five knots beneath it. Its most unusual feature was its gasoline engine. Holland had little faith in the primitive electric motors of the day and opted instead for an ingenious system of "valves and bottles" that allowed exhaust gases to be pumped out of the hull under pressure while fresh air for the carburetor flowed in from a series of compressed-air tanks. Despite the obvious dangers of combining an internal combustion engine with a small, unventilated space, there was only one mishap and that involved the pneumatic cannon rather than the boat's vital systems. Holland was conducting a submerged "attack" on a Brooklyn pier when the dummy projectile flew out of the water, soared several hundred feet into the air, and plunged to earth beside a solitary fisherman, who was left shaken but unharmed.

Just as Carroll had predicted, the chief problem concerned the Irishmen's inability to keep a secret. The boat was built at the end of Manhattan's West Thirteenth Street, in a marine foundry on the banks of the Hudson River, and quickly became an irresistible magnet for curiosity seekers of all kinds. A number of foreign military attachés traveled up from Washington to chat with the inventor and take notes. Holland rather liked the attention of fellow professionals, but he drew the line at newspaper reporters, whose incessant intrusions forced the Skirmishing Fund, as it was called, to spend even more money on a high wooden fence. One particularly persistent reporter hung around long enough to glean the basic bones of the story, which he filled out with a few speculations of his own, inadvertently christening the submarine in the process. The scribe took note of the inch-thick plate at the bow and theorized that the submarine was a "Fenian Ram" intended to punch holes in the hulls of its victims. The name stuck, making a total mockery of the "Jacobs and Company" strategem and forcing the Skirmishing Fund to tacitly admit its involvement.

Although the *Ram* was undoubtedly capable of performing the task envisioned for it, the opportunity was never allowed to arise. Instead, Holland tested it repeatedly through the summer and fall of 1883 in the waters in and around New York Harbor, on one occasion scaring the daylights out of a ferryboat captain when he surfaced unexpectedly off the ferry's port bow. Holland and his boat became one of the city's sideshows, receiving almost daily coverage in the press through the summer and fall of 1883 and even raising eyebrows in London, where accounts of the *Ram*'s performance led the Foreign Office to insist that the U.S. Government have the experiments stopped before events got out of hand. The complaint received summary attention in Washington, no doubt because Britain's concern was considered to be out of all proportion to the threat. As it happened, the British need not have worried, since the Fenians were doing everything in their power to snatch defeat from the jaws of victory.

As the excitement and unity engendered by the *Catalpa*'s success faded from memory, the old divisions and rivalries reemerged, to manifest themselves in a spate of petty feuds. One such issue reached a head in early fall of 1882, when the body of a recently deceased Irish official was sent back to Ireland for burial. The Clan-na-Gael man who accompanied the deceased comrade accepted the assignment in the belief that both the funeral bills and his own expenses would be met by the Skirmishing Fund. When he arrived in Ireland, to find the trustees had changed their minds, he threatened to issue an injunction freezing the fund and all its associated assets until a court could define the legal responsibilities of its guardians. The case was further complicated by a fresh outbreak of acrimony surrounding a number of other perennial sore points, not the least of which was the submarine. O'Donovan Rossa's radicals had renewed their campaign for an immediate attack, accusing Devoy and the others of cowardice. Holland was about to sue George Brayton, the inventor of the *Ram*'s gasoline engine, who was himself defended by Breslin on the grounds that any further legal action would serve only to widen the growing rifts. In a final twist, Breslin's pleas reached Holland via an intermediary because the two men were no longer on speaking terms. Something had to give, and late one night in November, it did.

In order to stop a rival faction from seizing the *Ram,* Breslin decided to take the boat himself. Late one night after Holland had retired to Paterson, Breslin and a small band of sympathizers slipped into the shipyard and convinced the night watchman to release the *Ram* in their charge. The boat was tethered to the stern of a waiting trawler, and together with a smaller, sixteen-foot model, which Holland had been using for advanced tests, the seagoing caravan left on the first leg of a long journey to the New Bedford brass foundry, the same spot from which the *Catalpa* had departed for Australia some seven years before.

If the rebels had chosen a warmer night, the trip might have passed without incident. Instead, the large quantities of liquor consumed to ward off the cold precipitated a minor catastrophe. Less than an hour out of the dock, someone noticed that the smaller boat appeared to be sinking. It went to the bottom a few minutes later, apparently because, in their haste to flee the dockyard, the Irish pirates neglected to close the turret.

Holland learned of the seizure the next day and his reaction was not anger but disgusted resignation. He was weary of his colleagues' interminable disputes and may even have been secretly pleased that the affair was over once and for all. In a letter to a friend, he explained that he planned to take no action other than allowing the boat "to rot on their hands." And that was precisely the fate that befell what was undoubtedly the most sophisticated underwater craft in the world to date. Unable to operate it themselves and yet unwilling to offer it for sale, the trustees simply abandoned it in one of the brass foundry's storage sheds, where it remained neglected and forgotten until the turn of the century. Today, it holds pride of place in Paterson's museum alongside the equally ill-starred *Holland I.*

Holland put the *Ram* behind him with high hopes that its well-publicized voyages beneath New York Harbor would prompt the U.S. Navy to reconsider its position. In this, as with so many of his later business decisions, he erred on the side of generosity in appraising the Navy's institutional intellect.

It would be hard to imagine a less suitable advocate to confront the enshrined prejudices of the American Navy. There was little of the firebrand in Holland's character and not a trace of the mercantile instincts that transformed Victorian arms inventors like Maxim and Nobel into some of the richest men of their generation. Nor was he particularly comfortable in the atmosphere of Byzantine politicking that even then dominated the selection of the nation's means of defense. Quiet, shy, and perpetually absentminded, he was an easy target for the infinitely more sophisticated businessmen and political operators who were to appropriate the fruits of his genius. The otherworldliness was best captured in a photograph taken about twelve years before his death in 1914. Peering at the world through his thick spectacles, he protrudes from the hatch of one of his later submarines like a timid burrowing animal ready to bolt back into its hole at the first sign of trouble. Never an entrepreneur or a lobbyist, he was an inventor, an innocent in a world dominated by forces and personalities beyond his ken.

Then, as now, a weapon's deficiencies on the field of battle were of secondary importance to its ability to win friends. When a branch of the armed services finds itself about to be saddled with an unwanted or insufficiently glamorous weapon, one of two things is likely to happen. The system itself may undergo a form of bureaucratic plastic surgery, which usually results in a piece of equipment significantly different in shape, function, and price from the one originally envisioned. That, however, represents the last line of defense, a course of action taken only when there is no longer any hope of scrapping the project entirely. In extreme cases—a proposal to buy a German tank instead of one from Detroit—the chosen instrument is apt to be a hefty tome of performance specifications conceived with the explicit intention of foiling any designer who sets out to meet them. Such willingness to do battle armed only with a variety of fine-print clauses was to be the bane of John Holland's existence for the next two decades.

The first overtures in his second campaign were rejected out of hand. After several fruitless rounds of correspondence, the inventor and the small band of supporters who had pooled their capital to establish the Holland Torpedo Boat Company were informed that the *Fenian Ram* had been nothing more than a novelty to enthrall the credulous masses. No right-minded seaman would serve in such a preposterous craft, various admirals opined, adding that the captain of a submarine would never know where he was going. Holland tried to point out that the *Ram* was designed to be "porpoised" through a series of shallow dives, breaking the surface for a second or so before returning to the obscurity of the deep. The argument fell on willfully deaf ears.

It was not until 1888, six years after his first contact with the Navy, that Holland was given a chance to refute his invention's detractors. Admiral Montgomery Sicard, head of the Bureau of Ordinance, had been monitoring official submarine research programs in France, Sweden, Russia, and even Austria for some time, and while he was unimpressed by the results, he believed it prudent for the United States to remain apace of the latest trend. Armed with the enthusiastic support of several relatively junior members of his staff, he succeeded in obtaining $150,000 from President Cleveland's Navy secretary, William C. Whitney, to finance the selection and construction of an American design. Sicard's defeat of the surface fleet was important but not unqualified. While the admirals could hardly dispute the wisdom of Whitney's decision, at least not in public, they did reserve the right to impose their own daunting list of performance specifications. More worrisome for Holland and his backers, they convinced Whitney to withhold the money until such time as the winning boat proved itself in a series of arduous sea trials.

When Holland emerged the winner, it was the absence of immediate financing that confounded his hopes. Although his design had been judged the most promising of four candidates by a Navy selection panel, the construction contract was declined by the shipyard assigned to build it when the owners concluded it could not be done, at least not within the stringent guidelines. That was all the defenders of the surface fleet required to have the project shelved.

Admiral Sicard proved to be a dogged campaigner, and within twelve months he had succeeded in launching a second competition, although this, too, was subject to the original conditions. Holland won the second round as well, only to see the prize once again plucked from his grasp. A new President, Benjamin Harrison, moved into the White House, bringing with him a new Navy secretary, Benjamin F. Tracy, whose sympathies resided firmly with the surface fleet. Tracy simply abandoned the competition and reassigned the money to the construction of surface ships. Now that the U.S. Navy had joined the ranks of his tormentors, Holland was coming to resemble an Irish Sisyphus, condemned to an endless cycle of strenuous achievements and crushing defeats. The first submarine, the diminutive *Holland I,* survived only a few days of cautious tests before the paranoid fancies of Breslin and the other rebel backers sent it to the bottom forever. The *Fenian Ram* remained in its inventor's charge a little longer but ultimately shared a similar fate, snatched away by events and personalities over which Holland could exert not the slightest degree of control. There was nothing he could do about this latest blow. Disgusted, dejected, and financially embarrassed, he turned briefly from his dream and accepted a job designing dredging equipment, occupying his idle moments with plans for a steam-powered flying machine. Even his closest friends were taken aback by his latest scheme, and the project died as quickly as it had arisen, a victim of the prospective shareholders' healthy skepticism.

The four years of frustration came to an end in 1893 with the Cleveland administration's return and the establishment of a third submarine competition committee. As usual, Holland won, and just as before, the defenders of the surface fleet made every effort to have the result ignored. This time, however, their task was not as easy. President Cleveland was himself a keen supporter of a submarine program, as were most members of his cabinet and a number of vocal congressmen, who argued that the modest expense was justified by the risk of being left behind by the march of naval technology. However, there have always been countless ways in which the inevitable may be delayed.

The first stalling stratagem saw the invocation of an entirely spurious technical argument. A number of senior naval officers expressed the opinion that while a submarine might survive the detonation of an underwater mine, its crew almost certainly would not. There was little basis for the theory, but the admirals insisted that it be put to the test before they would agree to authorize the contract award. A cat, a rooster, a rabbit, and a dove were sealed in a boiler and lowered into the sea off Newport, Rhode Island, where they were subjected to a series of large explosions, each detonated slightly closer than the one before. It was not until the final blast—one hundred pounds of gunpowder at a distance of just thirty yards—that the "crew" sustained its first casualties: the rabbit and the dove died. The cat and the rooster emerged no worse for wear.

When the theory of submarine concussion foundered off Newport, the battle swung from technical considerations to the corridors and offices of Capitol Hill, where Holland was opposed not just by the Navy but also by one of the two competitors whose entries had been rejected. A Chicago dentist, James Baker, had submitted a proposal that was inferior to Holland's in almost every respect. However, his deficiencies as a designer were more than compensated by his feats as a lobbyist. Through the latter part of 1893, Baker mounted a concerted bid to have the result suspended until a series of competitive trials could determine which design came closest to fulfilling the Navy's needs. There was an astonishing degree of bare-faced self-interest in the campaign, since

Baker alone possessed a working prototype, a decidedly impractical craft launched twelve months earlier on Lake Michigan.

If Holland had been representing his own interests, as he had always done in the past, there can be little doubt Baker would have taken and maintained the upper hand. He had enlisted the aid of influential Washington personalities like the Iowa senator and presidential aspirant William B. Allison and a notoriously wily capital lawyer, General C. M. Shelley. As for the Navy, the battleship admirals favored any development that would delay the introduction of this distasteful new weapon. Fortunately for the future of General Dynamics, the matter was no longer in Holland's hands. Some months earlier, a young New York lawyer and stock speculator, Elihu Frost, had taken over as the Holland Company's secretary and chief financial officer.

When a series of stories in the *New York Times* confirmed that the Navy was pressing to have Baker's proposal accepted, Frost brought his own extensive network of Washington contacts and family friends into play. The result was a stalemate that was to last almost two years. At one stage, Frost attempted to woo General Shelley away from the Baker camp with a sizable and unsolicited parcel of Holland Company stock. Shelley not only rejected the offer but turned the incident to his client's advantage by going public and deploring the Holland Company's devious tactics. Frost apologized, limply explaining that he believed Baker had recently dismissed Shelley and engaged a new counsel. It was hardly surprising that another of Frost's schemes—to lavish stock certificates on several of Baker's other supporters—was abandoned very quickly indeed.

Frost's next line of attack was a good deal more subtle. A number of Holland's friends were dispatched to Japan, South America, and various European capitals with instructions to take out patents on Holland submarines and components, and, wherever possible, to set up agencies empowered to negotiate foreign sales. As a business move, the acquisition of foreign patent rights was a sound investment in the future. In the more immediate context of the lobbying battles taking place in Washington, it was a form of gentle blackmail. With Holland's designs now available to any foreign power with sufficient funds to purchase them, the Navy found itself under increasing congressional pressure to make first use of them. In 1895, seven years after the first competition, the pieces fell finally into place when the legislators authorized the sum of $200,000 to finance the construction of the original winner.

The Navy had lost the war, but it remained resolute in its determination to be anything but cooperative in defeat. At the Baltimore shipyard where Holland's aptly named *Plunger* was finally taking shape, a succession of slight amendments to the original design became first a stream and then a flood, each wave of alterations prompted by an apparent desire to build a submarine that looked as much like a surface vessel as possible. Holland had specified one propeller, but the Navy demanded no less than five, three for forward propulsion and two horizontally mounted screws which, it was hoped, would permit the boat to "hover" at a fixed depth. The original design featured one forward-firing torpedo tube; the Navy's version sported two tubes and room to store five more. But the most bizarre piece of tinkering involved the *Plunger*'s power plant —not the combination of internal combustion and electric motors that Holland wanted but a totally impractical steam engine with a boiler so large it proved impossible for members of the crew to move from the forward compartment to the stern without suffering serious burns.

Like many modern weapons, the *Plunger* was redesigned to be the biggest and the best, a submarine that could do everything. Yet it proved incapable of anything. Its longest dive was a short dockside trial that concluded eight minutes after it had begun, when the heat-addled crew brought the ship to the surface and tumbled gasping onto the steaming deck. As for the seven torpedoes about which the Navy had been so insistent, they proved entirely useless, since the boiler generated so much heat the boat could never have remained submerged long enough to put them to use. Common sense should have led someone in a position of authority to cancel the entire project long before the slapstick finale. But having gained its own bureaucratic momentum, the *Plunger* plunged ahead despite Holland's oft-repeated avowal that it could never be made to work. There was a definite note of urgency in the inventor's protests: the *Plunger*'s failure would send many of his closest friends and fellow investors to the poorhouse. In one of their last bids to stymie the program, the surface admirals insisted that the company undertake to pay a penalty of ninety thousand dollars if the boat failed to meet the levels of performance promised in Holland's original submission. The fact that the final product of all those changes bore but a slight resemblance to its theoretical ancestor counted for naught.

There was, at least as far as Frost was concerned, only one solution. Canvassing Wall Street for what would be known today as venture capital, he approved Holland's plan to begin work on yet another submarine, this one to be built in a private shipyard beyond the reach and control of the Navy's Bureaus of Ordnance and Construction. The new boat, Holland's sixth, was to be powered by a gasoline engine on the surface and an electric motor beneath it. There would be just one propeller, the same system of self-adjusting ballast tanks, and a tapered, cylindrical hull that owed much to Holland's studies of dolphins and whales.

The *Holland VI* took shape rapidly and entered the water for the first time, as a partly fitted hull, within six months of the keel-laying ceremony. Frost made the most of the occasion, transforming the launching into a gathering of the best-known men in the country. Military attachés were brought up from Washington to add a touch of foreign glamour to the affair; a band played patriotic songs, while the Irish inventor produced just the right amount of blarney to guarantee a prominent display in the next day's papers. The submarine would be the most awesome instrument of destruction in the history of the world, he explained, adding that while he would like to see it go into service with the U.S. Navy, he was a businessman with shareholders to consider, and thus obliged to sell this and later boats "to whoever has the money." Working deftly behind the scenes, Frost put it about that competitive negotiations were already under way with several major, but unnamed, foreign powers. It was just what the press required to turn its gaze away from the stalled *Plunger* to the infinitely more colorful *Holland VI.*

"FICTION HAS BEEN OUTDONE," read the headline over a story in the *New York Herald,* which described in lurid detail the havoc such a boat might wreak against a future enemy. Holland himself went into print with an equally sensational piece of speculation, in which he described the destruction of Manhattan at the hands of a foreign submarine lurking beneath the waters of the Hudson River. Of all the papers, only the *New York Times* managed to maintain a sense of dispassionate understatement, informing its readers "that the submarine may or may not play an important part in the navies of the world."

Frost's flair for attracting attention developed in step with the *Holland VI*'s construction schedule, the high point coming less than two months after Saint Patrick's Day 1898, the date selected by Frost as the most appropriate occasion for the inventor to take his craft beneath the surface for the first time. Holland was led into a chamber at the company's offices on lower Broadway, where he was introduced to the waiting press as the man who would avenge the recent sinking of the *Maine.* If the Navy would agree to transport his boat, himself, and a volunteer four-man crew to a jumping-off point close to the Cuban port of Santiago, the Spanish fleet's home base in the Caribbean, Holland promised to clear a path through the minefields, enter the harbor, and sink every ship he encountered. His sole condition was that the Navy undertake to buy the boat if he returned alive. The event caused such a stir that Holland left for Washington the next day, explaining that he planned to present his offer in person to the Navy secretary and, if possible, President McKinley. The Navy, however, remained unimpressed and the two men returned to New York a few days later without having caught sight of Cuba, the Oval Office, or a contract of sale.

Whenever the public's appetite for submarines appeared to be waning, Frost always managed to manufacture some new stunt. The boat was put through a series of well-advertised sea trials in the waters around

Manhattan, where the spectacles were witnessed by throngs of some twenty thousand people gathered on hills, bridges, and flotillas of small boats. Celebrities also won press coverage, and Frost found no shortage of big names eager to take a dip. Teddy Roosevelt, then the under secretary of the Navy, accepted an invitation to "commune with the fishes" and returned to Washington one of the boat's most ardent advocates. Another passenger, Clara Barton, formed a decidedly different impression during the course of her underwater excursion: The first president of the American Red Cross berated Holland for squandering his genius on a machine that would make future wars even more deadly than the Victorian bloodbaths against which she had campaigned for more than thirty years. Holland's response deserves mention if only because it made him one of the first to float the doctrine Robert McNamara would later christen "mutually assured destruction." Submarines would preserve the peace forever, he asserted, since they would render impotent and obsolete the surface fleets of all major powers.

Frost kept using what little influence he possessed to subvert the Navy hierarchy's insistent opposition. At his instigation, a certain Captain John Lowe was given a twelve-month leave of absence to work at the New Suffolk shipyard, where the *Holland VI* was under development, while another known sympathizer, Lieutenant W. J. Sears, was moved from Manila to the headquarters staff of the Atlantic fleet, where it was believed his slight influence might advance the Holland Company's cause. Another tactic was attributed to Frost by Simon Lake, a contemporary of Holland's and a fellow submarine prophet, whose designs for wheeled submersibles capable of rolling across the ocean floor found even less favor with the Navy Department than those of his notoriously unsuccessful rival. According to Lake, who later prompted an indecisive congressional investigation into his charges, Frost had placed large quantities of Holland Company stock in trust for a number of senior navy personnel, thereby creating what his complaint characterized as "an unbreakable submarine monopoly."

A man possessed of Holland's narrow focus might have persisted with the fruitless campaign indefinitely. Disappointments had become the norm; every slight victory preceded the bitterness of yet another defeat. More than a quarter of a century later, the recently retired Rear Admiral W. W. Kimball recalled a chance meeting with Frost that took place several days after a series of semiofficial sea trials in and around Peconic Bay. The *Holland VI* performed admirably, fulfilling all the demands originally assigned to the *Plunger.* "Under your advice we asked for requirements. It took a pretty penny to meet them," Admiral Kimball quoted Frost. "You told me that an official report would give us status. We must have the approval of our own country before we can do business. We have the finest report whatever; we have the status you talked about—but we don't get any indications that we will ever get any orders for boats from the Navy. The Department is as much down on us as ever. . . ." The diatribe continued unabated for several minutes.

It was at this point that Frost decided to dispense with official channels altogether. Reasoning that nothing but the sheer weight of congressional opinion would force the Navy's hand, he informed Holland and the other stockholders that the boat was to be "sent down to Washington to lobby for an appropriation."

Taking a circuitous route through the network of inland waterways because no insurance company would accept liability for the boat on the open seas, the *Holland VI* reached Washington and the waters of the Potomac in late January 1900. As usual, Holland took the controls and put the boat through its paces, circumnavigating a series of obstacle courses laid out with flags and buoys near the foot of Mount Vernon. More trials followed, as did unscheduled demonstrations for the benefit of any congressman who expressed even the slightest interest in the machine. This time, Frost was to emerge the winner.

On April 23, Admiral Dewey told a hastily convened meeting of the House Naval Affairs Committee that Holland's boat was one of the marvels of the age. "If they had had two of the things in Manila I would never have been able to hold it with the squadron I had at my disposal. . . . With two of these in Galveston, all the navies in the world would not be able to blockade the place."

No politician could ignore the ensuing clamor. The idea of a submarine serving under the American flag—particularly one so warmly endorsed by the hero of Manila—captured the public imagination and swept away all but the last shreds of official opposition. Several of the most determined bureau chiefs did make one last attempt to scuttle the program, informing the Naval Affairs Committee that Holland's toy was useful only for impressing simpleminded city folk. For once, Congress was in no mood to listen; it immediately reallocated $150,000 from funds originally set aside for the construction of two additional *Plunger*-type boats.

On April 11, 1900, the Electric Boat Company, the new name for Holland's original firm, accepted an initial payment from Navy Secretary John D. Long. There would be many, many more payments over the years to come.

2. The Navy Blockade

Success changed everything. Holland was on his way out within months of the government contract, "buried with his glory," in the words of a close friend and fellow director, who witnessed the gradual eviction. The inventor was to be squeezed out of the company that had recently borne his name, stripped of his patents, and with scant regard for appearances or sentiment. It would be a dirty campaign of slights, demotions, petty insults, and dubious legalities exploited to the hilt. In view of the sometimes torrid scandals that would punctuate the future history of the Electric Boat Company and, later, General Dynamics, it was an entirely appropriate start to the ancestral line.

When Holland cast himself as the scourge of the Spanish fleet, a New York newspaper artist captured the element of madcap cheek with a cartoon that was to inspire several generations of imitators. Above the caption, "What, Me Worry?" the Irishman strode the deck of a submarine bound straight for Cuba and a clutch of looming battleships, of which he seemed totally oblivious. The depiction was apt in more ways than the artist knew. Holland had little reason to fear an actual encounter with a Spanish man-of-war; he did, however, have every reason to fear Frost.

Of the company's limited successes until the end of 1899, Frost was responsible for most. It had been his legwork and lobbying in Washington that turned aside James Baker's bid to have the third submarine competition declared void and ultimately confounded the Navy's repeated assaults. In New York, where the company found its popular and financial support, the lawyer's ability to attract investors in the face of so many setbacks was even more remarkable. He was, in the best New York sense of the word, an operator. By contrast, Holland remained an innocent in an era of carnivorous capitalism. Lacking the acumen to protect his interests, he became an easy and irresistible target.

Sentiment aside, his ouster was probably inevitable. Now that the pioneering was done, Holland's pursuit of perfection merely obstructed the real job of developing a standard design that could be sold at home and abroad, preferably in large numbers. The sale of the first boat to the U.S. Navy hardly qualified as a profitable transaction. With construction costs, later refinements, public relations expenses, and repair bills occasioned by a series of unfortunate accidents, the total investment amounted to some quarter of a million dollars. Yet the Navy was being asked to pay but $150,000 for the first boat and only a little more for each of the next five submarines for which orders had been placed. As Frost well realized, the real future for the Electric Boat Company rested not with the U.S. Navy but in the bloated naval budgets of Britain, France, Germany, and the other chief European powers. Winning the support of the U.S. government was an essential first step on the way to breaching the European market. Dispensing with the services of the aging Fenian was no less important, particularly if the company hoped to conduct business with Britain.

The inventor's slow ejection was such a deliberate process, it is possible to identify the exact day on which it began, July 4, 1898. Frost was keen to make the most of the holiday weekend and had arranged a full schedule of tours in and under New York Harbor for the benefit of several groups of potential investors. Though the itinerary was not unusual, some of the guests were. Along with the customary assortment of local and national political figures, Frost and Holland were to play host to one of the most successful stock players ever to make his mark on Wall Street, Isaac Leopold Rice. A year or so earlier, Rice had taken the spoils from one of his frequent raids on the market and bought out the Electro Dynamic Company of New Jersey, a small but innovative manufacturer of marine electric systems and the sole contractor for the design and installation of the *Holland VI*'s motors. By the time Frost and Holland brought him back to dockside at the end of the afternoon's jaunt, Rice had made up his mind not merely to invest in the Holland Torpedo Boat Company but to take total control of it.

It was only natural that Rice should look upon Holland's little company as a handy addition to a personal portfolio that included many of what were the high-tech stocks of their day. Throughout his career he had mined the potential of America's advancing frontiers, making his first fortune as one of the most prominent among "the 40,000 railroad lawyers" whom Jack London contemptuously described as "conspiring to defeat the people in the courts." Later, having left the service of the railroad barons for the independent career of a large private investor, he generated his next millions in the infant science of electricity. With the cash accumulated during his career at the bar, Rice bought control of the Exide battery company, a firm that had achieved modest success manufacturing what was then known as "chloride accumulators." The price was high but the deal came with one particularly attractive fringe benefit. Along with the plant, equipment, and goodwill, Rice obtained the basic battery patent. His timing was exquisite. In the space of the next few years, something of an electrical revolution took place as simple, direct-current appliances came into widespread use. Motors for Mr. Singer's new sewing machine, fans for the parlor, electric cars and buses—all required banks of rechargeable batteries. Exide was soon generating such a cascade of riches, Rice was able to indulge his instincts and secure the rights to hundreds of electricity-related patents. Until a few months before he accepted Holland's invitation to come aboard the submarine, Rice's investments included a large fleet of electric cabs, the Electric Launch Company, which built lavish, battery-powered pleasure boats and tenders, and Electro Dynamic of Bayonne, New Jersey. It was this last acquisition that was to set Rice on the path to his last and by far his greatest fortune.

One night, some months after the last of the wiring had been installed in the *Holland VI,* a dockyard hand went home without bothering to close one of the boat's seacocks. The next morning found the vessel beneath ten feet of water, its interior a waterlogged shambles of mud and sodden wiring. A technician sent out from Electro Dynamic in response to Frost's urgent plea for assistance inspected the recently raised hulk and concluded that it would be a waste of time trying to salvage the existing electrical systems. All the same, before the arduous chore of stripping and replacing the ruined circuitry was undertaken, the electrician, Frank W. Cable, decided to try a novel theory of his own. Reversing the field—running an overload of current backward through the wires—he managed to generate sufficient resistance to heat the wires and dry out the waterlogged insulation from the inside. Cable would shortly return to Electric Boat full-time as Holland's permanent replacement.

Rice was intrigued by Cable's reports about the submarine and the Holland Company's dire financial straits, including a recently discovered oversight that threatened to stretch the company well beyond its meager financial resources. Holland, having neglected to calculate the effect of the propeller's wash on the rear steering surfaces, had come to the reluctant conclusion that much of the aft section would have to be torn apart, redesigned, and reassembled if the boat was to be cured of a tendency to list as much as fifteen degrees when running submerged at top speed. The job was to cost thirty thousand dollars, a small enough sum in terms of Frost's ambitions for the company but sufficient to nudge it within a hairsbreadth of bankruptcy. Cable's intelligence reports fanned Rice's acquisitive instincts; with his usual blend of careful planning and diplomacy, Rice began to sound out his friends in Washington as to the Holland Company's chances of making a sale to the U.S. Navy.

His inquiries may or may not have led him to acquire the company without the unintentional incentive provided by two of New York's most influential citizens, Thomas Fortune Ryan and William C. Whitney, whom we last met as President Cleveland's Navy secretary. Between them, the two men controlled the city's largest and most powerful utility, Metropolitan Traction Company, which held franchises for streetcars, subways, and the elevated railway lines springing up in Manhattan and the boroughs. Like any monopoly, Metropolitan had a low tolerance for competitors, and it soon came to regard Rice's growing investments in electrical vehicles and patents as a threat in the making. Just as they had done with all previous challenges to their authority, Whitney and Ryan resolved to buy Rice out of business.

The pair moved first against Exide, buying a majority stockholding on the open market for what Rice himself described as "very high figures indeed." They next seized control of the cab company and were preparing for what would undoubtedly have been an assault on Electro Dynamic, when Rice counterattacked. Since he had had the foresight to transfer many of the most valuable patents to his own name, he was able to force Whitney and Ryan to accept a negotiated truce. If the takeover bids persisted, Rice explained, he would be obliged to increase his royalty rates or perhaps even deny their further use when the current licensing agreements expired. Since several of the patents applied to electrical equipment essential to Metropolitan's continued operations, Whitney and Ryan acquiesced without further protest.

Rice emerged from the aborted buy-out flush with cash and eager to find another investment. Cable's reports of the Holland Company's financial troubles provided the answer. If the submarine builder was to be consolidated into a new company together with Electro Dynamic and Electric Launch, there was less likelihood of any individual firm falling victim to a takeover bid of the kind Metropolitan had just agreed to abandon. Moreover, if Rice's reading of the situation in Washington proved correct, the Holland Company could shortly expect to record its first sale. Within days of meeting the inventor and making his first brief descent into New York Harbor, Rice agreed to finance the essential modifications to the submarine's stern. Seven months later, happy with the progress he had witnessed, Rice merged all three companies under the banner of a new corporate entity, Electric Boat.

For Holland, it was the beginning of the end. Like Frost and the few other directors who had been preserved on the new board, he received a bundle of preferred stock as well as a three-year contract that guaranteed a modest salary of ninety dollars per week. Holland was pleased with the financial package; it was the accompanying demotion that infuriated him. No longer general manager but chief engineer, he soon discovered the title was nothing more than a hollow courtesy. Old friends, some of them associates from the days of the *Fenian Ram,* were instructed to ignore Holland and take their orders from Frank Cable. Holland's indignation simmered for more than a year before being brought to the boil by two calculated affronts to his competence and intelligence.

The first followed a near-scrape with a tugboat when Holland, distracted by a minor malfunction inside the hull, allowed the boat to drift perilously close to the accompanying vessel's propellers. Several weeks later, the inventor was informed that he was no longer master of his own ship, that honor also going to Frank Cable. The former electrician celebrated the promotion by having a tailor run up an elaborate uniform of his own design, a trivial concession to vanity but an action that summed up the intensifying battle for prestige and preeminence within the company's rapidly expanding hierarchy.

Less than a year later, Frost and Rice delivered the crowning insult: Holland was sent off to Europe, ostensibly to make the acquaintance of several newly appointed sales representatives. Frost had reasons for wanting Holland out of the country and he scarcely bothered to contain his contempt for the simple Irishman. "I suggest you make the arrangements most explicit," the lawyer wrote in a letter to one of the company's European agents. "Holland has not the best memory for business arrangements as you well know." On the day Holland's ship was due to sail, Frost arranged to pay five years' taxes owing on the submarine patents registered in various European capitals. The settlement of Holland's personal accounts did not remove the patents from his possession, but it did represent Frost and Rice's first attempt to establish financial claim over them. Three months later, when Holland had returned to New York, Frost confronted him one afternoon with a sheaf of papers. He was tidying up some old bookkeeping, the lawyer explained, when he came across a few documents that required the inventor's signature. Perhaps uncharacteristically, Holland took them home and discovered that he was being asked to assign his European patents to the sole control of Electric Boat. Writing of the incident in a letter to a friend, he denounced the attempt at piracy and confided that he was off to see a lawyer. "Mr. Frost will not find that I am such a fool as he thought."

Holland proved to be every bit the fool Frost imagined. There is no record of his lawyer's advice, but whatever Holland learned, it failed to fire his fighting spirit. Evidently reluctant to reimburse the company for the substantial sums it had invested in renewing his patent rights, Holland meekly agreed to renounce all present and future claims on them.

Reduced to a sort of mascot useful only for impressing celebrity-conscious congressmen and reporters, Holland no longer had any reason to remain with the company that had once borne his name. In March

1904, he gave voice to his long-festering resentment in a letter of resignation memorable for its final note of subtle sarcasm.

> Dear Mr. Rice,
>
> As my contract with the company expires on the 31st instance and as it is proper that I should withdraw my directorship, I beg to offer my resignation.
>
> The success of your company can never be as great as what I ardently desired for it.
>
> Yours, very sincerely,
> John P. Holland

His departure was neither mourned nor protested, except by a few newspaper columnists and cartoonists, who expressed a measure of sympathy for the dispossessed inventor. One cartoon in particular summed up the attitude. Holland was shown about to be fired from the deck gun on a submarine, right into the mouth of a large and villainous fish that bore the label "The Electric Boat Co." The cartoonist was too late. With the exception of his dignity and a small parcel of stock, Holland had lost everything he had valued two years before. Rice and Frost were glad to see him go, happy to be rid of a potential source of embarrassment.

Some months after the first sale to the U.S. Navy, Rice had fulfilled the first stage of his plan to turn Electric Boat into a major provider of international arms by arranging a licensing deal with Vickers of England. Several of the British boats had since come to grief and there was much speculation at the Admiralty that the elderly Fenian had somehow tampered with the plans supplied to Vickers' shipwrights. The newly appointed head of the Royal Navy's submarine service, Captain Roger Bacon, suspected as much, as did Holland's onetime associate and chief patron of the *Fenian Ram,* John Devoy. For all his rebel sympathies, Holland was an unlikely culprit. It was Frank Cable who supervised the construction of the British boats, and trained their crews. If Holland had made any attempt to sabotage the plans, Cable would have noticed immediately and taken the appropriate steps. Nor did the Admiralty's courts of inquiry find, or even suggest, any evidence that Holland had been less than honest. In almost every case the accidents were attributed to the youth and inexperience of the boats' captains or, in one particular case, to the unadorned folly of a seaman who ignored a standing order and lit his pipe in the gasoline-saturated environment of the command cabin. Rumors persisted all the same, even finding their way into an obituary published by the New York *Gaelic American* several weeks after the inventor's death, in August 1914. Holland had opposed the licensing agreement with Vickers, the aged John Devoy asserted in the article, and had secretly taken steps to make sure that several of the most important valves were installed back-to-front. The deaths of the thirty British sailors who perished in the accidents provided absolute proof that Holland was "a true and loyal Irishman to the end."

Foreign sales had always been the key to Rice's plans for the company. Even before merging the Holland Company with his other possessions, Rice had made at least one attempt to establish a licensing agreement with Vickers. His offer was received with polite interest but no particular enthusiasm. How could any navy, particularly that of the world's premier sea power, be expected to invest its time, money, and prestige in an untried device that could not even win the support of its inventor's own government? The trials on the Potomac and subsequent congressional endorsement eliminated that objection but did nothing to improve the company's precarious financial position. America's naval expenditures were a trifle in comparison with the £26 million a year that sustained the Royal Navy. Access to that treasure trove would mark the difference between international success and slow starvation at home.

Rice set out to woo the Royal Navy through Siegmund Lowe, an English banker who was both a protégé of Lord Rothschild and the Vickers conglomerate's chief financial officer. The two men had met in New York not long after Rice's first cruise on the *Holland VI,* and remained in regular contact. Within weeks of the U.S. Navy's forced decision to buy its first boat, Rice wrote to request Lowe's help in obtaining an interview with Lord Rothschild and any other architects of the fleet who could be expected to lend their support. Rothschild's help was vital. Three years before, he had instigated Vickers' merger with the Maxim-Nordenfeldt munitions company, serving as both unofficial spokesman for the parliamentary cabinet and chief source of the financial backing needed to keep both parties at the conference table. If Rothschild could be persuaded to support the submarine's introduction, Rice knew that he would have little to fear from British admirals, who shared the view of their American counterparts in regarding underwater warfare as a dangerous and heretical indictment of their beloved dreadnoughts' supposed invulnerability. "A tin death trap no larger than a steamer on the Upper Thames," one admiral said in reaction to the U.S. government's recent decision. Another of the fleet's offended patricians, Admiral Jacky Wilson, VC, saw the submarine as a knavish device fit only for cowardly foreigners. It was, he said, "a damned un-English weapon."

Rice sailed for London in July 1900 and found Rothschild a good deal more receptive than he'd had reason to hope. Propelled through a series of meetings with increasingly senior officials at Whitehall and the Admiralty, the American soon found himself before First Sea Lord Sir George Goshen. Ignoring the protests of his uniformed subordinates, Goshen placed an immediate order for five boats on the understanding that they would be built by Vickers under the supervision of imported American technicians. A more sensible approach might have seen the boats completed in the United States and shipped across the Atlantic as deck cargo, but Goshen, like his avatars in the modern defense establishment, was obliged to abide by an unwritten law that demanded the naval budget be spent at home. The naval construction and supply industries had become the source of one job in six and existed as the unacknowledged mainstay of the domestic economy. Duty bound to offset the accelerating decline of the heavy export industries, Goshen was left with no choice but to hand the work to Vickers, even though the decision meant a 25 percent price increase. (The U.S. was paying $150,000 for the *Holland.* The Admiralty never paid less than $200,000.)

Lowe and Rice could not have been happier, each for his own reason. Since he was selling his company's technological expertise above all else, Rice had cause to dread the prospect of building more boats at home. Fulfilling the sudden rush of British orders together with those of the U.S. Navy would have required a considerable investment in an expanded work force and new equipment that the company could ill afford to make. It was much simpler, and almost as lucrative, to encourage Vickers' debut as a submarine builder in return for a substantial royalty on each boat.

Rothschild, Lowe, and Rice plotted the future of the Royal Navy's submarine force in a series of negotiations that proved a good deal more protracted than might have been expected, given Goshen's demand that the first boats be made ready for service as soon as possible. The talks consumed most of August and continued in fits and starts until a few days before Rice's departure for New York in mid-October, when they produced a finely tuned document that balanced Electric Boat's need to establish a regular cash flow with its chief stockholders' long-term ambitions. In return for Rice's promise that Electric Boat's patents would not be made available to any of Vickers' European or British competitors, the document promised royalties of 12 percent on vessels sold in England and variable rates of up to 12 percent for those destined for third parties. The two companies carved up the world like a piece of fruit. Lowe claimed Europe and the empire as Vickers' private domain, while Electric Boat walked away with Central and South America, the Caribbean, and parts of the Near and Middle East. Japan and Russia, countries where both firms held high hopes for future sales, were declared free trade zones and left open for competition or cooperation, as circumstances demanded.

Rice returned to New York the hero of the hour. For the first time in its short history, Electric Boat's future seemed reasonably secure. With Vickers' globe-trotting salesmen now able to cite the Royal Navy's decision as ultimate proof of the submarine's coming of age, large and small nations all over the world began to prepare for a coming era of underwater warfare. Germany was quick to follow Britain's example, obtaining much of its early expertise from Holland's less prominent American contemporary, Simon Lake. Even minor league nations far removed from the imperial rivalries of Europe began to invest in submarine fleets. It was becoming the fashionable thing to do and each sale dropped another windfall on Electric Boat.

In later years, Electric Boat was to acquire a reputation as one of the most unsavory of the infamous "merchants of death," the munitions firms that were to be vilified on both sides of the Atlantic by government inquiries into the international trade in weapons. Rice's next foreign sales certainly lent credence to the charge that even in an industry subscribing to no motive loftier than sordid, shameless greed, Electric Boat remained in a class of its own.

Within weeks of the Japanese attack on the Russian settlement at Port Arthur, Rice played host to a succession of purchasing missions representing the czar's navy. The Russians had already snapped up Simon Lake's decidedly odd *Protector,* designed to roll across the ocean floor on huge iron wheels, and soon added Electric Boat's *Fulton* to their fleet. Rice and Cable were glad to be rid of the boat. It had been long under construction, so plagued by mishaps and structural flaws that the U.S. Navy had already deemed it unfit for service. Now that the Russians had come to save Electric Boat's investment, the sole problem remained the best way to smuggle it out of the country.

Both Rice and the Russians realized the sale was in breach of U.S. neutrality laws, which neither party had any intention of observing. Determined that diplomatic niceties would not be allowed to cheat him out of a handsome profit, Rice chartered one of the largest floating cranes in the country and had it brought north from New York to a midnight rendezvous with a chartered steamer in international waters off the coast of New England. The submarine was towed to the site and slung on deck, while Frank Cable and a crew of dockyard hands maintained a nervous watch for a patrolling destroyer which loomed out of the darkness several times but passed without stopping to challenge the clandestine operation. Safely in Finland, Cable escorted the boat to Russia, where he trained a crew, received a decoration from the czar, and supervised the boat's storage on a specially designed railway flatcar that was to carry it three thousand miles across the steppes to Vladivostok and the war with Japan. Cable found the Russian excursion a frustrating and often unsettling experience. In his memoirs, he recalls being tailed by spies, offended by the haughty arrogance of the Russian admirals, and baffled by the internecine feuds of the imperial court, in which he claimed to have become an unwitting and unwilling participant.

Cable was glad to return home, but was denied the pleasure of an extended reunion with his family. Having just sold a submarine to one side, Rice turned to his first customer's enemy and disposed of half a dozen more. No more than a day or so after his return, Cable was dispatched to Tokyo. We can only assume he paused long enough in New York to remove the Saint Petersburg stickers from his trunk.

The Japanese and Russian sales added to Electric Boat's growing reputation for fast and shady dealings. When Cable departed from San Francisco on his way to Japan, he was dogged by reporters eager to learn how he had solved the dilemma of serving two mutual and mortal enemies. No doubt enjoying his newfound celebrity, the former electrician was tantalizingly vague in his replies, reluctant to send the press away but equally unwilling to reveal too much about his impending journey or the sealed cargo that had been loaded into the hold of the ship on which he was to travel.

However annoying it may have been, the publicity did serve to distract attention from a situation that was causing the major shareholders a good deal of grief. The sales to Russia, Britain, and Japan constituted outward signs of success, but financial statements and the first signs of a fresh scandal in Washington indicated that Rice and Frost would soon have to contend with several different varieties of crisis.

Frustrated in his attempts to sell the Navy what he claimed was an infinitely superior craft, Simon Lake accused Electric Boat of playing politics in the Navy Department and controlling "an unshakeable submarine monopoly." Rice did indeed hold a monopoly, although it was far from the lucrative money-spinner Lake imagined. Navy orders were few, far between, and placed with the greatest reluctance. The company was still doing its construction work in the yards of subcontractors, while its own plant and equipment investment was negligible. Albert Vickers, the crusty chairman of the English arms company and a major investor in Electric Boat, found the situation so distressing he took to haranguing Rice in a stream of transatlantic correspondence. Electric Boat had accepted "a tremendous amount of money without any means of paying it," he complained on one of many occasions, advising Rice to replace Holland and the other technically minded members of the board with "solid, sound businessmen." Evidently more vexed than usual, Vickers waited only a few days before firing off a terse second cable: "Have the dummies been retired and responsible directors named in their place?" Sir Albert's rudeness was understandable, since Vickers had been supporting Electric Boat with purchases of stock, unsecured loans, and advances on future British royalties since late in 1902. By the end of the following year,

Vickers and Rice held between them an absolute majority, which was to remain unchanged until 1909, when Electric Boat's rising fortunes permitted it to repurchase much of its lost equity. In the meantime, Rice was called upon to use all his diplomatic skills in deflecting Albert Vickers' incessant criticism, a task made even more difficult by the reappearance of a much aggrieved Simon Lake.

In response to Lake's charges that the Navy had demonstrated an unfair bias in favor of Electric Boat, Congress insisted that the next set of contracts be determined by a competitive trial. Electric Boat carried the contest without effort, only to discover that the resulting four-boat contract was a rather bitter victory. Convinced that the judges had not given him a fair deal, Lake accused Electric Boat of distributing stock certificates, cash, and sundry private favors to those who controlled the Navy and its budget. Together with other witnesses, representing Electric Boat's chief subcontractors, Rice was summoned before the Naval Affairs Committee to give his opinion of the charges. The lawyer refused to speculate, so the committee members made their own suggestions, dredging up several closets' worth of skeletons which did little to substantiate Lake's allegations but much to discredit Electric Boat. Under what circumstances had Holland relinquished his patents and for how much? Why was the Navy so reluctant to listen to the man known all over the world as the father of the submarine? Why were the latest B-boats so much more expensive than their Holland-type predecessors? Why did the Navy refuse to examine plans for an advanced, far-ranging boat designed by Holland after his resignation, opting instead for a clearly inferior model built by Electric Boat?

Rice insisted that he could not answer for the Navy, but did make a few guarded comments about his relationship with Holland. The Irishman was every bit the genius depicted so often in the press, Rice began. However, he was also a poor businessman, an inveterate and insubordinate meddler, a complainer, and on at least one occasion, a danger to his colleagues. Sounding like a man forced to reveal a distasteful episode from the past, Rice repeated a canard first flown and demolished in a series of court actions that followed Holland's resignation in 1904: Holland had avoided dismissal after one particularly acrimonious episode by surrendering all past and future inventions to the sole custody of Electric Boat. When the committee asked about the size of Holland's remuneration, Rice took refuge in a poor memory. No doubt embarrassment played some part in driving the figure from his mind, because in comparison with the $300,000 or so being paid for each of the company's latest submarines, Holland's ninety dollars per week amounted to no more than petty cash.

As for the charges of influence peddling in Washington, Rice referred his inquisitors to the company's financial statements as the final word on the Navy's low regard for submarines. If Electric Boat had done better than its competitors, this was due solely to the superior nature of its product.

Holland's version of events would have been very different, but he was not called as a witness and appears to have made no effort to present his case. He was sixty-six years of age, and his interest in underwater warfare was now entirely theoretical. For a time after his resignation he had tried to start a company of his own, but the job soon proved hopeless. Rice instigated a series of lawsuits that scared off most potential investors, while the Navy remained as resolutely beyond Holland's understanding and manipulation as it had always been. His one success was the sale of a single set of plans to the Japanese, and that venture was marred when the boat went down with all hands not long after its maiden voyage.

As for Electric Boat's attempts to develop upon the basic ideas enshrined in the *Holland VI,* the inventor found them laughable.

Rice had lured a young lieutenant away from his job at the Bureau of Construction and installed him as the company's chief designer and theoretician. Lawrence Y. Spear knew virtually nothing of submarines, while those few original thoughts he did possess were almost all at variance with Holland's standard formula. In supervising the construction of the ill-fated *Fulton,* soon to disappear forever off Vladivostok, Spear ordered brass valve housings and transmission components replaced with cheaper versions in cast iron. This resulted in a near-tragedy when a cast-iron clutch plate exploded, pelting the submarine's cramped interior with fist-sized lumps of metal that seriously maimed a chief mechanic. Cable took to referring to Spear as "a disadvantage not of our choosing."

It was Spear's less obvious attributes that made him a recruit of incomparable worth. He understood the system and counted many of the Navy's middle- and senior-level administrators as his personal friends. His most important quality, however, was an easy familiarity with the nuances of naval aesthetics.

John Holland once complained that the Navy resented submarines because they "lacked decks to strut upon." The jest contained a strong element of truth. The first generation of "tin cigars" were so unlike anything in service on the surface, the Navy refused to dignify the little craft with individual names, even replacing the Holland nameplate on the first boat with the uninspiring designation A-1. It would be Spear's job to balance the need for stability and underwater speed with the naval warrior caste's yearning for a boat that looked at least a little like a traditional warship. Under his guidance, the squat, windowed observation turret evolved into a soaring conning tower evocative of a surface ship's bridge, the beam was narrowed, and a deck materialized to spoil the extremely efficient hydrodynamic lines that had been an integral part of Holland's design philosophy since the *Fenian Ram.* The longer and narrower configuration allowed the new submarines to keep up with a squadron of surface ships, but only at the cost of considerable discomfort to their crews. Where waves had once washed over the Holland-type boat's circular hulls, they now battered, rocked, and occasionally swamped the newer models.

The major improvements that revolutionized the submarine during the first decade of the century were, almost without exception, the work of companies and individuals other than Spear and Electric Boat. Germany provided diesel engines to replace the unreliable and extremely dangerous gasoline models that had caused so many problems with the first British boats. And it was Vickers that solved the problem of visibility, by substituting a newly patented marine periscope for Holland's impractical observation turret. The final achievement, the high-speed torpedo, was the work of the Whitehead Company, a Vickers subsidiary, which increased the weapon's range from around one thousand yards in 1900 to almost six miles by the end of the decade.

Death spared Holland the dubious pleasure of seeing his predictions vindicated. On August 12, 1914, while Europe was marching to war, Holland finally succumbed to the chest ailments that been part of his life since childhood. A little more than one month later, a German U-boat surprised a British trio of heavy cruisers off the Dutch coast, sinking all three in less than twenty minutes and sending almost fourteen hundred seamen to their deaths. When news of the disaster reached Wall Street, Electric Boat stock began a precipitous climb, which was to continue unabated for the next four years.

3. Feast and Famine

On Wall Street, where irony is noted with no less enthusiasm than take-over rumors, talk of General Dynamics is apt to prompt a recitation of traders' quips. The most common is subject to many variations but usually goes something like this: "No matter how bright its prospects, General Dynamics always manages to get hit by a bus on the way to the bank."

Over the years, providence has had a way of taking a wry turn, usually at the most inconvenient moments. It may have been that a previously complacent congressman or committee took an unprecedented interest in a weapon's deficiencies or perhaps went even further and delved into the circumstances under which the original contracts were awarded. A presumed ally in the Pentagon may suddenly have emerged a bitter critic, complaining of shoddy workmanship or brandishing accusations of fraud. Even the most trusted executives have become instruments of fate's treachery, leading the company into ill-advised ventures or, in the most recent example, fleeing the country with several million dollars' worth of kickbacks extracted from subcontractors and the public purse.

This venerable tradition of mishaps and misadventures stretches back at least as far as November 2, 1915, when Electric Boat's directors were informed that Isaac Rice was dead, the victim of a sudden heart attack. His demise served as a cue for the first appearance of the now proverbial bus, unnoticed amid the mourning.

Though sudden, the lawyer's death could not have been entirely unexpected. He had never been a robust man, and his last few years witnessed a marked decline. Breaking with the habit of a lifetime, he took to spending his afternoons at home with a floating assortment of fellow chess and music buffs, delegating the more mundane aspects of the company's legal affairs to a number of newly recruited deputies. He was certainly well equipped to enjoy a lavish retirement. Over the previous three months, the financial press reported that he had sold most of his holdings in the company, to realize a profit of two million dollars or more.

On one of his relaxed midweek afternoons, the lawyer complained of chest pains and sent his wife to fetch a doctor. By the time she returned, he was dead.

Mrs. Rice immersed her grief in the glorification of her husband's achievements, both real and imagined. She donated a million dollars to a foundation for recuperating surgery patients, and another hundred thousand to the quaintly named Anti-Noise Society. In contrast to the quiet family affair that had marked John Holland's exit from the world little more than a year earlier, the lawyer went to his repose amid scenes of profuse formal grief. Hundreds of mourners filed through the family's newly acquired apartment in the fashionable Ansonia Hotel, paying their respects in tour parties organized according to mutual interest. The chess players and music lovers claimed the mornings and evenings, while the afternoons were devoted to the dead man's business associates and former colleagues from Columbia University Law School. For those unable to make it to New York, Mrs. Rice commissioned the artist John Sloan to produce a series of deathbed etchings, which were dispatched to destinations as far afield as Tokyo and London. Sloan's sketches, rediscovered in the early 1980s by a chess historian, display a talent for sentiment and subtle irony not found in the cityscapes and architectural renderings on which the artist built his reputation. In the largest and most moving of the series, the corpse rests upon an ornate chaise longue, the face all but hidden by a voluminous beard and soaring eyebrows worthy of an Old Testament prophet. Around the body, in a penumbra of suggestive detail, are numerous hints to the room's opulence and the dead man's wealth, both rendered irrelevant by the magisterial presence of death.

Lawrence Spear's reaction to the death is not recorded, but we can safely assume it must have been a jumbled mixture of elation and mild anxiety. The two men had always enjoyed an easy professional regard for each other's abilities, but the vast differences in their backgrounds, temperaments, and intellectual capacities precluded all chance of intimate friendship. Where Rice justified the arrogance of his ambitions with inevitable triumphs, Spear was an undistinguished plodder. Molded by nature and the formative influence of naval academy discipline, he was a born first mate but an unpromising skipper. Rice had always been the company's heart and soul, particularly during the years of spectacular growth that followed Holland's ouster. Without assistance he had transformed an organization overendowed with inept businessmen and amateur inventors into an increasingly profitable war supplier that would end the year with more than $30 million worth of business on the books. Now, quite suddenly, he was gone. His heirs were left with a robust legacy and a vacuum at the top, which would not be entirely filled until World War II.

The most eloquent testimony to Rice's achievements, his wife's effusive tributes notwithstanding, lay in the ledger of ongoing work. At the time of his death, Electric Boat faced at least a two-year backlog of work, while its specialist engineers and assemblers were laboring on the company's behalf in yards across the U.S. and Canada. Thanks to the world war, millions of pound's worth of royalties were flowing into New York, not just from England, where boats were under construction around the clock, but also from far-off corners of the world like Australia and New Zealand. Even Canada, a British dominion that had looked traditionally to the Royal Navy for the protection of its sea lanes, had bought a submarine from an Electric Boat licensee in Seattle. Electric Boat's naval business represented the most visible side of the company's mounting good fortune, but it was far from being the only source of joy in the Rice bequest. At the Elco Motor Yacht division's various facilities, the demand for pleasure craft remained undiminished by the distant rumblings of the war in Europe. And finally, there were the company's various merchant marine operations. Electro Dynamic faced a ten-month backlog in meeting orders for heavy maritime propulsion systems, while the New London Ship and Engine Company, a civilian shipbuilder established at Rice's instructions some years earlier, had moved to larger facilities in the town of Groton, Connecticut, where it was struggling to produce the first in a series of large cargo vessels.

Like many men whose achievements and personalities cast a shadow over their immediate associates, Rice devoted little time and less energy to grooming a suitable successor. Elihu Frost might have met the bill had he not been long gone. Struck down by a fit of disenchantment in 1909, he had sold off his holdings and retired to a quiet life with his family. Those who remained were a competent but unspectacular lot. Spear, the most prominent member of the new ruling clique, was invaluable for his appreciation of naval aesthetics and his old-boy links to the Bureau of Construction, but he remained a far from impressive figure in the boardroom. Henry Sutphen, Elco's chief executive, was little better. Another highly competent technician, he was often to be dragged beyond his depth by questions of finance, political considerations, and long-term corporate planning. The chief repository of business wisdom, such as it was, belonged to the new chairman, Henry Carse. He had joined the board in 1912, one of several "bright young men" whom Rice had recruited when his health began to fail. Carse had spent his earlier career in banking and knew next to nothing about submarines. At that moment, however, technical competence was not a qualification. Rice had laid down the basic path along which he believed the company should proceed. The pieces were in place; all Carse had to do was hold the line.

Even with their flaws, the members of the new regime would have been hard-pressed to make an immediate mess of their situation. Things were simply too good. The U.S. Navy alone had at least half a dozen boats on order at any one time; there was the flood of royalties from overseas, the insatiable demand for Electro Dynamic's motors and New London's diesel engines. The U.S. Navy's change of heart was entirely appropriate given the other good news. The admirals still had not quite managed to contain their distaste for submarines, but at least now they were ordering them without a fight.

And more tantalizing than any of the other cheerful portents, there was the war. Like most other major U.S. corporations, Electric Boat had already become a beneficiary of the conflict—the upturn in civilian shipbuilding being but one example of the war's largesse. Significant though this ripple effect may have been, it was as nothing compared to the limitless bounty available to any company able to find a way around the existing neutrality laws. It was something of a tall order. By any reasonable reading, the relevant statutes appeared to impose a blanket ban on the sale of all war-related goods to belligerents. The last of Isaac Rice's parting gifts was just what was needed—a plan hatched with Bethlehem Steel's Charles Schwab that promised to provide a loophole in the law through which it would be possible to sail a flotilla of submarines all the way to England.

The scheme was born not long after the first shots were fired, when Schwab called on Rice to discuss what amounted to an unofficial alliance between his own company and Electric Boat. Unlike many of his contemporaries, the steelman did not believe the war would be over by Christmas, nor was he prepared to accept the common wisdom that the belligerents' own immense arms industries would be able to meet the insatiable and prolonged demand for weapons of all kinds. As he saw it, the war amounted to an open invitation for Bethlehem to take its place alongside Vickers and Krupp as one of the world's largest industrial conglomerates. The hurried round of conferences with Rice was a small but key element of his master plan.

Although far from the largest American steel producer, Bethlehem was better placed than its larger rivals to make the most of the war. It was already the country's largest supplier of armor plate, and had gained much recent prestige from the sale of a consignment of large naval guns to the imperial German fleet, the first foreign-built weapons to break the Krupp family monopoly in more than twenty years. There was, in fact, only one area in which Bethlehem was ill-equipped to begin immediate, unrestrained war production: submarines. Electric Boat still controlled all of John Holland's principal patents, as well as several hundred innovations perfected since the inventor's departure. Since Schwab was certain submarines would play a much bigger part in the conflict than was generally predicted, Rice's pledge of cooperation and support was vital.

The meetings were brief and to the point; there was no need for preliminary pleasantries between two such hard-edged businessmen. Rice had known Schwab for at least ten years and recognized him as a man whose opinions were worth heeding. In 1907, less than four years after Schwab brought Bethlehem into being from the sound remnants of a bankrupt industrial conglomerate, Rice installed Schwab's Fore River shipyard, in Quincy, Massachusetts, as the company's chief subcontractor in charge of hull section fabrication and assembly. The work had formerly been handled by Lewis Nixon's shipyard and a change was definitely in order. Nixon had always been a true and loyal friend to John Holland and he took a dim view of events leading up to the inventor's forced retirement. Fore River was not only larger and better equipped to deal with the demands imposed by increasingly sophisticated submarine designs; it was also devoid of the bitter emotional baggage that had soured Spear's relationship with Nixon and his chief foremen. The union had been a complete success. Schwab reduced the construction period through the limited application of mass production techniques and established an enviable record for on-time deliveries. In view of their past successes, Rice had every reason to accede to Schwab's demands.

By the time the two men concluded their talks, they had established a relationship that was to endure until the war's end. By common accord, Schwab's Fore River shipyard and his West Coast facility, the Union Iron Works in San Francisco, were guaranteed the exclusive right to serve as prime contractors for the Royal Navy. In return for its pledge of cooperation and the promise of exclusive, unrestricted access to Electric Boat's patents, Rice signed on as Bethlehem's sole supplier of engines, electrical systems, motors, gearboxes, custom-fitted equipment, and specialist labor. As a further indication of his approval, Rice promised to promote Schwab's cause through Electric Boat's contacts with Vickers and its network of friends among the Admiralty's uniformed and civilian administrators. As for the problems at home, both parties agreed that the Wilson administration's inevitable objections would have to be dealt with as, and when, they arose.

Electric Boat's decision to become Schwab's junior partner represented a marked departure from all previous norms. As Simon Lake might have testified, Rice was a ruthless campaigner, seldom willing to settle for second best. In this case, however, there were any number of additional reasons why Schwab's appeal received such an immediate and sympathetic hearing. The first was a simple matter of physical stamina. Before the contracts with the Admiralty were finalized in January 1915, Schwab had crossed and recrossed the Atlantic six times in less than twelve weeks, pausing only long enough in New York between voyages to repack his bags for flying visits to Washington and Montreal. Increasingly infirm, Rice could not have stood the breakneck pace without risking a total physical collapse. Schwab's involvement also promised certain distinct advantages in both companies' future dealings with the United States government. The fact that he hoped to sell not just submarines but surface ships, artillery pieces, and armour plate provided his argument with the advantage of considerable economic clout, particularly among the many members of Congress and the senior bureaucracy who sympathized with the view that American industry was entitled to reap the maximum advantage from the military disease that held half the world in thrall. Armed with Rice's pledge of unconditional support, and his lawyer's opinion that the federal neutrality laws could be circumvented through the invocation of several obscure precedents, Schwab set sail from New York in mid-October aboard the liner *Olympic* for a meeting with Lord Kitchener and other senior members of the British war cabinet. It was a most eventful voyage.

Six days out of port, his ship was steaming through a rising swell off the southern Irish coast when the British battleship *Audacious* loomed out of the mist to escort the liner on the last leg of its journey into port. The warship's vast gray bulk must have soothed those passengers worried about a German attack, but whatever its palliative effect, the sense of security was short-lived. Some hours after the ships' rendezvous, the atmosphere of post-luncheon lethargy that had settled over the *Olympic*'s drawing rooms and salons was shattered by a tremendous explosion that sent many passengers racing to their lifeboat stations. There, no doubt to their immense relief, they saw that it was not their ship but the *Audacious* that had come to grief. The warship had struck a mine and now lay dead in the water, its bows peeled back like a stubbed-out cigar by the force of three thousand pounds of high explosives. Every attempt to take the stricken battleship in tow was foiled by the mounting seas, and two hours later it went to the bottom before the silent gaze of its own officers and crew, now jostling with the *Olympic's* passengers for vantage points at the liner's rails.

Any attempt to describe the Admiralty's reaction as one of intense embarrassment would be a gross understatement of the fact. The sea lords were horrified, less perhaps by the loss of a modern and formidable warship than by the certain expectation that the *Olympic*'s passengers would spread word of the sinking throughout the country.

Fate ordained that the man in whose lap the dilemma fell, Sir John Jellicoe, was temperamentally unsuitable for such a delicate task. At the later Battle of Jutland, the largest and most inconclusive of the war's set-piece naval stalemates, Sir John's fleet demonstrated a profound reluctance to close with the enemy, preferring to stand off from direct engagements and grope from a distance for openings that seldom materialized. His response to the problem posed by the *Olympic*'s passengers—many of them U.S. citizens immune to the strictures of British security laws—produced another, entirely characteristic bout of prolonged dithering. As if awaiting divine inspiration, Jellicoe ordered the ship to an isolated mooring station off Southend Harbor, where the astonished passengers and crew were informed they would be held under virtual house arrest for perhaps as long as a week or more.

The prospect of incarceration threw Schwab into a fury, not least because he had become the victim of his own taste for cloak-and-dagger melodramatics. Although he had an open invitation to call on Lord Kitchener at any time after October 20, no one in London knew that he was coming, since his tickets were booked under an assumed name and he had remained incognito throughout the voyage. Seething with rage at the ship's officers' refusal to pass on such an improbable tale to the captain, Schwab was forced to spend the best part of a day bombarding the bridge with a stream of notes whose tone reflected their author's impotent desperation. When the captain finally succumbed to his bluster and agreed to contact the appropriate authorities, the reply was both immediate and surprising. No less a person than Sir John Jellicoe himself came out to the liner, where he quizzed the purported tycoon before summoning an official car and driver to transport him to London. The less famous passengers were not so lucky. They were obliged to remain aboard for a further forty-eight hours until Jellicoe came reluctantly to the conclusion that nothing short of mass murder would guarantee their silence. As it happened, he need not have worried at all. When the Allies examined captured German naval intelligence estimates at the end of the war, they found that the *Audacious* was, at least in the minds of the Kaiser's admirals, still afloat. This was less a tribute to Jellicoe's ultimate solution—swearing the passengers to silence before allowing them ashore—than an indictment of the German espionage service. Although news of the loss was widely reported in the U.S., German intelligence remained skeptical, cautiously listing the ship as crippled rather than sunk.

Despite the dangers and vexations of the voyage, Schwab could not have asked for a more promising start to his sales campaign. As the millionaire and the admiral rode toward the shore in a navy launch, Jellicoe spoke candidly of the mounting U-boat threat. The mine that destroyed the *Audacious* had been sown by a submarine, he explained, adding that the Royal Navy was singularly ill-equipped to deal with this unexpected menace. The problem, at least as he saw it, demanded a twofold solution. There was first a chronic need for a new class of small, fast patrol boats, "submarine chasers," that could be armed with depth charges, torpedoes, and the other primitive antisubmarine weapons of the day, like steel mesh trawling nets and grappling hooks to haul the raiders to the surface. The second, and more important, of Jellicoe's remedies marked the first official endorsement of a doctrine still very much in vogue with modern naval strategists—the notion that a submarine's worst enemy is another submarine. The statistics of two world wars make this a dubious proposition at best, but at the time, the idea had a certain appeal. Winston Churchill, the Royal Navy's chief civilian overlord in the war cabinet, evidently thought so, because he had recently ordered a fivefold increase in the size of the underwater fleet. The trouble, Jellicoe continued, was that the war might well be over by the time the first of the new boats would be ready to put to sea.

This was another occasion on which Schwab may have felt inclined to mutter a brief word of thanks for the heaven-sent good luck that had marked every stage of his life and career. When still little more than a boy, he had gone to work as a dollar-a-day general hand in one of Andrew Carnegie's steel mills; within twelve years, he had risen to the second most powerful position in the company, with an annual salary in excess of $200,000 a year. He remained with Carnegie until 1901, when the Scottish industrialist astonished America by renouncing his business empire to concentrate on uplifting acts of international generosity. Schwab, his personal worth now well in excess of $20 million, struck out on his own. Bethlehem Steel and its family of eight principal subsidiaries engaged in shipbuilding and heavy manufacturing was the result. Now, as the British admiral continued a pukka, ever so proper recitation of the Royal Navy's woes, the former foundry assistant recognized another stroke of good fortune in the making. As usual, he moved quickly to exploit it.

Schwab began by suggesting that his shipyards might well provide an answer to the Navy's predicament. If the Admiralty was prepared to make the financial incentives sufficiently attractive, Schwab boasted that his workers would be able to build and launch a submarine in six months, and possibly less. Jellicoe was more than a little skeptical. Neither Vickers nor any of the minor British submarine yards had ever managed to complete a boat in less than fourteen months, the average construction period being closer to two years. By the time Schwab's bags were stowed in the Admiralty staff car, Jellicoe's initial doubts had been completely dispelled by the steel king's brash confidence. As they parted on the dock, Jellicoe urged Schwab to conclude his business with Lord Kitchener as soon as possible and then make haste to the Admiralty, where he could expect a private audience with the British Navy's aged but indominable first sea lord, Sir John (Jacky) Fisher.

Schwab did as he was bidden and found Fisher even more receptive than Jellicoe had promised. Rather than the ten long-range boats Jellicoe had predicated, Fisher placed an immediate order for twenty H-class submarines, a modern but thoroughly tested Electric Boat design that had already seen extensive service with the U.S. Navy. So delighted was the first sea lord, he raised not one objection when Schwab set the price of each boat at roughly half a million dollars, twice the usual price. Schwab's coup was further enhanced by a list of conditions that Fisher attached to the agreement in the mistaken belief that Bethlehem would prove unable to meet the accelerated deadlines. For every week a boat went undelivered after the six-month cutoff date, Schwab agreed to reduce his price by increments of five thousand dollars. The steelman replied with a counterproposal of his own. He pointed out that since Bethlehem would be obliged to honor the provisions outlined in the penalty clause, it was only fair to expect a corresponding system of bonus payments for early deliveries. Fisher innocently agreed, approving Schwab's demand for additional payments of ten thousand dollars per week prior to the six-month deadline. The deal was sealed with a handshake and Schwab was taken to the Admiralty's cipher department, where he encoded a message for his chief aides in the United States. Received and decoded at the British embassy in Washington, the messages were relayed by courier to the Quincy shipyard and Electric Boat's headquarters in New York.

The size of the agreement and the speed with which it had been negotiated buoyed Fisher's spirits long after his guest had departed. Several hours later, he penned a brief note to Jellicoe in which he praised the brash American's cocksure style and concluded with a gloating reference to his own part in the negotiations: "It is a gigantic deal, done in five minutes. That is what I call war!"

Schwab signed the first, formal documents on November 10, stopping off at the Admiralty on the way back to the ship that would return him to New York. The contracts differed only slightly from the basic agreement thrashed out with Fisher one week earlier, and those few alterations weighed heavily in Bethlehem's favor. As with all its wartime armaments orders, the British government reserved the right to cancel the agreement in the unlikely event that peace broke out. That liability was balanced by a generous concession approved by the civilian negotiators with whom Schwab had dealt after the first meeting with Fisher. In the event that "unusual circumstances or diplomatic considerations" forced either party to abandon the scheme, Bethlehem was to receive reimbursement of all costs plus 20 percent. Unique by Admiralty standards, the rider was a direct response to the very real possibility that the interpreters of America's neutrality laws would be unmoved by Schwab's jesuitical interpretation of the relevant statutes.

The obstacles would have daunted a less determined man. At a glance, the various federal laws regulating the international sale of arms in times of conflict gave the appearance of a monolithic doctrine: American ports were not to be used as bases or supply stations for belligerent vessels, nor were U.S. shipyards and manufacturers permitted to sell or supply arms likely to be used against a sovereign third party.

Among the many messages originating from the Admiralty code-room, the most important was dispatched to James H. Hayden, Bethlehem's Washington lawyer. Schwab ordered him to drop by the State Department and sound out reactions to a musty legal argument originally conceived by Isaac Rice during the Russo-Japanese War. It will be recalled that having sold the ill-fated *Fulton* to the Russians, Electric Boat was confronted with the problem of smuggling it out of the country. As his first step, Rice ordered the removal of the *Fulton*'s battery banks. This was done for no practical reason but as a matter of legal necessity. Rice reasoned that if the *Fulton* was not in full operational order, it could not be defined as a warship, at least not under the existing letter of the law. Moreover, since the boat would require at least seven days' preparation in Russia, Electric Boat could not be accused of outfitting a potential commerce raider. This last was a particularly important consideration, since it was just such a situation that had inspired Congress to draw up the first neutrality laws during the mid-1870s. The legislation followed the Treaty of Geneva, in which Great Britain agreed to pay Washington roughly $15.5 million in compensatory damages for shipping destroyed during the Civil War by the Confederate raider *Alabama,* a British-built ship that had been purchased and outfitted in England. Westminster agreed to the settlement in the entirely reasonable expectation that any future war might see Britannia's enemies wielding weapons manufactured in the United States. Ironically, on that early December morning, as Hayden strode down the marble corridors toward the office of Under Secretary of State Charles Lansing, the situation had been neatly reversed. Instead of prohibiting the flow of weapons and ships to Britain's enemies, the neutrality laws now threatened to stop the shipment of Schwab's submarines to the Royal Navy itself.

The fact that Hayden was calling on Lansing rather than Secretary of State William Jennings Bryan was yet another example of Schwab's habitual good fortune. The thrice-defeated Democratic nominee for the White House was resting at his home in Lincoln, Nebraska, after a campaign swing throughout the Midwest on behalf of his party's congressional ticket. It was just as well, because the once-dubbed "boy orator from the River Platte" would have taken a dim view of Hayden's argument. An isolationist and cracker-barrel populist who would later end a long and often laudable career on an ignominious note by leaping at the chance to lead the prosecution in the celebrated Scopes "Monkey Trial," he may well have chosen the occasion of Hayden's visit to deliver one of his messianic diatribes against the autocratic warmongers of the European nobility. Hayden could have expected no less. Like many of Washington's more refined citizens, the lawyer probably shared the opinion of Senator Henry Cabot Lodge, who had once likened Bryan to the Platte, "a stream 1250 miles long with an average depth of six inches and a wide mouth."

Lansing was likely to be far more sympathetic. He was widely regarded as a circumspect supporter of the British cause, although, in this particular instance, his leanings would have been insufficient in themselves to win Schwab a favorable verdict. As far as Hayden was concerned, Lansing's chief appeal lay in his reputation as an impartial and incisive arbiter of complex legal arguments. Since Schwab's case rested equally upon legalistic hair-splitting and the inadequacy of existing definitions, it was best presented to an adjudicator whose opinion would not be weighed down by personal and political prejudices. Xenophobic in his view of the world beyond America's shores, Bryan would have been unable to consider the case on its legal merits alone.

The meeting, brief and to the point, matched Schwab's brightest hopes. After no more than a few minutes' quiet reflection, Lansing opined that the sale and shipment of submarines to a belligerent would be permitted if the boats were exported in sections. This was precisely what Schwab wanted to hear. Although the initial contract had been vague, promising only that the submarines would be handed over to the Royal Navy as soon as they were launched, Schwab had since introduced several refinements. Now, instead of launching the boats in the United States, he planned to have them completed, cut into sections, and shipped to a British port in either England or Canada, where they could be easily reassembled. This fell within Lansing's personal interpretation of the law, and Hayden, thanking the under secretary for his time and trouble, returned to his office, where he informed Schwab's aides in New York that there was now no obstacle to be overcome or circumvented. Schwab himself, returning to America aboard the liner *Adriatic,* received word of the decision in a coded cable within hours of the meeting's conclusion.

Lansing's interpretation of the law galvanized Electric Boat and its Groton work force. The next day, Lawrence Spear informed his foremen that the company had received "a very substantial order" for submarine component parts from an unnamed purchaser. Forty large-capacity marine diesel engines had to be built within the next four months—two for each of the twenty boats on order—as well as a corresponding number of marine electric motors, battery systems, and circuit boards. This staggering volume of new work, the largest single order in the company's history, was rounded out by a final list of expensive extras like the torpedo room fittings and countless precision valves. Farther north, at

Schwab's Fore River yard, the consequences of Lansing's decision were even more apparent. Preparations were made to lay the keels for four submarines, as many as the yard could accommodate comfortably at one time, and a recruiting campaign was launched throughout New England to find the workers essential to the plan's success.

One of the more peculiar aspects of the affair was the ill-fitting cloak of secrecy in which Schwab and his cohorts at Electric Boat sought to conceal the details of an expanding scandal. Of course, it was a forlorn hope from the beginning. In London, Schwab had been tailed by a permanent corps of British and American pressmen, who noted his frequent trips to the Admiralty and Whitehall and drew their own conclusions. By the time the *Adriatic* docked in New York, the story had been plastered over every front page in the country. Many of the reports were comically wide of the mark, some even speculating that Schwab had agreed to devote his entire industrial empire to the production of war matériel for Britain. These wild flights of fancy were more than balanced by some very accurate reports in the *New York Times* and the *Wall Street Journal.* The *Times* gleaned the bare bones of the deal from an unnamed executive in the steel company's hometown of Bethlehem, Pennsylvania, as well as its reporters' own observation of the frenzied activity taking place at Groton, Quincy, and many other manufacturing towns along the New England coast. The *Journal*'s accounts filled out the gaps with explanations of the political ramifications and a surprisingly accurate estimate that put the value of the deal at $25 million. The only thing missing was an official confirmation, and on November 16, the day the *Adriatic* docked in New York, a throng of reporters gathered at the foot of the first-class gangway to get it.

The steel tycoon admitted the substance of the submarine story without saying very much else. He refused to confirm the *Journal*'s essentially accurate appraisal of the sums involved, nor would he even identify Great Britain as the purchaser. Silence would have been infinitely preferable. Certainly, Schwab was entitled to take heart in Lansing's opinion that the sale did not contravene any laws, but in the more emotional court of public opinion, the future of the deal was fraught with uncertainty.

Bryan's reaction to the news was predictable. Fulminating against the machinations of lawyers, financiers, and the British, he rumbled out of his hometown aboard the first eastbound train. As with most of Bryan's undertakings, this latest assumed the tone of a moral crusade. Lansing made repeated attempts to explain the legal technicalities, all to no avail. For more than thirty years, Bryan had considered himself the anointed mouthpiece for righteous, right-minded, God-fearing America, and he was not about to let legal chicanery propel his country into Europe's squalid war.

Late in November, after two weeks of impassioned intramural debates, Bryan stalked into the White House for a private meeting with President Wilson. Unlike his under secretary, Bryan made no reference to precedents; indeed, he scarcely mentioned legal arguments at all. Instead, he appealed to the President's sense of fair play, concluding with a reminder that the administration was bound to defend the spirit, as well as the letter, of the law. The next day, having given Lansing his chance to put the opposing view, Wilson declared Bryan the winner. Schwab's $25 million deal would have to be scrapped—or so it seemed.

At this point, it is necessary to break away from Washington and shift the focus of attention to Wall Street, where early rumors of the submarine sale had pushed Electric Boat's common stock to a new high of just under thirty dollars. That changed dramatically when Wilson handed down his verdict. Electric Boat dropped more than five dollars in a day, while Schwab's Bethlehem lost three times that figure. Even worse, in Electric Boat's case the damage seemed likely to spread.

When the initial outbreak of frenzied bidding for munitions and war-related stocks subsided a month or so after the outbreak of hostilities, many analysts took a closer look at Electric Boat and concluded that the stock price had risen well beyond its true worth. With the exception of its patents, the company had little in the way of hard assets. Most heavy construction was still done on the premises of subcontractors, and much of the machinery was leased. Shareholders had not seen a dividend since 1909, and while there was reason to believe the situation was about to change, the consensus was that the figure would not exceed 2 percent. As for the near and long-term future, the analysts' perceptions of Electric Boat were clouded by the same misconceptions that had led many Europeans to conclude the war would be over in a matter of months. Food, chemicals, shell casings, horses, and boots were the hot items being hawked and haggled over in the better New York hotels. By the time the first of the war-ordered submarines hit the water, the fighting would be long finished.

Electric Boat's junior partnership with Schwab postponed further critical scrutiny. But now that the relationship appeared to have foundered, the norms of the marketplace indicated an imminent outbreak of profit taking that might drive the stock down to levels only a few dollars higher than at the time war was declared. That nothing of the kind took place testifies to Schwab's determination, not to mention his complete disdain for diplomatic niceties. By the time the British submarine affair had run its course, the conspirators' resolve to ignore the White House had prompted an acrimonious exchange between Canada and Britain, and honed Germany's misgivings about American "neutrality" to a bitter cutting edge. Along the way, Electric Boat became the hottest investment in the country, with stock price gains amounting to an increase of roughly 2,000 percent over twelve months.

If Bryan had been capable of learning from past defeats, he would have been wary of such an easy victory. Only a few months had passed since he won, and lost just as quickly, a battle to prohibit American banks from issuing loans to belligerents. On that occasion, the bankers simply sidestepped the ban by replacing conventional loans with "short term commercial credits." Now, as Washington prepared for the festive season, it was Schwab's turn to outflank the ponderous Bryan.

Meanwhile, the secretary of state rejoiced in the success of his crusade. Instead of ordering an aide to relay the news to Schwab in New York, Bryan spoke personally with the millionaire by telephone. It must have been something of an anticlimax. There were no protests from the other end of the line, nor even the slightest indication of annoyance. Schwab conceded defeat with sportsmanlike grace and patriotic dedication. If the President himself believed the proposed sale would work against America's best interests, who was Schwab to dissent? Bryan had no cause for further concern, Schwab continued. Both Bethlehem and Electric Boat would henceforth abandon all plans to build foreign warships in the United States.

Untroubled by Schwab's easy and uncharacteristic capitulation, Bryan convened a press conference to put the President's decision on the record. "This closes the submarine incident," he noted at the foot of his official statement. Had he known of Schwab's activities over the previous three days, he would not have been so confident.

Like any businessman with extensive government contracts, Schwab maintained a well-oiled intelligence network in both the White House and the upper levels of the public service. Long before Bryan had left for his appointment in the Oval Office, the steel magnate realized that the deal, at least as originally negotiated, could not now survive the inevitable White House opposition. There was, however, an alternate solution, and on the evening of December 3, 1914, Schwab left New York on the overnight express for Montreal to pursue it. Midmorning found his five-man party, two of them technical advisers on loan from Electric Boat, surveying the city and the Saint Lawrence River from a rocky outcrop on the outskirts of town. Directly beneath him, in a grimy sprawl of noise, dirty buildings, and antlike human figures, the largest shipyard in Canada was coming to life with the arrival of the morning shift. Schwab liked what he saw. Most of the slipways and construction areas were unoccupied, and the one vessel under way, a large icebreaker on order to the Canadian government, appeared to have reached the stage where it could be pushed to one side in deference to more urgent business. However, it was the name over the front gate that was the most important of the yard's attributes. Canadian Vickers was a wholly owned subsidiary of the British arms firm, and Schwab realized its managers would do as they were instructed by the head office. All he had to do was win the Admiralty's approval, and total control of the shipyard would be as good as his.

Having stayed in Montreal only long enough to swear the yard's senior executives to secrecy, Schwab returned to New York, where he arrived just in time to take Bryan's call. The millionaire was not trifling with the truth when he promised no British submarines would be built in the United States. They were to be built in Canada instead.

Schwab probably never stopped to consider the diplomatic consequences of his scheme, and the terse exchange of official cables between London and Ottawa that followed would have left him unmoved. For more than ten years, successive Canadian political figures had been urging Britain to help with the establishment of a Canadian shipbuilding industry. All the pleas went unanswered. When Canadian Prime Minister Robert Borden heard about the Schwab plan, he was furious. Without his government's knowledge or permission, the much-needed icebreaker had been moved to a corner of the yard. And that was not all that riled him. In the worst insult of all, the task of prefabricating the submarine sections had gone to a syndicate of U.S. firms. The British were apologetic but unmoved, and the submarines continued as before.

The following day, again traveling incognito, Schwab left New York aboard the *Lusitania* for another round of meetings with Fisher and several senior Vickers executives. The Englishmen accepted his plan without reservation, approved some minor changes in the contracts' penalty provisions, and granted a two-week extension on the original six-month deadline. Four days later, Schwab was back at sea. Arriving in New York on December 23, he was, as usual, surrounded by pressmen. Was it true that the submarines would now be built in Canada? "Nonsense," Schwab replied, adding that he had no facilities in Canada capable of performing that kind of work. As for questions about the "secret" trip to Montreal and the possible involvement of Canadian Vickers, Schwab professed total ignorance. If Canada and a Canadian company wanted to build submarines, they had a perfect right to do so. After all, the country was at war.

Once again, though few people would have believed it, Schwab was sticking pretty much to the truth. At the London meetings, the Admiralty and Vickers had agreed to transmit all payments through the Canadian subsidiary, which in its turn would obtain all the essential component parts through Bethlehem and Schwab's family of U.S. manufacturers.

The days following Bryant's announcement and Schwab's subsequent departure for London had seen some curious goings-on at Groton,

Fore River, and the Union Iron Works in San Francisco. Far from slowing down, Electric Boat's work on the forty diesel engines had picked up pace and was now proceeding around the clock. At Fore River and San Francisco, the activity was even more pronounced. Six submarine keels laid in the weeks that followed Schwab's first meeting with Fisher were dismantled, ostensibly because the sale had been canceled. In reality they were taken in sections to a large enclosed shed and packed into crates emblazoned with the lettering "Prefabricated Bridge Girders—Canadian Vickers."

As with most of Schwab's deceptions, this latest was widely regarded as a ruse. The Navy inspector of machinery who was on permanent duty at Groton sent a long, detailed memo to the office of naval intelligence, in which he passed on the rumor that one hundred or more workers and supervisors would soon be taking up indefinite residence in Canada. The yard was also subject to regular visits by British naval personnel, some of them in uniform, who took a proprietary interest in work under way in the machine assembly plant. On the West Coast, another U.S. naval inspector peeked inside one of the crates and wondered, tongue in cheek, what new school of architectural design dictated conning towers on bridges. The papers, too, were not long in catching on, splicing together reports from Montreal with information gleaned from Schwab's workers and executives. Yet despite this apparent violation of Bethlehem's promise to withdraw from the deal, nothing was done. Once again, Schwab had Lansing to thank.

After returning from Montreal, Schwab sought the help of Paul Cravath, a New York lawyer whose pervasive influence in the worlds of politics and business warrants comparison with the omniscient Mr. Jaggers of *Great Expectations.* The cofounder of Cravath, Swaine and Moore, still one of America's best-known and most expensive law firms, he was the only man Schwab trusted to promote this latest variation on the original submarine agreement. Cravath advised Schwab to continue his Canadian exports while the lawyer prepared the legal groundwork and waited for an appropriate moment to unveil his argument in Washington. The opportunity came three weeks later, when Bryan returned to Nebraska to recover from influenza and left Charles Lansing to carry on in his place.

No doubt aware that he might be required to tread on very thin ice, Lansing dispensed with the customary stenographer and arranged for Cravath to call at the State Department late on a Saturday morning. The lawyer had come seeking approval for his own, inspired interpretation of the law, and once again, the argument revolved around a question of definitions. Like his client, Cravath conceded the President's right to veto the export of prefabricated submarines. What troubled him, he said, was the degree of completion that constituted prefabrication. While it was wrong to export hull sections with their internal fittings already in place, would the same strictures apply if the hull pieces were dispatched to some foreign assembly point in one crate and the engines in another? Unlike Rice's earlier removal of the *Fulton*'s battery banks, the absence of an engine, a periscope, or any other integral component would render a vessel totally useless. And if a submarine had no military value at the time it left U.S. territory, common sense decreed that Bethlehem could not be accused of outfitting a belligerent warship.

Lansing agreed, although he appears to have had mixed feelings, because he declined to put his opinion in writing. All the same, from Schwab's point of view, the verbal assurance was more than enough.

At this point, Schwab must have been expecting Bryan to come thundering out of the West, a reasonable assumption after the success of his November offensive. Nothing of the kind took place, and that remains the episode's most mysterious feature.

It may have been that the secretary of state was preoccupied on other fronts and decided to direct his energies where they would do the most good. His position and his influence within the administration were both deteriorating (less than eighteen months remained before Lansing would become his permanent replacement). Or perhaps, as Disraeli once said of Gladstone, Bryan was merely a spent volcano spluttering through his final years in a series of irregular explosions punctuated by longer periods of exhausted dormancy. Whatever the reason for his compliance, it stretches credulity to assume he had been won over by the logic of Cravath's argument. Any man who could reject all Darwin's theories and observations in preference to the Book of Genesis would seem incapable of appreciating the academic subtleties of such a finely drawn point of law.

The absence of opposition from the White House is a little easier to understand. After receiving a series of protest notes from the German and Austrian ambassadors through the last weeks of January and early February, President Wilson turned to Lansing for enlightenment. With Cravath directing the legal architecture of the defense, the facts and documents upon which Lansing was able to draw provided a near-seamless shield. Schwab had obtained a letter from Canadian Vickers' general manager, P. L. Miller, that piled lies and half-truths one upon the other. "We are making no purchases from the Bethlehem Steel Company and our purchases from its subsidiaries are comparatively insignificant," the letter said in part. Lansing's own explanation of the legal issues was also most persuasive. After making a diplomatic suggestion that recent events might warrant the introduction of fresh legislation, he concluded that Schwab's venture did not now contravene any existing laws.

Of course, Wilson could have pressed the matter, but he chose to push it from his mind. It was typical of the ambivalence that marked his approach to the question of American neutrality up until the declaration of war in 1917. At the conflict's onset, he had urged his countrymen "to be impartial in thought as well as action," yet his own sympathies lay with Britain, which he regarded as the least offensive of the chief protagonists. Other examples of Wilson's selective inconsistency are not hard to find. He had approved the ban on loans to belligerents while permitting "short term commercial credits" that gave the U.S. a vested interest in a British victory to the tune of more than two billion dollars' worth of outstanding debts. Wilson's advisers were similarly disposed toward Britain. His close friend and roving diplomatic adviser, Colonel Edward House, certainly favored England, remarking at the time of the original submarine agreement, "Bryan might cause trouble" and that "he should be treated like a small child." Now that Lansing and Cravath had cleared away the obstacles and provided at least a pretense of respectability, Wilson evidently found it more convenient to accept their advice without raising any fresh arguments of his own.

Electric Boat played very little part in the active maneuvering. The legwork, the lobbying, and the logistics of production were all supervised by Schwab and associates like Cravath—"my boys," as the millionaire sometimes called them. But now, as news of the favorable turn of events began to reach Wall Street, it became the submarine builder's turn to take center stage. In short order, Electric Boat became first a stock to watch, then an impressive performer, and finally a self-fueling phenomenon. At the start of 1915, a single share could be purchased for around $20. By December it had gone as high as $420, been subjected to a ten-for-one stock split, and prompted such frenzied trading in the New York Stock Exchange's old Curb Market that one observer wrote of "persons of large means and much knowledge" shouting each other hoarse in bidding battles "for parcels of no more than ten shares." Along the way, in what was to be his last major decision, Isaac Rice instigated a new and short-lived change of name. For the next few roller-coaster years, Electric Boat became the Submarine Boat Corporation.

Rice was reputed to have made two million dollars when he began liquidating his holdings in the old Electric Boat back at the start of the spring. It was an extraordinary profit in an era when a respectable apartment in Manhattan could be rented for thirty dollars a month, but it was far from the gain he might have realized had he hung on a little longer. On the day the *Lusitania* went down, May 7, 1915, Electric Boat had risen to a healthy figure in the mid-seventies. Six weeks later, carried along by what can only be described as an outbreak of submarine mania, the price went surging past $275, with no indication that the peak was anywhere in sight.

It was the *Lusitania* more than anything else that started the stampede. Submarines went from a vague threat to a public fixation. They were denounced as the weapons of cads and cowards, as the murderers of women and children. And of course, the entire topic of underwater warfare became all the more fascinating as a result. The papers were full of stories about submarines—how they might be built, avoided, or destroyed, depending on the author's point of view. The mania even reached to the upper levels of the Navy Department. In June, for example, Navy Secretary Josephus Daniels took the evening train to Menlo Park, New Jersey, for an evening conference with night-owl inventor Thomas Edison. Daniels told reporters that he had turned to Edison in search of guidance. "Bright, intelligent innovators" were needed to set Navy policy and oversee procurement decisions. Who better than the great inventor, he asked, to identify the weapons that would protect American sovereignty and the safety of its ships on the high seas. The newspapers were quick to speculate about the other marvelous machines and secret weapons that might have been discussed. Predictably, submarines figured prominently among the more colorful fancies. Secretary Daniels was said to be considering a fleet of twenty titanic submarines that would carry cargo across the Atlantic beneath the German surface blockade. According to the press reports, Edison went one better by suggesting a heavy cruiser that could submerge and surface at will. Given Edison's fine technical instincts, it is safe to assume that idea originated elsewhere.

Daniels did obtain two pieces of good advice from his trip to Menlo Park, one of which proved to be a boon for Electric Boat. Edison advised that the Navy buy a standard seaplane, presumably to scout for submarines, and he also gave top priority to an accelerated U.S. submarine program. "We should develop the best submarine engines in the world," he said. Electric Boat's stock surged again.

Five years earlier, Isaac Rice solved one of the Holland boat's worst flaws by replacing the original gasoline engines with less volatile diesels. In true Rice style, rather than simply buying the engines, he purchased the American rights from the engine's German inventor and set up a company near the village of Groton in Connecticut to manufacture it and other submarine and ship components. By mid-1915, when Edison made his suggestions, the payroll was up to fifteen hundred men and growing constantly. Electric Boat's rising star twinkled ever brighter.

The facts of Electric Boat's happy situation were responsible in part for its newfound glamour, but they were far from the only forces behind the meteoric rise. As with any runaway bull market, rumors were of prime importance. There were some, for example, who insisted Electric Boat was the target of a ham-fisted takeover bid mounted by British agents who showed their hand too early and started the stampede. There was a strong line of logic, and quite possibly some truth, to the theory. Electric Boat had always enjoyed a particularly close if not always harmonious relationship with Vickers and it did control the world's largest collection of submarine patents. With America still neutral, it would have been entirely reasonable for the Admiralty to consider the company a desirable purchase, if only to block any German moves in that direction. The *New York Times,* among others, was convinced that ownership had passed into British hands, and informed its readers that "a majority of stock is believed to be held in a voting trust by English capitalists." Other reports from Wall Street pursued the same line, in some cases asserting that Rice had sold his holdings to an Admiralty nominee. If the company was being run from Whitehall, there was no obvious evidence other than the lack of evidence. Although he had liquidated most of his and his family's stock, Rice continued to serve as the company's chief executive, while all his old directors retained their seats on the board.

The other, most frequently reported rumor about Electric Boat concerned its purported reorganization, an event speculators had been anticipating since early June. In just six months, Electric Boat had taken in more than $20 million worth of business, more than six times the volume of work it had handled in all of the previous year. For the first time since 1909, there were plans for a dividend, and there was also a lot of ambitious talk of branching out into other war-related areas. Elco was experimenting with a variation on one of its larger and faster motor yachts, a prototype for what would become the PT boats of the Second World War. The stories about the diesel plant at Groton were even more intriguing in their speculations that Electric Boat would use the division to oversee work on a large civilian shipbuilding venture.

Rice lived just long enough to see his company reach its zenith. On August 4, 1915, three months before his death, Electric Boat's secretary announced the long-awaited stock split: ten shares in the new Submarine Boat Corporation for every one in the old Electric Boat. There would never be a higher moment in the company's history. Inundated with work and free from scandal, it was, however briefly, the toast of Wall Street. At the end of the year, gross earnings totaled $5.6 million—five times greater than for 1914—and the next year they rose again, to around $7 million. And that was when the Rice magic began to disappear.

All the pieces Rice had put into place performed perfectly until 1917, when America entered the war. By any reasonable expectation, it should have been the start of an even more lucrative chapter; ironically, it was the start of the end. Tempted by the further riches to be obtained from Washington's war chest, Rice's heirs began to make a few alterations to the chess fanatic's grand strategy. Pretty soon, almost before anyone in authority noticed what was happening, the boom days slipped quickly into the past.

The seed of this, the company's first brush with disaster, was sown by a government authority called the Emergency Fleet Corporation, which had been set up at the onset of the war to replace the merchant ships going down in the North Atlantic and the Irish Sea. It was the opportunity to enter the civilian shipbuilding market for which Henry Carse, Rice's successor, had been waiting. With a surfeit of zeal and little thought for the consequences should hostilities cease, the Submarine Boat Corp. took out a lease on a large shipyard in Newark, New Jersey, and began work on the first order for thirty cargo vessels. Like the Liberty ships of the Second World War, they were built to a common, 5,500-ton design, usually with prefabricated hull sections that were mated each to the other on the construction slipways. The work was not complex but it was labor intensive, and the cost of all those new employees was soon reflected by the company's financial statements. In 1916, the various enterprises within the Electric Boat family of companies grossed $7 million in profits; the next year, weighed down by the mounting expense of the shipbuilding program, the figure fell to $2.6 million, and by the end of 1918 and the declaration of peace, the sum had dropped to a scant $1 million.

More sober minds might have paused for reflection in the weeks that followed the armistice, but Electric Boat chose to plunge ahead. The Newark shipyard had already robbed the company of millions of dollars in potential war profits, compromising what would otherwise have been a splendidly productive conflict. Now, despite a surplus of wartime cargo ships, Carse and his chief deputy, Lawrence Spear, decided the time was ripe to establish an international shipping line.

The invitation to embrace disaster originated with the Emergency Fleet Corporation, which celebrated the war's end by canceling its orders for all prefabricated cargo vessels. Electric Boat could have accepted the proffered option and simply walked away from the Newark yard, its disappointment mitigated by the lump-sum payment offered as compensation for the inconvenience and expense incurred by the sudden cancellation. Instead, they chose to gamble on the shipping line, and very nearly drove the company to bankruptcy.

There was a good deal of support for the scheme within the firm. In addition to Carse and Spear, there was the company's Paris representative, former Dutch submarine captain Paul Koster, who wrote to New York with predictions of an impending shipping crisis. Koster was a man whose opinion was treated with immense respect, at least on some issues. Among his other achievements, he was rumored to be a freelance intelligence agent, a man "known all over Europe as an international spy," in the words of a U.S. senator. It was the Dutchman's opinion that the Europeans would shortly be in dire need of new ships. In England and France, most industrial plants had been devoted to supplying the land war, and he reasoned it might be years before shipyards that had turned to building tanks and guns would return to maritime pursuits. If Electric Boat moved quickly, it might well grab a large section of the market for itself.

Carse and Spear, who both believed there was untapped shipping business to be had out of South America, moved quickly. Rather than accepting the Emergency Fleet Corporation's compensatory offer, they agreed to buy the unassembled hull sections for thirty-two cargo ships and have their own workers finish the job. A new subsidiary, Transmarine Corporation, was set up to operate the ships as they became available, and by 1919, the first were already at sea. Carse and Spear were effusively optimistic about the scheme, but their public pronouncements about its prospects did little good. Transmarine began operating in the red and never came close to declaring a profit. The ships had been designed to make three or four trips across the Atlantic at most, before being either sunk by the Germans or scrapped. They were uneconomical in peacetime and unpopular with the seamen who served on them. In 1923, after four years of mounting losses, the enterprise was wound up officially. By that stage, most of the ships had been taken off the high seas; cut down to barges, they were being used to haul low-paying cargoes up and down the Mississippi.

The collapse of the shipping venture would have been bad enough in itself. Unfortunately, it was only one of many problems. In a preview of the troubles that would beset the company a decade later, when U.S. senators responded to the disarmament movement by scrutinizing Electric Boat's international sales, Peru announced that it lacked the money to pay for three ships bought from Electric Boat immediately after the war. All manner of compensatory deals were discussed in talks between the company's representative in Lima and government officials, but none came close to producing the $2 million Electric Boat was owed. The Peruvians promised to levy additional taxes on telegrams, tobacco, and cocoa, and offered Electric Boat a concession to sell guano in the United States. The idea was ultimately rejected. Although the company had enough boats of its own to transport the potent fertilizer in large quantities, an investigation revealed that the potential profits amounted to no more than $36,000 a year. "It was such a small amount of guano," Carse later remarked.

The final blow to Electric Boat's darkening prospects was delivered by the U.S. Navy, which informed the company not long after the war that it would be devoting what was left of its much reduced budget to the surface fleet. The company would be free to continue working on the boats already under way, and it might even get one or two new orders to test the innovations observed in captured German U-boats. In 1923, at the same time the shipping venture finally went under, the final U.S. boat on order was commenced. It was to be the last for eleven increasingly grim years.

Faced with this endless series of setbacks and close to total financial collapse, Carse did the only sensible thing, and instituted the second reorganization in eight years. The Submarine Boat Corporation once again became Electric Boat, its operations now centered around the Groton shipyard. While the twenties roared by, the company suffered the privations of peace.

4. Dirty Laundry

By the standards of the day, the news broadcast aired by the Columbia Broadcasting System on the evening of September 1, 1934, must have struck its audience as grimly typical. The chief story concerned a national strike by textile workers and gave a brief description of the rioting that had broken out in several southern mill towns, where tear gas, dogs, and state militiamen had been loosed on picketers. The announcer quoted a former President and leading figures in previous administrations, who blamed the violence on "Communist agitation," usually concluding their denunciations of the strikers with obligatory attacks on the current occupant of the White House. President Hoover was said to have described Roosevelt, in the latest edition of the *Saturday Evening Post,* as "a tyrant," and the New Deal as "a foe of human liberty." Even some of the President's own men appeared to be turning against him. Another story reported that the director of the budget, Lewis Douglas, had threatened to resign unless Roosevelt agreed to slash relief spending and balance the budget without delay. There was only one item of "good news," and even that must have struck many members of the Depression-wearied audience as a bitter joke. After winning the right to defend the America's Cup, the millionaire yachtsman Harold Vanderbilt announced that his impending victory over the British challenger would make a splendid tonic for the national spirit.

On any other Saturday, listeners could have taken refuge in the escapist serials and knockabout comedies that had helped CBS to become the largest, and most frequently heard, radio network in the country. But on this particular evening, events in Washington took precedence over *Amos 'n' Andy* and the network suspended its regular schedule to relay the live broadcast of a much-awaited speech by a senator from North Dakota. The speaker was Gerald P. Nye, "an isolationist of the deepest hue," and what he had to say amounted to a declaration of war against the U.S. munitions industry.

For two years, Nye had stumped the country, railing against the sins of the armament makers, and the Saturday night address was more of the same. This was, however, one occasion when timing was of much greater importance than substance. On Tuesday morning, the senator was scheduled to convene the first in a series of far-ranging hearings that would place him in the unique position of judge and prosecutor. He left his audience with little doubt that the investigation of the "insane racket" would be pursued with unrestrained partisan zeal.

It had taken a lot of hard work to get the inquiry off the ground, and Nye was determined that it would commence amid a blaze of publicity. Like Harry Vanderbilt, at that moment celebrating his selection with a lavish party at the family mansion in Newport, Nye was brimming over with enthusiasm for the task at hand. The time had come to settle the score with those who dealt in death and misery. The arms companies were about to be laid open to public scrutiny and their sins revealed for all to see.

The senator began with a series of rhetorical questions: "If we defeat the Depression caused by the World War, and we accomplish national recovery, what good will it do if we are moved to another war." Next, he asked if the American people were aware of the profiteering that had been part and parcel of the war, and went on to announce the names of ten companies whose business had increased by more than 500 percent between 1914 and 1918. The gallery of corporate rogues began with Ford and General Motors and reached its climax with Du Pont, which, Nye asserted, had lifted its profits from a prewar average of $6 million a year to $59 million in 1918. His point, implied but never stated in so many words, was clear. Could anyone doubt that those same companies would not again leap at the chance to boost their earnings?

Nye was no shining orator, but in his own church hall style, he managed to present his argument as a reasoned appeal to his listeners' sense of morality. The existence of an organized arms industry at this point in the twentieth century was an affront to all right-thinking people, a blot on modern civilization which he believed his committee would do much to remove. The United States should have a Department of Peace, equal in status and provided with a budget at least as large as that of the Department of War; there should be a blanket ban on the transportation of weapons cargoes in all American merchantmen, and the munitions industry itself should be placed under strict government control. Finally, he promised that his inquiry would demonstrate the clear need for an amendment to the revenue laws that would "take the profit incentive out of war." As one possible solution, he suggested an act of Congress that would oblige the President to authorize new, and much higher, rates of corporate taxation at the first sign of hostilities.

When he stepped out of the sound booth, Nye was met with congratulations from a small group of admirers, and a barrage of questions from a much larger assembly of newspapermen. Whom would he call as his first witness? Did the senator think that his hearings would lead to criminal prosecutions? Would he care to name some of the worst offenders? Nye calmed them down like a scoutmaster with a troop of rowdy children. "Boys, I can promise you that we'll be going through the weapons Who's Who," he said. "I think we are all about to see one of the most amazing chapters in the history of American arms and ammunition."

Explaining that it would be wrong to preempt the inquiry by releasing information ahead of time, Nye refused to name names or describe in any but the most general terms the evidence unearthed by his eighty full-time investigators. He was not, however, averse to fine-tuning public anticipation with a few tantalizing promises. The panelists would not defer to influence, prestige, or wealth, he said. His staff had drawn up a list of more than one hundred witnesses representing some of the world's largest corporations. There would be no favoritism and no closed sessions. If the inquiry was to succeed, every scrap of the industry's dirty laundry, however embarrassing, had to be washed in public.

Only when the reporters' notebooks were closed did the senator become more candid. There was one company, he said, that epitomized all that was wrong with the weapons trade, and he planned to begin the hearings by placing its three most senior executives on the stand at the same time. The company had bribed officials in countries all over the world; it had entered into secret agreements with the arms companies of Europe, and it had conspired to increase international tensions, particularly in South America. During the World War it had made money from the weapons used by both sides. Within America, the company's activities were even more objectionable, Nye continued. One of its vice-presidents was a well-known Washington lobbyist who, judging from his own boastful letters, appeared to have half the city at his beck and call. By his own admission, the man had rigged the appointment of senators and representatives to congressional committees and meddled constantly with the internal affairs of the U.S. Navy, often using senior officers as unpaid salesmen at home and abroad.

There was no need for Nye to spell out the names of Electric Boat or its Washington representative, Sterling Joyner, and he refrained from doing so. The congressional investigations of fifty years ago were every bit as prone to leaks as those of today, and the committee's intense interest in the submarine builder had been common knowledge for several weeks. The press conference dissolved, the reporters converging on the nearest telephones.

American history is punctuated with outbreaks of reformist excess, moments when enthusiasm smothers prudence and the perceived solution to any particular problem is apt to be an act of Congress. Prohibition is the classic example. Launched with the best intentions, it collapsed amid the most sordid of consequences, Al Capone and the Saint Valentine's Day massacre springing most readily to mind. By 1932, the year Roosevelt made the repeal of Prohibition one of the main planks in his platform, the nation's crusading spirit had fixed upon a new cause—the end of wars and the abolition of the weapons that made them possible. It would be wrong to claim that the isolationists and arms abolitionists were the direct descendants of the hatchet-wielding veterans of the war on drink, but the two movements had much in common. Each shared the same faith in legislative remedies for ancient vices and both managed to inspire support among people who might normally have been apathetic at best.

Religious organizations were among the first to call for the reform of what one church leader described as "this dreadful industry of death and suffering." They were soon joined by an informal coalition of diverse groups that spanned the length and breadth of American society. In the months leading up to the first Nye hearings, the Federal Council of the Churches of Christ, the National Education Association, several unions, and at least one farmers' group called for government control, if not outright state ownership, of the weapons industry. Even Eleanor Roosevelt endorsed the committee and its objectives, an expression of support that must have occasioned a moment or two of extreme annoyance in the Oval Office, since her husband regarded Nye's brand of evangelical isolationism with considerable dismay. Secretary of State Cordell Hull later recalled that Roosevelt always believed that the affair would get out of hand, quite possibly saddling the White House with numerous fresh constraints on the conduct of its foreign policy. By Hull's own estimation, the hearings were a disaster. In his memoirs, the secretary maintained that no other congressional committee "ever had a more unfortunate effect on our foreign relations." Just as Roosevelt had feared, the hearings did generate legislation that barred the U.S. from an active part in world affairs. Moreover, the revelations infuriated many friendly governments, whose confidential, and frequently scandalous, arrangements with the arms companies became front-page news around the globe. In Argentina, just one of the half-dozen South American countries where arms dealers had lavished hundreds of thousands of dollars' worth of "special commissions" on politicians and members of the naval staff, the resulting scandal took a particularly Latin twist when a group of leading citizens demanded that the nation's most senior admiral commit suicide to redeem the nation's lost honor.

The real wonder is that the storm was so long in coming. There had been an earlier wave of revulsion, but that had been very different from the reformist fervor that motivated Nye and his fellow committee members, as different as the writings of Robert Graves and Siegfried Sassoon from the Budd family novels of Upton Sinclair. Those who had fought the war and survived—Graves, Sassoon, and others—tended to write about the nightmare's human face, recounting their experiences from the perspective of tiny, easily replaced cogs in a vast, impersonal machine. When they mentioned the stupidity of the general staff, or reflected on the greed and patriotic smugness of those at home, it was only in passing. Such aberrations were symptoms of the disease, not the cause. The reaction in America was rather different. Sinclair's villains—and he was but one of the leading propagandists—were those for whom war was a source of power, "depraved minds" as Nye called them, who wrung dollars out of human misery. Coming as it did in the wake of an economic collapse that had shaken public faith in established institutions, the idea that many of the world's ills could be traced to a shadowy guild of master puppeteers exerted an irresistible appeal.

The hunt for culprits received a major boost in 1929, when evidence brought to light during the course of a sensational court case appeared to confirm many of the most serious accusations. William Baldwin Shearer, a former nightclub owner, boxing promoter, and convicted bootlegger, brought suit against three of the largest naval shipbuilders, claiming that he was owed more than a quarter of a million dollars in unpaid fees. Shearer explained that he had been engaged to lobby for increased naval construction budgets and confessed that his contacts with assorted admirals and congressmen had resulted in millions of dollars' worth of additional business for his patrons.

The most shocking revelations, however, concerned the arms companies' attempts to thwart their government's peaceful intentions. Two years earlier, Shearer had been sent to Geneva with instructions to sabotage an international conference on naval disarmament. The howls of public outrage could be heard even over the roar of the collapsing stock market.

When the Geneva conference was first announced, the arms companies of Europe and the U.S. met the news with a degree of apprehension out of all proportion to the slim chance that the talks would produce any meaningful restrictions on the size of the major powers' fleets. Most dispatched their own representatives to observe and, whenever possible, exert whatever influence they could to steer the negotiations toward a deadlock. Shearer, a relatively new arrival in Washington, who had somehow managed to acquire an extensive network of contacts at the highest levels of the Navy Department, approached the Bethlehem Steel Company and offered his services to its proprietor, Charles Schwab. The tycoon enlisted two other shipyards, Newport News and the New York shipbuilding companies, and concluded an unwritten agreement under which the trio agreed to pay Shearer $25,000 a year and provide him with an unlimited expense account. Shearer also approached Henry Carse at Electric Boat, but was told that the submarine builder would be making its own arrangements. Only during the Nye hearings almost seven years later did it emerge that the lobbyists and confidential correspondents of whom Carse spoke included at least one member of the official U.S. delegation and an unknown number of senior officers attached to various South American contingents.

Described by one of his contemporaries as "brash, good-natured, cordial to all, back-slapping, hand-shaking, a true salesman type," Shearer left for Geneva, where he put his backers' money to good use. He rented a palatial apartment within walking distance of the conference center, took on a small staff, and began to distribute a daily torrent of press releases and information kits.

To the reporters, Shearer was both a blessing and an enigma. His credentials proclaimed him the official observer for the Daughters of the American Revolution, yet his apartment was renowned for boisterous parties, an open bar, and high-stakes card games. He also claimed to be on a first-name basis with William Randolph Hearst, in whose papers his dispatches appeared under a variety of by-lines. The *New York Times* correspondent, Wythe Williams, was astonished by Shearer's instant access to the most important figures at the conference. On the first day, the lobbyist left his seat in the public gallery and walked calmly to the delegates' table, where he sat down, called for copies of the conference papers, and remained for several hours, taking copious notes. To Williams, a veteran reporter who was not easily impressed, Shearer's masquerade as a delegate was "no mean feat," every bit as baffling as the silence that surrounded the source of the lobbyist's funds.

For all the mystery that surrounded him, Shearer was an invaluable source of hard news, too valuable for any reporter to ignore. The conference took only a few days to degenerate into a series of fruitless wrangles about gun bores, varieties of armor plate, displacements, ordnance weight, and fleet tonnage. Many of the American reporters had not the foggiest idea what it was all about and turned to Shearer for enlightenment. He was always happy to oblige, often arranging background briefings with U.S. Navy officers, which usually concluded with the distribution of what Williams described as "vast amounts of elaborate data, all of it violently and tactlessly anti-British."

Shearer's intention was to have the American papers portray the talks as a British ruse that would force the U.S. Navy to relinquish its position as a front-rank naval power, and his basic argument was not very much different from the bargaining-chip theories of the Reagan administration.

Only when the U.S. Navy was equal in size and sophistication to that of Great Britain should a treaty be drawn up. Until that imbalance was corrected, he said, any attempt to restrict the size of the American fleet would be "a treasonous betrayal of the American people."

When the conference failed to reach a consensus, the lobbyist was quick to claim that it was all his doing. This was another of his exaggerations, since the conflicting interests of the chief protagonists were more than enough in themselves to assure failure. However, many observers, European journalists in particular, were inclined to agree, and blamed Shearer for all manner of mischief, including several disruptive episodes that were probably the work of others. Shearer was delighted and passed on their opinions in a series of letters to Schwab aide Eugene Grace. *The Guardian,* he wrote, had taken to describing him as "the man who broke up the conference," and to prove the point, he enclosed a bundle of press clippings with the most scathing comments underlined in blue pencil. In another letter, he explained that his attacks on Britain were intended to exploit ethnic divisions within America by inflaming traditional suspicions. It was nothing more than a cynical sales pitch, he said, one that he hoped would "fool the simple Irish" in particular.

The publicity that accompanied the disclosures created such a furor that the Senate's Naval Affairs Committee was soon holding a series of special hearings, at which Shearer proved to be an eager witness. Still waiting for a decision in his civil suit, he appears to have regarded the trip to Capitol Hill as a splendid opportunity to turn the screws on his former employers. Schwab, his "boy" Grace, and executives of the two other companies were called to give evidence; all found the experience acutely embarrassing. They agreed that Shearer was a rogue, yet none was able to explain how such an obvious scoundrel came to be in their employ. Schwab, bristling and indignant, provided the best entertainment. Asked if he had hired Shearer to protect lucrative Navy contracts, the steelman replied with a note of weariness that he did not care "whether we ever do any of that work again." The millionaire further maintained that Shearer had not been acting on his behalf to influence the outcome of the talks. All he had wanted, Schwab said, was someone who would keep him informed on the ebb and flow of debate. If Shearer chose to pursue his own interests at the same time, what right had Schwab to stop him?

Shearer gave a very different account of his activities in Geneva. He stressed that he had been in constant contact with his masters and added that they approved of his strategies and supported him at every turn. He also boasted of his many well-placed friends and told of being given free rein to browse through the Navy's most sensitive reports and statistics. The Naval Intelligence Service certainly held him in high esteem. During one of his weekend excursions to Rome, the service arranged for his train to be met by a reception committee that included the American ambassador and his wife. He was driven to the embassy, where the resident military and naval attachés spent the afternoon explaining the tactical balance of power in the Mediterranean. Shearer's only regret was that his business forced him to spend vast amounts on entertainment and cigars. The preparations for his trip to Geneva, for example, entailed the wining and dining of twenty-four admirals and senior captains, as well as every commander and lieutenant commander in the Navy's Washington operation. The guest list was so extensive, one panelist could not resist asking if Shearer was running a benevolent society for hungry sailors.

The inescapable conclusion was that Upton Sinclair and the other activists were correct. It was all there in black on white—the lies, the corruption, the constant aroma of bribery and conspiracy. The weapons companies really did control the armed forces, perhaps even entire governments. With predictable speed, a champion emerged to lead the fight for reform. She was Dorothy Detzer, a onetime prohibitionist and the current secretary of one of the largest disarmament groups, the Women's International League for Peace and Freedom. Late in 1932, she turned her attention to Washington in search of a senator prepared to sponsor an unrestricted inquiry into every aspect of the arms business. The group's president, the equally energetic Mrs. Hanna Clothier Hull, summed up the political advantages for sympathetic politicians: "We intend to focus the attention of the electorate on the munitions racket and make disarmament an important issue in Congressional elections." It was a threat to be taken seriously. The League's membership was approaching 100,000 and its endorsement could make or break a campaign—as had been the case in California, where Upton Sinclair won the Democratic gubernatorial nomination despite strenuous opposition from the party leadership. The result was announced, appropriately enough, on the first day of the inquiry.

Of those Mrs. Detzer approached, Nye was by far the most receptive. He not only agreed to introduce a resolution calling for an inquiry but also asked his fellow senators to appoint him its chairman. Nye quickly assembled a staff of fifty legal aides and researchers, which included a bright young man who would later have troubles of his own with a Senate committee, Alger Hiss. Subpoenaed documents, internal memos, and financial records began to arrive in such quantities Nye was forced to ask for thirty additional investigators. By the last weeks of August, their work was finished, the evidence indexed and cross-referenced under a variety of potentially explosive headings.

Nye conducted a preliminary review of his own and decided to lead with the strongest, and most sensational, evidence. It was an easy choice. Of all the companies investigated, Electric Boat appeared to offer the richest haul of headlines. The committee secretary, Stephen Rausenbush, was told to inform Henry Carse and his two senior vice-presidents, Lawrence Y. Spear and Henry Sutphen, that they were required in Washington.

The hearings began with a theatrical flourish in the caucus room of the Senate Office Building. Just as he had promised, Nye asked if the Electric Boat witnesses would take the stand together, explaining that he wanted to be sure his questions received immediate answers. The trio made a sorry sight. Carse, arms crossed resolutely across his chest, was flanked by Sutphen on the right and Spear to his left, their faces equally devoid of expression. Jammed into chairs that had been placed just a shade too close for comfort, the three men sat shoulder to shoulder like a knot of huddled schoolboys waiting for a reprimand from the headmaster.

Nye dispensed with the customary greetings accorded most congressional witnesses, and called instead for a list of the banks with which the company did business at home and abroad. Carse provided six or seven names off the top of his head, only to be asked if he had forgotten to mention his Paris bankers. Apparently at a loss to understand the chairman's motives, the witness conducted a brief conference before he felt free to reply that all European transactions were handled through the London bankers. It was not the answer Nye had hoped for, so he became more specific, asking if Electric Boat had paid commissions to any European representatives other than Lieutenant Paul Koster, a former Dutch submarine commander who administered the Paris office. Once again the witnesses put their heads together, only to have their discussion cut short by a further question, one that drew a surprised gasp from the public gallery: "Did you pay any commissions to Basil Zaharoff?"

If the senator was trying to establish guilt by association, he was off to a good start. Basil Zaharoff was the merchant of death himself; it was much better than any of the reporters could have hoped. If Electric Boat was mixed up with someone of that ilk, there must be no limit to its depravities.

The stories about Zaharoff were legion, all of them scandalous and many true. Greek, Turkish, or Russian by birth—his ancestry was no less mysterious than his dealmaking—Zaharoff had risen, by one account, from a day laborer's job on the Piraeus waterfront to control one of the greatest fortunes of his day. He made his first millions selling machine guns to the armies of Europe on behalf of the Nordenfeldt company during the 1890s. When Nordenfeldt was absorbed by Vickers, Zaharoff rose quickly to become the British firm's chief salesman, and in the opinion of several biographers, its largest shareholder. During the war he profited from foundries and powder factories located in England, France, and Russia, as well as others in Germany and Austria. There were rumors that he had ordered political assassinations and others that depicted him as the root cause of the war itself. Even the details of his sex life were the subject of lurid speculation. One of the most fanciful stories was repeated often during the late twenties and thirties. Enamored of a young English girl, he was said to have floated a specially constructed black velvet bed in the flooded basement of his Paris mansion. Some accounts insisted that it was built to resemble a swan while others insisted it was inspired by a Venetian gondola.

Admitting that Zaharoff had received certain payments, Carse was adamant that they were not commissions, at least not in the accepted sense. What, then, did these sums represent? Nye persisted. Surely Zaharoff must have done something *to earn* commissions totaling more than two million, almost half stemming from the Spanish business alone. Sounding rather like a missionary who must explain some obscure theological tenet to the savages who are preparing to eat him, Carse insisted that the moneys were "transmissions," not commissions. The senators were eager to have the distinction explained.

"We do not pay him but, under an old agreement, a certain percentage is paid to us and we transmit it to Sir Basil," he began, adding that the agreement applied "to the Spanish business only."

Semantics proved an inadequate defense against Nye's probing, and Carse was soon forced to concede that the agreement to which he referred was nothing more than a simple kickback scheme. Under a hail of questions from Nye and his fellow committee member Senator Homer T. Bone, it emerged that Zaharoff all but controlled Sociedad de Construcción Naval, the chief supplier of ships and submarines to King Alfonso's navy. The Spanish vessels made extensive use of Electric Boat patents, for which the company received royalties that sometimes amounted to more than 10 percent of the boats' total price. These sums were sent first to New York, where half was converted to French francs and remitted to a Paris merchant banking house in which Zaharoff also held a major interest.

The panelists were curious to know why Zaharoff did not avoid a lot of pointless paperwork and simply write himself a check on the shipyard's Cádiz bank. Carse professed to have not the slightest idea. Sir Basil was a much misunderstood man, he said, "the victim of a lot of slurring." As far as he was concerned, the arms dealer would always remain "one of the greatest men I have ever had the honor to meet." One gathers that Carse would have made every effort to forward the "transmissions" via the moon if Zaharoff had so wished. "After all," he said, "he is the man who brings us the business."

The suspicion that bribery was the mainstay of the arms trade was deepened by several letters in which Zaharoff mentioned the payment of "discreet commissions" to unnamed Spanish officials. The correspondence also did much to illuminate the curious arrangement that existed between the American submarine builder and "the most mysterious man in Europe."

Zaharoff's contacts with the Spanish Navy ensured that most foreign submarine builders faced an insurmountable disadvantage in bidding for construction contracts. The one exception was Electric Boat, whose control over several hundred vital patents gave it the power, at least in theory, to make the construction of submarines within Spain a prohibitively expensive endeavor. Zaharoff had solved the problem at a meeting with Isaac Rice in London during the summer of 1912. In return for Zaharoff's promise to split the profits on each boat down the middle, Rice agreed to stay out of the Spanish market and make his patents available at a nominal cost. Ever the realist, Rice understood that he had no choice but to agree. Had he rejected the proposal and dispatched salesmen to compete directly for the Spanish business, it was far from certain his company would have emerged the winner. As for the patents, both men realized they were a hollow threat at best. There was no way Rice could stop them being used without permission, and little chance that Electric Boat would win compensation from a Spanish judge and jury. The considerable advantages, on the other hand, were obvious. In return for a little cooperation and a minimal investment, Electric Boat could look forward to a handsome profit on each new boat. And with Zaharoff controlling operations in Madrid, there was reason to hope the windfalls would continue well into the future.

There was a provision for Zaharoff's remittances in the original contract, but it was a relatively small amount, seldom more than 3 percent. Not long after the war, certain elements in the Spanish naval high command began to agitate for an end to the old monopoly and, as Zaharoff made clear in one of his letters to Carse, many more "special commissions" were necessary to quiet the clamor. The "transmissions" were Electric Boat's contributions to the stepped-up fighting fund.

The committee took a particular interest in an invited observer at the London meeting, Albert Vickers, the chairman of the arms company that bore his family name. The inference that three rich and powerful men had entered into a conspiracy to split the world into mutually profitable chunks was inescapable. Moreover, it would soon be confirmed by fresh quantities of documents taken from Electric Boat's correspondence files.

The licensing deal gave Vickers the first rights to sales in Europe and the empire, while Electric Boat claimed the Americas and much of the

Near and Middle East. In Japan, one of Groton's oldest customers, both companies were free to pursue their own best interests.

The witnesses made several vain attempts to put matters in what they maintained was the correct perspective. Spear, the technical expert, presented the company's more questionable practices as unfortunate but inescapable facts of life. However distasteful bribery may have seemed in Washington, it was just another ancient custom in other parts of the world, he said. Every business had its unique norms and codes of behavior, and the arms trade, for all its peculiar ways, was really "no different from any other industry." The senators were no more impressed by Spear's argument than the members of the Church committee forty years later, who heard, and rejected, a similar plea of expiation from executives of the Lockheed and Northrop Aircraft companies.

Nye, his question laced with veiled sarcasm, asked if Spear was being flippant. Did the witness really believe that there was nothing unusual about graft and corruption? Furthermore, was Mr. Spear aware of any other companies that conspired to supply their customers with second-rate goods? To support the allegations implicit in his questions, Nye held aloft a sheaf of correspondence concerning Lieutenant Koster, the unfortunate Paris representative, whose activities were rapidly becoming a source of considerable embarrassment.

Koster was an eternal optimist, whose promises of impending orders usually came to naught. Carse once grumbled that the Dutchman "had not brought in a single dollar's worth of business" and blamed him for helping to get Electric Boat involved in the disastrous merchant ship construction program during the last years of the war. Zaharoff always referred to him with the undisguised contempt of a consummate professional for a bumbling and vexatious amateur. One of Koster's sales promotion schemes—an essay competition that asked Dutch schoolboys to suggest uses for submarines in the country's East Indian colonies—was so far removed from the sordid reality of the arms trade that even the senators appeared surprised at the eager innocence of its sponsor.

Koster's biggest mistake, however, was to try to encroach upon Zaharoff's Spanish bailiwick. Unaware of the confidential agreement, the Paris agent learned that several submarines under construction in Cádiz lacked many modern features. No doubt expecting a pat on the back for his initiative, Koster wrote to a number of Spanish admirals and politicians, criticizing what he termed the boats' antiquated design and all but asserting the navy was being sold a bill of goods. The Electric Boat Company, he declared, would be pleased to quote on future orders and might even be able to remedy the glaring deficiencies in the boats already under way.

Koster's meddling prompted what Zaharoff described as "a real mess in Madrid." Zaharoff had put it about that Electric Boat was intimately involved in the construction program, claiming that he had personally negotiated the use of the most advanced patents, some so modern they had yet to be adopted by the U.S. Navy. Koster's revelations destroyed this carefully nurtured fiction in an afternoon and obliged Zaharoff to add many fresh names to his payroll. Embarrassed, out of pocket, and beside himself with rage, the arms dealer attacked Koster in a letter to Carse that concluded with a demand for his immediate dismissal. "It has always been an exceedingly difficult problem to create a monopoly for Electric Boat Co. products," he complained. "The Spanish authorities say we have always assured them they are getting the very best that exists; yet your agent in Paris must surely be in possession of improvements we are keeping back from them. This intervention has caused serious friction."

That the Dutchman survived for another five years before leaving of his own accord was largely due to the protective influence of Lawrence Spear, who recognized Koster's failings as a salesman but valued him as an irreplaceable source of information on the shifting political and military currents in Europe.

In 1929, for example, Koster reported that Germany was preparing to build submarines in Dutch and German shipyards, a clear violation of the Treaty of Versailles. He provided the names of those in charge, detailed the construction schedules, and identified the German investors who had put up the working capital. The senators wondered why the information was not passed on to the State Department or some other government agency. There was no ready answer save the suspicion that this was yet another example of cynical self-interest.

The man identified as the head of the German operation, a certain Herr Techel, also happened to be the chief witness in a legal action that Electric Boat had launched against a number of German shipyards. The suit demanded some $17 million in compensation for alleged patent infringements, and Techel, a chief architect of the wartime construction program, was the one man in Europe who could answer the most important technical questions. All but two of the Kaiser's 441 U-boats had been scrapped or used for target practice and no trace could be found of the plans the Germans claimed to have surrendered to the Allies. In the absence of any physical evidence, the case rested upon Techel's willingness to cooperate, and until Koster began his sleuthing, the German had suffered from a very poor memory.

There were strong indications that Koster's findings had been used to win Techel's cooperation. The panel learned that the ten-year-old dispute was concluded within months, and while the settlement was trivial, less than 1 percent of the amount requested, the accounts in the papers the following day depicted it as a significant victory, a shot in the arm for a company so hard-pressed for cash that all but one of its directors had just agreed to a 25 percent pay cut. With the case settled,

Koster's proposal that the League of Nations be alerted and military sanctions imposed on Germany proved to be totally unnecessary. The covert construction program was left free to continue work on what would become the first of the Third Reich's new U-boats.

The news that Electric Boat had profited, however belatedly, from the German war effort produced the most unfair smear of them all, the emotional and quite erroneous charge that Carse and his fellow directors "had dealt with both sides."

Electric Boat's ties with Germany stretched back to the turn of the century, when Isaac Rice concluded a series of licensing arrangements with the major European shipyards. His German partner, the Vulcan shipyard of Hamburg, was forced to renounce the agreement a few years later, when the Kaiser let it be known that companies with extensive foreign ties would not be asked to build ships for his navy. The work went instead to the Krupp yards at Kiel, and so, too, did the patents, which were appropriated from the Berlin commercial registrar's office without acknowledgment or apology. From 1909 until 1913, Electric Boat battled with Krupp in the German courts, eventually winning a settlement that would undoubtedly have led to an appeal if the war had not intervened. Seizing the opportunity, Krupp announced that all royalty payments would be suspended for the duration. Electric Boat was back where it had started.

Other British, French, and American companies suffered similar thefts and many sought compensation before the various war claims tribunals during the early 1920s. Electric Boat's claim was typical yet perceived as somehow different because it concerned armaments. All but the most ardent abolitionists were prepared to concede that a company had a right to a decent profit on work done for the U.S. Navy; reaping the same benefits from a recent enemy was, however, quite another matter. The fact that the royalties were owed on submarines—the weapon that sank the *Lusitania*—merely served to make Electric Boat's pursuit of its blood money appear even more reprehensible.

Carse made a vain attempt to point out that the $125,000 award was a drop in the bucket, insufficient to meet what amounted to twenty years of legal costs. Nye refused to listen. He was out to make a point, as the following exchange illustrates:

> **NYE:** Doesn't it pretty nearly come down to this: The only thing that is left to be honored in time of war is a patent on war machines?
> **CARSE:** The recovery was only $125,000!
> **NYE:** All the same, where there was a recovery on the rights to manufacture a machine, there has been no right to recovery for any of the lives or other property destroyed.

Generally missing from the newspaper accounts of the hearing was the real reason for Herr Techel's newfound candor.

Koster recalled that France had asked the League of Nations for permission to keep two of the U-boats seized at the war's end. French dockyards and marine foundries had been devoted almost entirely to the production of weapons for the trenches, and since the country's naval strength was well below treaty limitations, the request was approved. A little checking revealed that both boats were still in service and available for inspection at a training base near Marseilles. Koster next persuaded the French Navy to convene a board of inquiry to inspect one of the boats and notarize his sketches of the main ballast tank system. He went even further with the second U-boat, arranging to have several compartments stripped of their fittings in a painstaking search for lost and forgotten copies of the shipwrights' instructions. Just such a set turned up in a forward compartment. They were the plans for the torpedo room and included a plumber's chart that described the operation and installation procedures for a small but vital set of compensatory flood tanks that maintained the boat's underwater equilibrium whenever a torpedo was fired. The maze of valves, ballast chambers, and pipes was a refinement of an old idea first used in the *Fenian Ram* to absorb the recoil of John Holland's pneumatic cannon. It was these plans, rather than any threat of blackmail, that carried the day, since the system was an obvious copy of the designs stolen from the Berlin patent office.

Koster's enthusiastic quest for new business for Electric Boat remained a source of embarrassment even after his departure in 1932. One year later, he wrote to Spear on behalf of his new employer, the Bergmann Gun Company of Berlin, asking if the American would obtain a large quantity of machine guns. Koster told his former protector that the weapons had been ordered by "certain organizations in Germany," adding that this information should be kept "strictly confidential at all costs." Neither the senators nor the press required any prompting to identify the clients as the Nazi party's armed militias.

Spear let the request pass without taking any action, but he was given no credit. The members of the committee evidently believed that anyone on intimate terms with Basil Zaharoff would feel completely at ease with Hitler's thugs.

The hearings' progress could be monitored by the exertions of the committee's junior aides, who shuttled into the chamber with successive cartons of documents, each representing a fresh source of embarrassment. One box was devoted to a series of letters in which Carse advised Zaharoff about the best ways to avoid paying income taxes on some $2 million worth of commissions.

Carse's attempts to portray the incident as an innocent courtesy exposed one of the few cracks in his stony reserve. Scrambling desperately to distance himself from his earlier comment that Zaharoff was "one of the greatest men in the world," Carse now tried to persuade the panel that he knew next to nothing about the arms dealer. They were unimpressed. After wading through literally hundreds of intimate, gossipy letters, Nye and his colleagues were loath to believe that Carse, Zaharoff, and Spear were anything but the best of friends.

Several other boxes produced evidence that two U.S. admirals had pressured the Turkish government to buy a large package of armaments from Electric Boat that included two submarines, a surface tender, and several hundred antiaircraft guns. Admiral Hilary P. Jones, the former commander in chief of the fleet, and Admiral A. T. Long attended a meeting in Washington where Electric Boat's representative Sterling Joyner thrashed out a deal with the Turkish ambassador that established the size, and the recipients, of the special commissions that would have to be paid before the sale could be finalized. Together with the volumes of other evidence that indicated that U.S. Navy and embassy personnel often worked as unpaid lobbyists for American companies, the Turkish disclosures prompted Nye to make one of the inquiry's most durable observations: "It makes one wonder whether the Army or the Navy are just organizations of salesmen for private industry paid for by the American government."

It was, however, the letters to and from Sir Charles Craven of Vickers that did the most to blacken Electric Boat's reputation. With its web of secret agreements, enormous wartime profits, and what the committee evidently regarded as de facto control of British foreign policy, Vickers epitomized all that Nye opposed, and he attacked the latest bundles of documents with an avenger's determination to see justice done. The mere mention of Craven's title was enough to provoke one of the chairman's little sermons. It would be preferable, he said, if the witnesses deferred to the committee's egalitarian sensibilities and referred to their British associate as "Mr. Craven" rather than "Sir Charles."

The committee was both fascinated and appalled by the tone of amoral ruthlessness that underscored the otherwise chatty and informal notes. In one round of correspondence, Spear passed on a Chilean request for two submarines with the explanation that Electric Boat could not handle the business for fear of losing its contracts with Peru. Nevertheless, Spear was prepared to accept a cut of the final price and gave no hint that he saw anything unusual or unethical in his position as a procurer of arms for two traditional rivals. In bidding for the work, he advised Craven to bolster the total price by at least 10 percent. "My own experience is that, at the last moment, something extra is always needed to grease the ways," he wrote. "We all know that the real foundation of the South American business is graft." Spear concluded by placing Electric Boat's Chilean representative at Vickers' disposal, suggesting that "it may well be he knows the proper people to pay in Santiago."

The most damaging, and widely publicized, of all Spear's observations was a mere aside tossed off in another letter about the South American prospects. "It is too bad the pernicious activities of our State Department have put the brake on armament orders from Peru by forcing the resumption of diplomatic relations with Chile," he grumbled. If the committee had any lingering doubts about the justice of its cause, this indiscreet comment dispelled them forever.

The final, ironic twist in the investigation of the relationship with Vickers came when Carse was asked if he knew whether the Admiralty was aware of the royalty payments to Electric Boat. Acutely embarrassed, he floundered, struggling to find the right words before allowing his reply to trail off into a series of incoherent mumbles. Nye leaped to the obvious conclusion, interpreting the silence as a sign that this was another kickback scheme, similar to Zaharoff's little scam in Spain.

In fact, as the official history of Vickers explains, Carse was the victim of Craven's calculated deceptions over more than two decades. The Admiralty was not only aware of the licensing agreement; it had helped to set it up back in 1902, when Isaac Rice first journeyed to London with copies of John Holland's patents. We can only assume that Carse chose to say nothing because he really believed that his company was involved in something underhanded. If this was the case, it was exactly what Craven wanted him to think. Vickers' official historian, J. D. Scott, explains the Englishman's motive: "Carse had forgotten, or had never known, that Rice had gone to Vickers with the Admiralty's blessing and evidently Craven had never enlightened him, perhaps with the idea that if Electric Boat believed themselves to be involved in a secret transaction that they might be more receptive to Craven's perpetual pressure to reduce their royalties." It seemed there was no honor whatsoever in the armaments business, not even among the players themselves.

Craven's deceptions reaped their own reward. When accounts of the Washington hearings appeared in the British press the following day, there were immediate calls for a royal commission along similar lines. Craven's initial response was to deny having written anything at all, but it was to no avail. Public pressure became irresistible and Parliament was soon obliged to investigate not only the letters to Carse and Spear, but also the details of secret agreements and wartime profits whose exposure did much to confirm the decades-old stories of Vickers' perfidy.

The atmosphere in the car that carried the witnesses back to their suites at the Washington Hotel must have been oppressive with gloom.

Electric Boat had taken a terrible mauling and the three men were well aware that the worst was yet to come. With the exception of its few minor, domestic digressions, the committee had confined its attention to shady dealings in other parts of the world. The evidence made wonderfully colorful front-page copy—but that was all it made. For all the scandals and ethical breaches, the fact remained that no laws had been broken, at least no laws of the United States. Indeed, on the strength of the foreign evidence alone, it was still possible to feel some sympathy for Spear's assertion that Electric Boat was a victim of circumstance. Since the U.S. Navy had seen fit to withhold all major construction contracts for a period of ten years, the company was forced to scramble for whatever work it could find. It was, Spear said, a simple matter of survival. If the directors wished to honor the investments of their stockholders and protect the jobs of thirteen hundred workers, they had to play the game by the same dirty rules that applied to their competitors. Without the foreign business and the royalty payments, he said, "there would not now be any such organization as Electric Boat."

A few minutes before the witnesses' first day on the stand came to an end, Nye had asked if there was any news of a conspicuous absentee, Sterling Joyner, the Washington representative. A clerk replied that the gentleman was confined to bed, where he was likely to remain for quite some time to come. Suffering from an an ailment variously described as "a heart lesion" and "cardiac fatigue," Joyner was said to be "in grave danger of a complete collapse" unless he avoided all exertion and excitement. It was Nye's first piece of bad luck. One of his chief and most vulnerable witnesses had been stolen away by a doctor's exemption.

If the chairman was disappointed, as indeed he must have been, he showed no sign. There were certain matters he had intended to discuss with Mr. Joyner, and he would now be obliged to ask the witnesses to return in his stead. As an ominous parting warning of the ordeal ahead, Nye suggested that the witnesses speak with Joyner during the evening to refresh their memories of his activities.

Joyner had come to Electric Boat nine years earlier, after an undistinguished early career as a middle-ranking aide to the industrialist Alfred Du Pont. With the exception of a few errands that brought him to Washington as a confidential messenger, his most notable achievement appears to have been overseeing the construction of a summerhouse on one of the chemical king's Delaware estates. At some point—the details are far from clear—he made the acquaintance of Sir Basil Zaharoff and, later, Sir Charles Craven. The Vickers knights were impressed by the cocksure American and decided that he would make a useful replacement for Mr. A. S. Roberts, one of Vickers' two nominee representatives on the board of Electric Boat. Roberts had been talking about resigning to concentrate on the management of a thriving real estate portfolio that had just grown to include the Washington Hotel, Carse and Spear's favorite home away from home.

As it happened, Roberts remained on the board for a further two years, while Joyner made a series of trips to the other side of the world in a frustrating and ultimately unsuccessful attempt to sell the Japanese Navy what would have been the largest foreign order for armaments ever received by an American manufacturer. It was Joyner's first major job for his new employers, and it gave the lobbyist a free hand to indulge all the indiscreet and foolhardy tendencies that would later incur the wrath of the Nye committee. The chief source of wonder in the Japanese episode was that no one at Groton or New York realized what a self-inflating, egotistical, arrogant, and devious viper they had taken into their midst.

Electric Boat had been dealing with Japan since 1904, when Frank Cable assembled and tested the Navy's first five submarines in and around Yokohama Harbor. Twenty-one years later, a delegation of Japanese naval attachés and technicians from the Japanese embassy in Washington journeyed north to Groton to discuss a proposal for a massive increase in the size of the Imperial fleet. In a series of talks that must have made Carse's mouth water, the delegation spoke of plans to acquire several hundred additional submarines, minesweepers, patrol boats, and destroyers, and indicated they would like Electric Boat to become the prime contractor in charge of those boats to be built in the United States.

The proposition was irresistible, but it called for kid-glove treatment. The advantages—a financial mainstay well into the next decade and an end to the company's dependence on the capricious purchasing policies of the U.S. Navy—had to be weighed against Japan's deteriorating reputation in the West. Having bid unsuccessfully for several of the Navy's latest, long-range submarines, Carse and Spear must have known the new boats were specifically designed to take any future war deep into Japanese home waters. Yet no one seems to have noticed, or cared, that a commercial relationship with Japan might lead to the same sort of complications that had marked the earlier dealings with the Kaiser's Germany. The chief concerns were speed and secrecy. It was to be one of the greatest prizes in modern naval history, and Carse was determined that it would not be snatched away by a rival or jeopardized by premature publicity. Joyner left for Tokyo within days, taking with him some hastily prepared construction estimates, a variety of technical submissions, and a final piece of advice—the lobbyist was not to breathe a word until all the papers were signed and safely back in America.

The weeks that passed while Joyner's steamer crawled across the Pacific must have been an agony of anxious waiting. Carse had no idea if Japan had made similar overtures to rival firms, and Joyner, the new boy on the team, was still an unknown quantity. His first reports must therefore have occasioned considerable excitement, if not outright jubilation.

His tidings seemed too good to be true, as indeed they were. He claimed that Japanese officials were well disposed toward Electric Boat and had promised they would not hold talks with any other firms. In another rumor, the most tantalizing of them all, Joyner maintained that millions of yen had been transferred to a New York bank, in anticipation of being applied as a down payment on a slate of orders many times larger than the one originally discussed. There were only a few details remaining to be settled, and Joyner was confident they would be sorted out within the month.

The fact of the matter was that while the Japanese were interested, very interested, in examining the latest plans and technical advances, they had no intention of writing any checks. Domestic production had been an integral part of Japanese naval strategy since the great modernization program in the last years of the nineteenth century, and there was no reason to believe the policy was about to change. Enraptured by the prospect of a secure future, Carse and those at home failed to notice that they were being played for suckers; the submarine negotiations were part of a systematic Japanese quest for inspiration in the patent files and sales proposals of their future enemies. If Joyner realized, or even suspected, what was happening, he chose to say nothing to the head office. Instead, as he would do so often in the years to come, he sustained the excitement, and his own prestige, with a stream of optimistic rumors that found their origin in his own fertile imagination.

Joyner's compulsion to sing his own praises also doomed any hopes Carse may have harbored about the talks being kept under wraps. Word of Electric Boat's impending windfall was soon radiating out from Tokyo via diplomatic pouches, embassy cables, and the grumblings of resident representatives for Vickers and several other manufacturers, who resented the cocksure newcomer and his self-proclaimed successes.

Long before there were any official announcements, Sir Basil Zaharoff, every bit as well informed as usual, dropped Carse a friendly note of congratulations in which he demonstrated a comprehensive knowledge of every minor detail under discussion. The news also reached Wall Street, where the company's stocks rose precipitously. Since Joyner was both an adept trader and a major shareholder, we can only assume he did well from the market's reactions to the two years of on-again, off-again talks that followed.

Like a good many others, Senator Nye wondered why, when everybody from embassy cipher clerks to Wall Street speculators knew of the deal, the U.S. government had been kept in the dark. An official request for information was rejected, with the blanket denial that Carse, or any of the other directors, knew anything of value. The only man who could speak reliably on the matter was Mr. Joyner, and he always seemed to be out of town.

Carse attempted to put the best face on his conduct by hinting that he would have been more cooperative if the deal had gone through. "It would have been a very nice piece of business for the United States," he said, explaining that he had hoped to provide his government with an intelligence bonanza. "These things are always beneficial, not only for labor, but also for the knowledge of what is going on." So passionate was his defense, observers might have been excused for thinking it was every firm's patriotic duty to shower weapons and sophisticated technology on potential adversaries.

The most sensational of the second day's disclosures concerned Joyner's activities in Washington, or rather what the junior vice-president claimed were his activities. Like the letters from Tokyo, his weekly dispatches to the New York office were a mixture of fact and fiction.

Late in 1928, for example, Joyner wrote that he had engineered the election of two Republicans to the House Rules Committee. In staccato sentences, he described how he had marshaled the support of his other contacts in the House:

> Successfully managed campaign for candidates for the Rules Committee which is most important to us when any legislation is brought up.
>
> Brought in some western states, New England states, New York, Pennsylvania and Michigan in Fort case. And New Jersey, Michigan, New York, Illinois, South Dakota and Pennsylvania in Martin's case.
>
> Candidates successfully elected to the Rules Committee: Hon. Joseph Martin, Massachusetts; Hon. Frank Fort, New Jersey.
>
> The rules committee is the most important committee in Congress. It absolutely controls legislation.
>
> Thanking you with kind regards,
>
> Sincerely,
> S. J. Joyner

Once again Carse found himself compromised by the deceptions of an intimate associate. For all his talk, the truth was that Joyner had done nothing to influence the outcome of the election, as the two men whose careers he claimed to have advanced soon made clear. Frank Fort, by then a former congressman, conducted a lighthearted press conference at his Newark home, where he told reporters that he had never even met Joyner and chuckled openly at the gullibility of Electric Boat's senior executives. "Apparently Joyner was trying to fatten his prestige with his own employers by claiming to have accomplished something with which he had absolutely nothing to do." Representative Joseph Martin, a future House Speaker, was the picture of outraged dignity. His election, he said, followed a unanimous endorsement by Republican representatives from the other New England states, who had formed a committee to manage his campaign. As for being a pawn in the service of Electric Boat, Martin pointed out that he had voted consistently for arms treaties as well as restrictions on the international traffic in U.S.-built weapons.

Carse was in no position to dispute the letter's authenticity, but he did deny ever having seen it. If that was hard to believe, his account of Joyner's relationship with the head office was even more so. The lobbyist's "legislative activities" were not directed by anyone in Groton or New York, Carse asserted. "Speaking personally, I do not know any members of committees nor have I sought to influence the decisions of Congress," he said.

The words were scarcely spoken before the committee turned its attention to another letter, one that skewered Carse as a liar and mocked the claim that Joyner always acted as a free agent.

In March 1929, Joyner reported that he had engineered his greatest coup to date, orchestrating the passage of a bill that awarded Electric Boat roughly $3 million as a final settlement to a long-running legal dispute with the U.S. Navy. The affair dated back to America's entry into the war when Groton agreed to an official suggestion that it minimize lost time by taking a softer line on wage and overtime demands. There was a vague promise that the excess labor costs would be reimbursed at some future date, but when an official claim was lodged in 1920, the Navy balked. The dispute eventually went to the Court of Claims, which handed down a favorable verdict after almost eight years of oft-delayed deliberations.

The court's decision was a moral victory, but it was far from an end to the affair. The Treasury Department, the body responsible for actually releasing the funds, placed Electric Boat's name at the bottom of a long list of over three thousand companies and individuals, all of them waiting for Congress to authorize payments in a series of deficiency bills. The achievement in which Joyner took so much pride was the company's leap to the front of the line. The latest piece of correspondence proved that Carse was intimately informed of his agent's activities. The letter read, in part:

> I sincerely promised you the day we lunched together in New York that we would manage, after overcoming a variety of handicaps and jumping some hurdles, to get the Second Deficiency Bill through . . . and we expect to receive payment at 2 o'clock this afternoon or early tomorrow morning.

> I am not prepared to write to you; however I shall be glad to verbally tell you what really happened with reference to the Deficiency Bill and the part this office played in getting that bill through.
>
> I absolutely and positively believe and feel safe in making the statement that if it had not been for the action taken in this office on the day before the bill was passed, the Second Deficiency Bill would have gone over until the next session.
>
> My reason for putting this in writing is out of respect for those who helped and who were so powerful and friendly. The code of honor between men makes it unethical to name persons.

Waxing lyrical, Joyner concluded by promising that his close friends at "the Army and Navy, the Treasury, the Commerce Department and the Shipping Board" would make sure the company enjoyed "a bigger future with less sales resistance and pleasant hours free from past worries and cares."

Who were Joyner's connections, these people who were so "powerful and friendly"? Once again Carse professed ignorance. Nye decided to try a little prompting. Flourishing another sheaf of papers, these latest taken only days before from Joyner's Washington office, Nye asked if the names of Ernest Lee Jahncke, a former assistant secretary of the Navy, and John Q. Tilson rang any bells. Carse made a noncommittal reply, so Nye ordered one of the letters read into the record. It was one Jahncke had written to Tilson, suggesting ways in which the pair might direct several upcoming submarine contracts to Electric Boat. Tilson's concern was understandable. Although best known as the Republican leader of the House, he also represented many of Groton's workers and was keen to keep his constituents at work. Jahncke's interest in Electric Boat's well-being was, however, open to conjecture. Many concluded, perhaps unfairly, that it had something to do with Joyner's courteous concern for the assistant secretary's wife and daughter.

Some four years earlier, Mrs. Jahncke had taken her daughter on the Grand Tour, and thanks to Joyner's intervention, the pair appear to have had a splendid time. One of Joyner's titled friends in London introduced the ladies to an assortment of knights and duchesses. In Paris, Koster's replacement attended to their hotel and railway reservations as well as squiring them through the city's tourist haunts. Even Basil Zaharoff made a contribution to the success of the trip. Joyner asked the head of the Monte Carlo–based United States-International Hydrographic Survey, Admiral T. Long, to call on the aging arms dealer and remind him that he had promised to open his clifftop villa to the women for the duration of their stay. The old rogue swept the ladies off their feet, whirling them through casinos, receptions, balls, and garden parties attended by the motley remnants of European nobility and assorted Riviera celebrities.

Like any father, Jahncke said that he was glad that his family had a guardian angel throughout the European trip, but he was adamant that the performance of his official duties had not been influenced by any sense of gratitude he may have felt. In evidence given several months later, he justified his support for Electric Boat on the narrow grounds of national security, the same rationale that underscored all his letters to Tilson and Joyner. "As Electric Boat is at present the only private company in America specializing in the design and construction of submarines, the department considers it necessary to do everything possible, so far as is permitted by the Laws of Congress, to give that yard its reasonable share of submarine construction. Therefore, you may rest assured that I will do everything in my power to further that result," he said.

Other, less senior but no less influential naval officers and officials also figured prominently in Joyner's contact book. His Monte Carlo messenger boy, Admiral Long, was just one member of the old boy network that linked the Navy to its principal suppliers. An old Annapolis classmate of Spear's, Long had been nominated by Congress to represent his country as a governor of the International Hydrographic Survey, an offshoot of the League of Nations. Since America was not a member of the League, his ratification by member countries might have been more difficult had it not been for Joyner's behind-the-scenes support. Taking up his pen once again, he begged Zaharoff to use his influence with Spain, France, England, and Greece to assure their governments' endorsement of the American candidate. Long won without serious opposition and went on to revel in a plum job whose only disadvantage was the enforced isolation from the Navy rumor mill. Always a good friend, Joyner set out to remedy the situation with a series of reports that demonstrated a gossip columnist's flair for breaking news. In one letter, he mentioned over fourteen transfers, promotions, and petty quarrels among upper-echelon admirals and captains, as well as a good many tidbits about the wives and children of Long's best friends back home.

In another letter, one that must have come as an eye-opening shock to Spear and Carse, the company's two longest-serving executives, Joyner informed Zaharoff that he was thinking about resigning. The company was burdened with "a good deal of dry rot," he wrote, adding that he had given up all hope of convincing his superiors to adopt a more aggressive sales policy. "My voice is the call of one in Babylon, lost in the confusion," he said. The witnesses' reaction to the dry-rot metaphor was not recorded, but it is safe to assume they recognized the lonely voice as that of Babylon's whore propositioning a potential client.

However, like so many of his other promises, Joyner's letter of resignation failed to materialize. It was nipped in the bud, so he claimed, by "a rather strange development. . . . The Secretary and the Assistant

Secretary of the Navy, Admiral Jones and other admirals, Colonel Tilson, the leader of the Republican Party, the White House and other people brought me into conference and asked me to remain that they might have faith in the honest construction of any program they favored us with, and also that their dealings might be without conflict or confusion, enmity or doubt," he said. In what was undoubtedly another self-prompting fiction, he added that it was the first time in the history of the United States that the Navy had asked an officer of a private company to further the service's interests by remaining at his corporate station. Closer to fact were the services Joyner had rendered in several past battles for increased naval appropriations. At one point he threw his lobbying resources behind a funding bill for two additional heavy cruisers and received in return the hierarchy's support for an increased submarine budget. Even today such a scheme would raise eyebrows; fifty years ago, the idea that the composition of the fleet might owe more to backroom politics than to strategy was deeply shocking.

And so the hearings continued, ranging through scandals large and small. For three days, Electric Boat's directors were subjected to probing examinations of deals, letters, conversations, and agreements which the transcripts indicate the witnesses had often long forgotten. There was only one final scandal of major proportions, the allegations that Carse and Spear had conspired to float and promote a dishonored issue of Peruvian bonds. Once again the witnesses denied the charges, maintaining they had acted in good faith. All suggestions that Electric Boat had pledged its support in order to speed the payment of money due on submarines already in service with the Peruvian Navy were dismissed out of hand. The proof, Carse pointed out, could be found amid the columns of red ink that dominated the company's annual reports. Not only did the Peruvians fail to pay for their boats; they also tried to ensnare Electric Boat in an extremely unprofitable deal to sell guano in the United States. If the senators really believed that the fruits of corruption amounted to a lease on a mountain of bat manure, Carse asserted the panel's idea of a fair profit must differ markedly from his own.

The hearings concluded with an unfamiliar note of praise from the chairman, who thanked the directors for their cooperation and apologized for any inconvenience. One gathers Nye would have enjoyed delivering one of his sermons about the evils of the armaments trade but was stymied by the scarcity of evidence. Despite his pre-hearing comments that the press would see the greed of the munitions kings laid bare, the first three days of the proceedings had produced nothing of the sort. True, the revelations had been plastered across every front page in the country, but in and of themselves, they represented nothing illegal. In fact, if the evidence was examined calmly and without partisan passion, the picture that emerged was not one of insidious corporate corruption but of ham-fisted bungling. If Electric Boat's directors were the crooks their many critics supposed, they represented some particularly incompetent specimens of the breed since the documents indicated the company had been fleeced by everyone from Zaharoff to Joyner, their deceptions and misrepresentations cheerfully accepted by Electric Boat's trusting executives. Joyner could boast all he wanted about having Washington in his vest pocket, but the actual results of his lobbying campaigns indicated otherwise. Not once did Electric Boat win a contract as a direct result of his efforts, nor did the evidence support his claim to controlling the destinies of elected political leaders.

As for the other allegations—the claim that Electric Boat had dealt with both sides during the Great War, for example—there were only disappointments. The one thing that resembled German profits was the settlement of the long-running patent infringement action, and that, as Carse noted several times, was insufficient to cover the company's excessive legal fees and research expenses.

The witnesses' polite dismissal from the stand was not a clean bill of health or an endorsement of the company's peculiar code of ethics. Nor was it what Nye had been hoping for. Carse, Spear, and Sutphen went on their way while the panel turned its attention to Du Pont, Bethlehem Steel, the Ajax Powder Company, and representatives of the major banks.

Of course, there was never a real chance that Nye and the other reformers would prompt any reduction in the size of the armed forces. Many things had changed since the Washington Conference and there were now several good reasons why the U.S. was both unwilling and unable to engage in another outbreak of battleship scrappings and war surplus sales. The first was practical: There were very few battleships left to scrap. The second was more pragmatic: However loathsome they may have found it, even the most ardent crusaders for disarmament were forced to admit that the arms industry kept people at work, and no one, elected officials least of all, was prepared to prolong the misery of the Great Depression. Even at the height of Prohibition, that other flawed and failed campaign to improve the world through legislative decree, the legal production of potable alcohol had not been entirely abolished. So now would it be with the implements of war. The federal government would assume not only the responsibility for establishing the moral fitness of prospective purchasers but also the burden of whatever guilt accrued to the weapons' eventual use.

No less than Prohibition, the attempts at reform were destined to become some of this century's better examples of the treachery of good intentions. Far from stopping the arms trade, they now mark the first small steps on the road to today's acronym-infested system of export controls, which have made arms sales some of the foremost instruments of international persuasion and coercion, a diplomatic device we know as weapons diplomacy.

II MIDDLE AGES

5. War and Redemption

For all its noise and scandal, the turmoil of the mid-1930s affected Electric Boat scarcely at all. With the exception of the smears on its escutcheon, the company emerged from 1936 in better shape than it had been for more than a decade. The Second World War was about to begin. At Groton, Elco, and Electro Dynamic, workers laid off during the early years of the Depression were recalled and new ones taken on to replace those who had moved to fresh jobs in other parts of the country. By the year's end, the situation had improved enough for Carse to celebrate the festive season by announcing a profit of seventy thousand dollars, the first in seven years.

What made the turnabout so remarkable was that it had been achieved without recourse to foreign sales of the kind Senator Nye and the Women's League for Peace found so objectionable. It was not, however, a victory for disarmament; far from it. The simple fact was that the newly implemented restrictions were largely meaningless. The predators —Italy and Japan, for example—had their own well-established naval construction industries, while their victims, nations like Abyssinia and China, whose cause might have benefited from infusions of modern weaponry, lacked the money to buy. The few foreign orders on Electric Boat's books were strictly nickel and dime requests for spare parts, refurbished equipment, and technical advice.

The immediate source of Carse's yuletide cheer lay in an unfamiliar quarter, the U.S. Navy. Groton had just finished work on the last of three long-range boats originally ordered in 1931, and was well advanced on three more boats scheduled to enter service over the next two years. Even more encouraging, there were firm indications that the Navy would soon be asking for at least four more boats and possibly as many as seven or eight. As the only submarine specialist to survive the Depression more or less intact, Electric Boat was about to extract compensation for the slights, insults, and neglect it had been forced to endure over the course of almost two decades.

The naval budget had been rising by slight degrees since 1933, when Roosevelt reacted to Germany's withdrawal from the League of Nations by demanding that Congress agree to pledge $1 billion for big-ship construction, to be followed by subsequent amounts of $100 million every year to finance gradual improvements in the quantity and quality of the fleet. Given the growing strength of the disarmament crusade, Congress was surprisingly generous in allocating roughly $250 million for four heavy cruisers, the submarines, and some general purchases of guns, stores, and ammunition. The marine industries did not receive as much as they might have hoped, but there was still room for considerable consolation in a legislative afterthought that was tacked on to the appropriation in deference to White House assertions that America's ability to defend itself was in grave peril. The naval budget, Congress decreed, should continue to expand through the next ten years, with the object of bringing the U.S. fleet up to permissible treaty strength by no later than 1944. In a more significant but less successful move, Roosevelt also sought to divert $238 million from funds marked for emergency relief to the construction of thirty-two warships. That proposal was rejected out of hand, but even in defeat Carse must have been cheered immensely. For the first time in the country's history, an administration had embraced the notion that defense spending can have a stimulating effect on the national economy. In the mid-1980s, with the Pentagon now consuming roughly one third of the federal budget, it is fair to argue that the economy has been stimulated to the point of unbalanced intoxication. At the time, however, even the relatively small amounts Congress was willing to approve had an immediate effect.

Electric Boat came back to life with a quickening vitality that rippled through Groton and its neighboring villages. A few years earlier, yard manager O. Pomeroy Robinson had roamed New England in search of any work that would keep the machine shops open. An oft-repeated tale, perhaps apocryphal, maintains that Robinson did his rounds on an old bicycle, once returning with a contract to repair beauty parlor equipment. By 1937, with the stink of corruption dissipating as the clouds of war came rolling across from Europe, Robinson was able to retire his bicycle and return to the production of the weapons of war. Employment, down to a few hundred during the yard's darkest days, soon reached 1,300 as a spate of fresh Navy orders added to a growing backlog.

Each of the new fleet-class boats taking shape at the mouth of the Thames River was worth roughly $3.2 million, almost twice the price of the American subs built during the closing stages of the Great War. Despite the substantial increase they remained, however, something of a bargain. The old S-series boats had been rolling, reeking, unreliable torture chambers for their officers and crew. Between 1920 and 1927 five S-boats were sunk or damaged in accidents and collisions, a chain of events that promoted a number of congressmen to demand that the Navy abandon submarine warfare forever. The new models were quick to the helm on the surface and a good deal more responsive than their predecessors when running submerged. Crew accommodations were vastly improved, the top speed higher and the crush depth much greater. As with most modern weapons the evolutionary steps that led from the so-called pigboats of the First World War to the fleet-class boats was determined by an equal measure of politics and science.

At the conclusion of the Washington Conference in 1922, the naval strategists of Britain and the United States found that their elected leaders had saddled them with a major dilemma: how best to enforce an effective naval blockade of the Japanese home islands in the event of war?

Under the terms of the agreement, France, Italy, Great Britain, and the United States agreed to forgo the construction of any new naval bases in the Pacific, a decision that ceded the ocean to Japan, the fifth party to the talks. The U.S. was limited to two bases, Subic Bay and Pearl Harbor, while the British had only one, Fortress Singapore. Since none of the three was close enough to serve as a practical offensive base for the short-winded and lightly armed boats then in service, both the U.S. Navy and the Admiralty made the development of long-range submarines one of their principal goals. Britain's decision to favor long-range boats was more than a little ironic, since its delegation opened the Washington Conference with a demand that submarines be banned altogether as "immoral weapons." The politicians' and diplomats' perception of morality no doubt owed much to the country's experience as a victim of the German U-boat blockade, but once the measure had been rejected, all qualms were quickly and quietly forgotten as the architects of Admiralty policy adopted the concept of unrestricted underwater warfare as their own.

In the United States, the twin themes of politics and technological innovation produced what one submarine officer later described as "the most compact and complicated man-of-war ever conceived by man's inventive hand . . . the most complex war engine of them all." While overall length varied from class to class, none was less than three hundred feet long and each required a crew of between sixty-five and seventy-five men, nearly every one a specialist. Though often next to useless, torpedoes had grown to number more than two dozen, while the deck guns, which had been removed during the early 1930s, were replaced and increased to a three-, four-, or even a five-inch bore. Theodore Roscoe, a wartime submarine fleet operations officer in the Pacific, conveyed some of the complexity in his book *Pigboats:*

> Tucked into a space approximately twice as large as a six room house, there are living accommodations, a control room, diesel engines

and electric motors, fuel and water tanks and 252 battery cells, each weighing in excess of one ton. There are air compressors and high pressure air banks for blowing tanks and charging torpedo flasks. There are torpedo rooms fore and aft, 10 torpedo tubes and storage space for 24 torpedoes. Crammed into the remaining nooks and crannies are refrigerated and dry stores, stills for manufacturing fresh water, air conditioners and air purifying equipment, ice machines, shower baths, main ballast tanks, variable ballast tanks, electrical equipment for operating bow and stern planes as well as wells for the periscopes, lazaret, chain lockers, ammunition magazines, the galley and the maze of oil, water and air lines that are its veins and capillaries.

After so many years of relative inactivity, Electric Boat was caught off guard by the sudden rush of fresh orders. There was a critical shortage of trained workmen, not mere tradesmen but artisans skilled in arcane disciplines like welding with mirrors and prepared to work in conditions that were always unpleasant and often dangerous. The need for skilled labor was never entirely met, not even at the height of the war years, when total employment stood in excess of fourteen thousand people. Lawrence Spear and Pomeroy Robinson attempted to solve the problem by establishing two in-house welding and construction courses, as well as an unofficial "apprentice" scheme that saw gangs of locally recruited fishermen, farmers, and lobstermen assigned to ten-man work teams, each under the supervision of a veteran hand. A former employee who served at Groton throughout the war years recalled Electric Boat's renaissance as an exercise in controlled chaos. "I don't think anyone really knew how the place worked from day to day," he recalled with evident pride. "We could never tell if the materials would arrive on time, if there would be enough men, enough steel, enough time." Yet, somehow, it did work. In August 1940, President Roosevelt spent a morning at Electric Boat, where he inspected four new slipways and praised the management's recent pledge to cut five months off the average construction time of a typical boat. "Twice as many submarines as before are being constructed there now," the President told an audience of Groton citizens later that day. "Soon they will be turning them out at a rate of one a month." Roosevelt was wrong—the figure was much, much higher.

In 1936, at the start of its renaissance, Electric Boat had the capacity to build perhaps three boats every year. By 1944, thanks to some $20 million worth of Navy-financed investments in new equipment, renovations, and extensions, the yard was turning out that many boats every month. Nor were the boom times limited to Groton. At Elco's headquarters in Bayonne, New Jersey, Henry Sutphen's team of woodworkers, engineers, and designers was developing the first in a series of almost four hundred PT boats and submarine chasers, worth in excess of $120 million. Not far away, at Electro Dynamic, the last of the divisions to benefit from the growing naval budgets of the prewar years, fresh employees were being taken on in response to an insatiable demand for marine electric systems, auxiliary motors, battery banks, and the innumerable other fittings required by a burgeoning fleet of naval and merchant marine ships. The war remains General Dynamics' finest hour, and the company's success, a tribute to the workers, management, unions, and New England residents who turned Electric Boat into one of the most productive cogs in the entire Allied war effort. The man chiefly responsible was Lawrence Y. Spear, the new president, who had stepped up to replace the aged Henry Carse.

Even the most charitable reading of Spear's first forty years with Electric Boat provides no hint of the organizational abilities he demonstrated during the scramble to meet its wartime obligations. Back at the very beginning, it had been Spear who overruled Frank Cable and insisted on cast-iron parts in the ill-fated *Fulton* of 1906, all but killing the test crew when the cheap replacements disintegrated during a drydock trial. Later, he figured as an ardent supporter of the disastrous plan to break into the global shipping business, and most recently, it had been some of his letters to Zaharoff that prompted many of the Nye hearings' most embarrassing disclosures. We can only conclude that he had been hiding his light under a bushel, because he rose to the challenge of war with an unanticipated degree of aggressive confidence. So great were his achievements that the Navy later bestowed on him its greatest accolade, by naming a class of nuclear submarine tender ships in his honor.

In 1940, Spear repeated an earlier suggestion that the Navy accept submarines assembled out of prefabricated hull sections. When he had advanced the idea three years before, it was scotched on the suspicion that uniform welding standards would be impossible to achieve. The second time around, Spear was more forceful in his talks with the offices of both the secretary of the Navy and the Bureau of Ships, pointing out that Electric Boat could not hope to meet its commitments if it was bound by the Navy's insistence on old-fashioned and unnecessarily detailed construction techniques.

No technical obstacles threatened the proposal, but as Spear no doubt expected, the Navy demanded proof before agreeing to change its ways. In an experiment reminiscent of the test that took place off Rhode Island in 1893, Spear prepared a sealed prefabricated section that was lowered beneath the surface replete with a barnyard menagerie of small animals and birds. Depth charges were detonated at various distances throughout the day without rupturing the welded seams or harming the section's occupants. Pomeroy Robinson, who had to disrupt an already tight schedule to indulge the Navy's whim, later quipped that the tests produced nothing more than an expensive omelet. "Every time we set off a depth charge our duck laid an egg. Our engineer later figured up the ratio of cost to production and estimated that the duck eggs cost us exactly $7,000 a dozen."

Jokes aside, the new construction methods revolutionized Groton as had nothing before. First the keel was laid and then it was matched with prefabricated hull sections lifted into place by a railway crane. Main sections for a three-hundred-foot boat could be welded into place in less than four days, the electricians, plumbers, and internal fitters moving into the empty hull as the metalworkers departed to assemble another boat in the next slipway. So well organized was this astonishingly complex operation, Electric Boat never failed to bring in a boat by the specified date and often managed to make deliveries as much as six months early.

In 1942, for example, Groton's supervisor of naval construction, former submarine skipper Commander D. K. Day, asked a visiting reporter to guess how long it had taken to build the P-class submarine they were touring at the time. Calculating that the boat was about one third complete, the reporter guessed at three months. Day laughed out loud. The truth was that work on the boat had begun less than one week earlier.

Spear also served as an unofficial consultant to the Navy Department and the Office of Production Management, suggesting ways in which small, independent yards could be brought into the submarine effort. One such firm, the Mantiowoc Shipbuilding Company, on the banks of Lake Michigan, became one of the nation's least likely submarine builders when Spear dispatched a team of Groton engineers, foremen, and production executives to set up the operation in mid-1942. Although the wartime production records credit them to another firm, the twenty-eight boats built in Wisconsin over the next three years were Electric Boat's in everything but name. Even the solution to the apparently intractable problem of moving the finished craft from the Great Lakes to the sea was the product of Spear's newly revealed genius. Under his instructions, and in accordance with plans of his own design, Mantiowoc's engineers built a number of huge pontoon docks, on which the submarines were borne first along shallow inland waterways and then down the Mississippi to a temporary submarine base near New Orleans.

When the war actually broke out, the problems confronting Spear and Robinson increased exponentially. On the morning Pearl Harbor was attacked, Electric Boat held firm orders for forty-one submarines. Within days, Spear was instructing his foremen to ignore all previous construction schedules and prepare to build as many boats as they could in the least amount of time. Any tactics, innovations, or refinements, however unorthodox, that promised to speed deliveries would receive the management's and the Navy's wholehearted support. Originality became a crowning virtue, as Spear informed a VIP guest in May 1942. Explaining that chronic shortages of nickel, copper, lead, and quality steel plate often left the plant with less than two days' supply of raw materials, Spear praised his purchasing department for its initiative and perseverance. His "boys," he said, have to "go out and dig it up, and then work like the devil to make sure the War Production Board doesn't give it to someone else."

The shortage of raw materials largely disappeared as American industry and the bureaucracy that supplied it adjusted to a total war economy. However, the dearth of manpower was never entirely solved. Driven by Electric Boat's insatiable demand for workers, Groton's population reached a new peak of fifty thousand people, yet it was still not enough to meet Spear's needs. In desperation, he took to advertising in New England papers and ordered handbills distributed as far afield as New Jersey and northern Vermont. To make up the shortfall, Groton's unions allowed their members the option of exchanging vacation time for war bonds and agreed to management's demand that average weekly working hours be increased to fifty-six hours. By 1943, the labor shortage was so acute Spear and Robinson had to forget their reservations and take on several hundred townswomen as trainee welders, fitters, and general yard laborers. They were also obliged to hire a good number of workers who spoke English only as a second language, a development the Navy feared would make the yard an easy target for Axis secret agents. Spear's remedy, although hard to imagine in the multicultural America of the 1980s, was both simple and absolute. In a warning that won praise from the press, Spear announced that anyone overheard conversing in a foreign tongue would be subject to instant dismissal and a lengthy interrogation at the hands of the FBI. The French-Canadians, Hispanics, and European refugees on the payroll presumably learned to speak English or kept quiet, because there is no record of such action being taken.

Amid all the other changes, Spear's most enduring contribution to the company's future went largely unnoticed. During the run-up to the war years, both Spear and Carse realized that the administrative side of the business needed an infusion of fresh blood. Their own talents, like those of Sutphen at Elco, were essentially those of front-office managers, jacks-of-all-trades who were familiar with the legal, technical, and political aspects of naval construction but masters of none. A new man was required who could oversee contract negotiations in Washington, permitting the old guard to concentrate its energies on the more immediate problems associated with day-to-day operations. The Nye hearings had shown that an imposing presence and an intelligent layman's grasp of the technicalities were no longer enough. The rash of prewar orders merely added to the urgency.

The prospective recruit was a forty-four-year-old corporate lawyer, financier, and a former assistant Treasury secretary during the Hoover administration. He was a self-made man, a millionaire, and, by all accounts, one of the more adept players on Wall Street. Carse offered him the position of chief legal adviser and contract negotiator, but the lawyer was unmoved. He had amassed his fortune representing oil, mining, and railroad interests, and while he had also served as a U.S. Navy ensign during World War I, he knew next to nothing about submarines and little more about the defense business. Carse and Spear persisted, and early in 1937 the lawyer agreed. His name was John Jay Hopkins and his enlistment was undoubtedly the smartest move that Electric Boat's directors ever made. By the time the war was over, Carse was gone and the two surviving members of the triumvirate, Sutphen and Spear, had lapsed into virtual retirement. The new boy ran the show with a flair for the bold stroke not seen since the final days of Isaac Rice. In 1938, however, Hopkins' appearance as senior midwife at the birth of the military-industrial complex lay almost a decade in the future. The immediate task was to make sure that Electric Boat's performance matched its obligations. For once in its history, the company met with spectacular success.

By V-J Day, Groton had built seventy-four submarines, more than the combined production of the five other submarine shipyards and an incredible average of one launching every two weeks. The total becomes even more impressive when the job lots of component parts produced for submarines under construction in other shipyards is taken into account. Groton's boats were also a favorite with their crews, who liked to believe they were stronger and better built than those from nonspecialist yards. It is impossible to know if any were lost due to structural flaws or faulty workmanship, but there is every indication the sailors' faith was not misplaced. Among the countless action reports compiled by U.S. skippers, there are literally hundreds of accounts of boats being taken past their official crush depths, often while damaged or under depth charge attack, and their crews living to tell of the experience. In fact, it was the Japanese Imperial Navy that paid Electric Boat the greatest compliment. The Japanese entered the war with three-hundred-pound depth charges that exploded at a maximum depth of three hundred feet; by 1945, both the standard weight and the maximum operable depth had doubled, so much stronger had the U.S. boats become.

One of the privileges of victory is the right to mold history into the curves of popular myth. And one of the most widely held misconceptions to arise from the Second World War concerns America's response to the challenge of Pearl Harbor. Often quoted by supporters of large peacetime defense budgets and advocates of "containment," the cautionary tale maintains that the U.S. was left open to attack by soft-minded isolationists who muted America's disapproval of tyranny in other parts of the world while simultaneously emasculating the nation's armed forces. However, once the clarion call of war roused the country from its slumber, the sheer strength of American willpower brought about a transformation worthy of Charles Atlas. There is a certain amount of truth in this summation, but like all simplifications, it ignores the obstructions and eddies in the current of events, the conflicts that blemished the birth of Roosevelt's "arsenal of democracy."

Some weeks after the *Wehrmacht* cakewalked through France and the Low Countries, the President went before Congress to ask for an increase in the defense budget that staggered even the heads of the armed forces. He began with a call for an Army Air Corps of fifty thousand planes and one million men to keep them flying. Next, there was a demand for similar, though smaller, commitments to the Army's ground forces and the Marine Corps, and finally, a priority request that the Navy receive $4 billion to finance a two-ocean fleet. Shocking as they seemed at the time, the figures were trivial in comparison with the actual amounts poured into the war effort before the end of 1945. More than $50 billion worth of military supplies went overseas under the auspices of Lend-Lease alone, while the exact amount spent on U.S. forces is almost impossible to calculate with any degree of accuracy, since the figure includes the cost of factory extensions, tax breaks, and direct grants from a score of government agencies. As for the statistics of production, they remain awe-inspiring. Within days of Pearl Harbor, industrial organizations representing some eleven million workers voluntarily surrendered their members' right to strike, and by December 1942, U.S. industrial output had grown to match the combined total for all the Axis powers. The next year, war production would double, before reaching its peak in 1944, when it doubled again.

What the statistics fail to mention is the very thing that made the transformation possible, the strained alliance between Hoover-style conservatives like Jay Hopkins and Roosevelt's New Dealers. Every bit as important to the outcome of the war as any major battle, the union was achieved only in return for what many White House sympathizers saw as the abandonment of the administration's agenda for social and economic reform. The Civil War historian Bruce Catton, then a Washington newspaper reporter, later wrote that "oddly matched rivals had made a truce, but it was only a truce and not a treaty of enduring peace."

The potential obstacles to a rapid industrial mobilization had been identified three years before by a congressional committee investigating the concentration of economic power. "Speaking bluntly," the report began, "the government and the people are over a barrel when it comes to dealing with business in time of war or other crisis. Business refuses to work, except on the terms it dictates. It controls the natural resources, the liquid assets, the strategic position in the country's economic structure and its technical equipment and knowledge of processes. The experience of the First World War, now apparently being repeated, indicates that business will use this control only if it is paid properly. In effect, this is blackmail." The report concluded with a question that was soon answered: "What price patriotism?"

The peace had been struck, wrote the celebrated William Allen White of the *Emporia Gazette,* at the cost of placing war production in the hands of "absentee owners of amalgamated wealth," men who were determined to "come out of the war victorious for their stockholders." A report compiled by a Senate committee in 1946 confirmed his prediction. More than 70 percent of all war contracts went to the top hundred industrial companies, which also absorbed roughly $22 billion worth of government-financed renovations, extensions, and investments in new equipment. The general attitude of the business community to the wartime bonanza was best and most bluntly expressed by Lammont Du Pont, chairman of the board of Du Pont de Nemours and a member of the resolutions committee of the National Association of Manufacturers. "Deal with the government and the rest of the squawkers the way you deal with a buyer in a seller's market. Nineteen hundred and twenty-nine to 1942 was the buyer's market—we had to sell on their terms. When the war is over it will be a buyer's market again. But this is a seller's market. They want what we've got. Good. Make them pay the right price for it . . . and if they don't like the price, why don't they think it over?" There was, as Du Pont well knew, no time for consideration, since the U.S. government was indeed "over a barrel." While the resolutions committee sat formulating the association's attitude to war production, U.S. Marines were fighting on Guadalcanal and Hitler's eastern army was sweeping all before it on the road to Stalingrad.

Electric Boat's attitude to the war was a good deal less combative than Du Pont's. Unlike the airframe industry, which journalist I. F. Stone described as "a front for the rest of business in its fight for special tax privileges on defense contracts," the submarine builder maintained a pliant profile in its dealings with Washington, never refusing to accept new contracts or complaining about inadequate compensation. Instead, Spear was content to accept the fringe benefits accruing to a successful war producer. When *New York Times* reporter Sidney Shallett visited Groton in mid-1942, Spear was able to take him on a tour of the old iron foundry that had been Electric Boat's next-door neighbor on the granite banks of the Thames River. The site had been largely cleared by gangs of workmen and demolition experts, who had just begun blasting ten new slipways out of the exposed rock. It was hard, demanding work, Spear explained, adding that the renovations would cost the Navy between $10 and $12 million. However, the cost was entirely justified by the expectation that production would double when the new facilities became fully operational. Nor were the slipways to be the last gift from Washington. By war's end, Electric Boat was the beneficiary of more than $20 million worth of government-subsidized improvements and capital investments.

The Groton complex has changed much over the years, expanding and evolving in response to each of the progressively larger generations of nuclear-powered boats that have followed the *Nautilus* down to the sea since 1952. But beneath the current clutter of dirty sheds, cranes, floating tenders, and sawtoothed rooflines, the bounty of World War II is still apparent. When the fighting stopped, the company continued to operate the Navy-financed extensions as its own. There was nothing unusual or untoward about the situation; it was simply one of the peculiar advantages enjoyed by defense producers over their counterparts in civilian industry. The Fort Worth plant, the home of the F-16 fighter, was put up originally to build heavy bombers for the war against the Axis. Hitler and Mussolini are long gone but the plant remains government property to this day. So, too, are the Land Systems division's tank factories, where it is the U.S. Army that controls the lease. The question of ownership helps to explain why, in the years that followed the war, none of these facilities has ever fallen completely idle. At risk of oversimplification, it represents an example of irreversible expansion: once established, the U.S. munitions industry could not withstand more than a slight contraction without risk of total and catastrophic collapse. Even more than the windfall improvements to its physical facilities, Electric Boat's new position as one of the main props supporting the national economy was the war's greatest gift.

For all the light they shed, the production statistics are also misleading, because they imply that battles are won by the captains of industry rather than the men in the field. This is simply not true. Wars reflect the personalities, gifts, eccentricities, and blind follies of the men who prosecute them. Popular myths to the contrary, the evidence indicates that heroic efforts on the home front were often undermined by the blinkered stupidity of military commanders. Electric Boat's submarines make an excellent case in point, a still horrifying example of the reasons why sophisticated hardware, and the ability to produce it in unlimited amounts, are not enough in themselves to assure victory.

American submarines sent some six million tons of enemy shipping to the bottom, a figure that is impressive only so long as it is viewed in isolation. In comparison with the results achieved by the Japanese, Italian, or German submarine commands, the total is paltry. Over the course of just six months in 1942, German U-boats sank three million tons of shipping in the North Atlantic alone. It is difficult to understand why this should have been so. The boats and crews of the rival submarine fleets were evenly matched in terms of numbers, sophistication, and competence. Yet the Axis boats proved so much more effective. Why?

The blame lay partly with an American naval high command that had little idea what to do with its submarines. This was not a case of oversight but a product of the official view that the next war would be fought primarily on the surface. The admirals were prepared to admit that submarines would have their part to play, but there was no formal consensus about what that role might be. The theory most favored depicted titanic surface battles that would see submarines scouting the fringes of the engagement like an underwater cavalry corps. Incredibly, the lessons of the Great War were ignored and the possibility that merchant shipping might become the submarines' chief target was given formal but scant consideration.

The irony of the situation was brought home with the force of the bombs that fell on Pearl Harbor. With its surface fleet in tatters, those in charge of Pacific operations found that the only battle-ready force at their disposal was a seven-vessel flotilla of submarines. The battleship fixation that had dominated prewar strategy led Clay Blair, Jr., one of the most energetic chroniclers of the "Silent Service," to comment that there was "a failure of imagination at the highest levels." The Navy moved its submarines about the Pacific with the uninspired flair of a small child pushing some little-favored vegetable about an empty plate. The boats seldom encountered potential targets, and when they did, the norms of peacetime training placed their skippers at a considerable disadvantage. "The U.S. force was divided and shunted willy-nilly on missions for which it was not suited while the bulk of the Japanese shipping sailed unmolested in Empire waters through the bottleneck in Luzon Strait," Blair wrote.

The inadequacy of theater tactics during the early stages of the war was but one part of a much broader failure. Throughout the 1920s and '30s, submarine officers were taught that caution was the best, indeed the only, approach. They were to remain concealed, rising to periscope depth every so often and risking a torpedo attack only when conditions were near-perfect. The tactics worked well in peacetime naval maneuvers, where they guaranteed that the carefully orchestrated game plans would not be disrupted by unplanned sinkings. Under the very different conditions of war, they robbed submarine captains of their chief advantage, the ability to combine judicious aggression with the element of surprise.

Even in today's era of so-called "smart" weapons, submerged torpedo attacks are difficult at the best of times—something that was proven most recently by the captain of an Argentine submarine who is thought to have fired between three and five torpedoes at various British ships during the Falklands war without scoring a single hit. In the early days of World War II, the difficulties were infinitely greater thanks to an official policy that banned the use of periscopes lest they be spotted by eagle-eyed enemy observers. Instead, the captain was supposed to launch his salvo with the aid of a primitive and highly unreliable acoustic range finder. The consequences of this white-cane approach are easy to imagine. A submarine groping about in the vicinity of an enemy convoy dispatches its torpedoes without any idea if they will even come close. The salvo misses its mark, and quite suddenly the hunter becomes the hunted. Tracing the torpedoes' wakes to the point of origin, the enemy captain calls in an escorting destroyer and sails safely away while the submarine and its crew are left to fight for their lives.

The idea that caution was the submariner's greatest virtue was destined to go on causing trouble long after the Navy repudiated it. Many captains of the old school proved incapable of producing the newly required degree of aggressiveness and had to be relieved of their commands. In fact, America's war was almost eighteen months old before a new generation of combat-trained skippers began to take the kind of risks so often depicted in Hollywood movies. People like Captain Thomas Klakring, the future head of Electric Boat's postwar testing and trial team, were typical of the new breed. Klakring himself once took his boat so close to the Japanese mainland he was able to amuse his crew with periscope descriptions of the passengers on a coastal train. Another skipper, Commander Dudley (Mush) Morton, adopted such an aggressive attitude he would almost certainly have been tried as a war criminal had he fought on the other side. Whenever he sent a ship to the bottom—something he did quite frequently during 1942—he made a point of coming to the surface to machine-gun survivors. Slowly, the submarine fleet's deplorable record became merely dismal.

The newspapers made wonderful propaganda out of Klakring's exploits in sinking six enemy ships in the course of a single patrol. What they failed to mention was that the figure should have been much higher. When he arrived back in Pearl Harbor, the newly famous skipper lodged an official complaint about the quality of his torpedoes. More than half his "tin fish" failed to explode, he wrote. Some leaped like porpoises from the water, while others veered off on unpredictable arcs. Of the shots that actually found their targets, most struck home with nothing more than a dull thud. The official response was a lecture and a warning. American torpedoes, he was told, were the best in the world. If they failed to explode, then the cause had to be human error. Klakring was firmly advised to stop making waves and sent back to sea with fresh supplies of the same torpedoes. Two other, less celebrated skippers who made similar complaints were summarily stripped of their commands and sent home to dockyard jobs as an example to potential troublemakers.

In equipping its boats with defective weapons, the Navy's Bureau of Ordnance fell victim to a malady familiar to students of the modern Pentagon. It placed too much faith in technology and too little in the opinions of its own line officers.

When the victors divided up the pitiful spoils of the Great War, few trophies excited more interest than Germany's magnetic mines, surprisingly simple devices that seemed at times to possess an independent intelligence. Unlike earlier contact mines, with their prickly external spikes, magnetic mines relied upon a trigger mechanism that was based upon a compass needle and a bank of batteries. Whenever an iron ship came close enough to disturb the magnetic field, the needle swung away from its true heading until it closed an electric circuit that detonated two thousand pounds of concentrated high explosives. The Germans made some preliminary efforts to adapt the exploder to their torpedoes, but soon gave the effort away when it was decided conventional detonation systems were good enough to deal with the thin-hulled steamers that were the U-boats' chief targets. The idea went back on the shelf to await the coming of peace and a sponsor with the time and determination to overcome the formidable development problems.

That patron turned out to be the U.S. Navy, which began the 1920s with a frantic search for an effective battleship-killer. The men in charge of the Bureau of Ordnance appear to have learned nothing from the war, since they clung to the notion that the next war would see submarines pitted against not merchantmen but large and heavily armored warships, a task for which the existing torpedoes were considered next to useless. Magnetic-influence exploders appeared to present the perfect solution. If they could be made to detonate beneath the keel, where the layers of armor plate were least thick and the concussive effect was most intense, a single shot might be enough to sink the most formidable enemy.

Research progressed quickly, given the diminished peacetime budgets of the day, and by 1926 the Bureau of Ordnance was confident it had perfected an infallible weapon. The new torpedo was certainly sophisticated. It could be adjusted to run at speeds of up to forty mph over short distances, or slowed to three-quarter speed for long-range attacks on targets up to nine thousand yards distant. All carried a standard five-hundred-pound warhead and two independent exploders, the first an improved version of the old-fashioned contact fuse, and the second the magnetic-influence sensor, which the Navy conspired to keep wrapped in a veil of impenetrable secrecy.

The Navy's faith in the new torpedo was curious, to say the least, since it had been subjected to a realistic test only once, when an ancient submarine was struck by a single shot launched from a moored barge. The target boat was lifted almost out of the water by the blast and sank in less than a minute. For all its spectacular appeal, the demonstration proved very little. The submarine in question was old, top-heavy, and thin-hulled. Even if the test torpedo had carried nothing more than a dummy warhead, the mere impact of the collision would have been sufficient to breach the thin layers of plate and flood the unsecured interior. A superannuated battleship or even a large cargo ship might have provided a more telling target, but the Bureau of Ordnance refused to authorize what it regarded as an unwarranted second test. Its sole concession throughout the late 1920s and the 1930s was a series of "theoretical appraisals" that saw a handful of torpedoes fired beneath the keel of the cruiser *Indianapolis.* Each shot was painstakingly prepared and all carried a magnetic sensor that was supposed to indicate if the magnetic detonator circuits lived up to their reputation. Not surprisingly, the torpedo emerged from the tests with a clean bill of health. Cynics at the Newport Torpedo Station who queried the procedure were warned to keep their objections to themselves. The absolute perfection of the Mark XIV torpedo had become a basic tenet of Navy dogma, to be challenged at the peril of an officer's career. It was a creed that would claim more than its share of martyrs before being laid to rest.

The new torpedoes went into production surrounded by a concern for security that now seems laughable in view of the weapons' many flaws. A booklet that explained how the magnetic exploder was to be installed and operated was set for distribution, when the Bureau of Ordnance intervened and ordered the entire shipment withdrawn from circulation. As for the exploders themselves, after leaving the factory they were shipped off to guarded warehouses, where they were kept under lock and key at all times. Captains who sought to learn more about the alleged superweapon were given wooden dummies and told that all would be revealed when the time was ripe. It was a policy of unmitigated stupidity and the consequences were appalling. More than a year after Pearl Harbor, most skippers still had no clear idea how their main weapon was supposed to work, nor any theoretical grounding that might have permitted the more astute to figure out what was going wrong.

Nothing better illustrates the dangers of the Mark VI torpedo than the fate of the old V-series boat *Argonaut,* which was lost with all hands off the coast of New Britain in the early months of 1943. When Captain J. R. Pierce spotted a small convoy with an escort of three Japanese destroyers, he staged a quick-fire, copybook-perfect attack that followed Navy teaching to the letter. The torpedoes were set to run deep to make maximum use of the magnetic exploder's reputed ability to blow up beneath the keel and "break" the victim's back. Unfortunately, they did nothing of the kind. Two of the torpedoes either struck their targets and failed to explode or passed harmlessly beneath them. The third detonated prematurely, doing little more than superficial damage to the third Japanese warship. The two undamaged destroyers peppered the sea with depth charges, quickly forcing the *Argonaut* to the surface, where the crewmen were gunned down as they tumbled onto the deck and attempted to surrender. As fate would have it, the entire incident was witnessed by the crew of an Army bomber returning to base after a raid. By any reasonable expectation, the aircrew's account of the defective torpedo shots and the resulting massacre should have prompted the Bureau of Ordnance to revise its position. Incredibly, the official reaction was nothing more than an irritated bristling. Despite conclusive evidence to the contrary, the bureau continued to insist that its torpedoes were perfect, implying that the *Argonaut* was somehow responsible for its own demise. Meanwhile, the toll of lost ships and men continued to grow.

Every skipper had a different horror story, and each seemed to concern a different component part. If they set their weapons to run deep and explode beneath the keel, there was a better than even chance the projectile would simply continue on its way, oblivious of the target; if, on the other hand, they opted for a shallow-running contact fuse, there was still no guarantee of an explosion, since this device was no more reliable than the first. Nor were the problems confined to the warhead. Many torpedoes shed their steering vanes as they left the tubes, while others ran about in circles, occasionally turning about completely, to bear down on the boats that launched them. To base commanders like Admiral Robert Lockwood, the plethora of complaints was simply unbelievable. Like all submarine officers of his generation, he had been taught that the Mark XIV torpedo was the acme of naval weaponry, and while he could accept that it occasionally ran several feet deeper than anticipated, he would not at first concede that so many mistakes could be crammed into such a small and revered package.

One quality that set Lockwood apart from his peers was the ability to change his mind. He did just that when the captain of the submarine *Tinosa* told of pumping one dozen torpedoes into a solitary tanker from a variety of angles without the satisfaction of a single detonation. The *Tinosa* brought its remaining torpedoes back to Pearl Harbor, where Lockwood ordered it fired into the base of a cliff along with several others selected at random from the ordnance store. Much to everyone's surprise, the first pair of shots exploded on cue. The third did not, and was soon being dismantled by a crew of very brave technicians who had no clear idea what they were looking for and no reason to believe the smashed, twisted, and temperamental projectile would not blow up in their faces.

It took less than an afternoon to isolate the problem and another half day of testing to confirm the diagnosis. After a fourth torpedo was dropped on its nose from the top of a thirty-foot tower, it was established that the pins holding the contact exploder in place were simply not strong enough. Whenever a torpedo struck its target at any but the most oblique of angles, the exploder was torn loose from its mountings and crushed before it could close the electrical circuit. A couple of additional bolts and a makeshift bracket of angle iron were all that were needed to set things right.

Confronted for the first time with unimpeachable evidence, Admiral Robert English and the Bureau of Ordnance's other presiding desk officers were forced to begin the development of a replacement. Ironically, the quest for a new mechanical exploder also marked the end of the magnetic device. No doubt recognizing an opportunity to end the constant stream of complaints, the bureau omitted any trace of magnetic sensors from its replacement design, offering neither apology nor explanation for the change of heart. All the same, it remained the best news the submariners had heard since Pearl Harbor.

The disappearance of the despised exploders signaled a major shift in the submarine fleet's fortunes. Early in 1943, the bureau released the first of its new all-electric torpedoes, to immediate and instant acclaim. The new weapon was slower than the one it replaced but infinitely more reliable. It was also much safer since, unlike its steam-driven predecessor, it left no wake of telltale bubbles to reveal the attacker's position. Only when the war was over did it emerge that Admiral English and his staff had made a strenuous effort to cripple that torpedo as well. The new model was based upon a captured German weapon, a fact the bureau resented, apparently because it did not originate from one of its own research programs. The bureau's procrastination became the subject of a scathing report in April 1943, when the Navy's inspector general concluded that "the torpedo station had its own electric torpedo, the Mark II, and the personnel assigned to it appear to have competed and not co-operated with the development of the new Mark XVIII. . . . Failure to provide experienced and capable submarine officers to the Bureau for submarine developments has contributed largely to the above deficiencies."

It is fair to say that the U.S. submarine fleet achieved its limited successes despite the naval hierarchy rather than because of it. When the problem concerned neither tactics nor weapons, it was likely to be one of the obscure feuds that split the Pacific submarine fleet's headquarters. At one point the force was divided between the east and west coasts of

Australia, not because the deployment promised any strategic advantages but because two senior staff members could scarcely stand the sight of each other. Tucked away in the southwest corner of the island continent, the submarines in the second group were required to sail thousands of miles to the north before reaching their operational patrol areas among the islands of the East Indies. There was only one enemy against whom the submarine fleet's leaders presented a unified front, and that was not the Japanese but General MacArthur, who was universally suspected of an attempt to establish his own private Navy. MacArthur's requests for submarines to support espionage and commando raids always met with entrenched opposition. That the military commander managed to obtain the little cooperation he did remains a tribute to his forceful personality.

It is easy to dismiss the fumbling, culpable stupidity of forty years ago as another of the war's quirky incidents, a dismal but essentially unimportant chapter in a much larger story that still somehow turned out for the best. Easy, but shortsighted. The world and the weapons that dominate its uneasy peace have both changed much since World War II, but the outlines of old follies are still evident beneath the high-tech distractions of today's strategic scenarios. The theoreticians of the 1930s gave little thought to preparing for anything other than titanic sea battles; today, when many would argue that surface ships are more vulnerable than ever before, the Navy has said it will send aircraft carrier task forces steaming straight into the teeth of the Soviet Union's coastal defenses. Similarly, plans for the ground defense of Eastern Europe demonstrate a faith in untested technologies and tactics that puts the exponents of the magnetic exploder to shame.

However, of all the areas where history appears to be repeating itself, the most telling concerns the evolution of individual weapons. Once again, the F-16 jet fighter provides the perfect example. It is hard to imagine any two things with less in common than a World War II–era submarine and a supersonic aircraft. Yet the comparison remains apt for a variety of reasons, many not immediately obvious. When America's submarines went to war in 1941, there was nothing wrong with the boats themselves—those of the fleet-type were among the most sophisticated, battle-ready items in the Navy's inventory. Their initial lack of success stemmed not from any inherent flaws but from the poor training, faulty tactics, and useless weaponry that was forced upon them by a naval bureaucracy as far removed from reality as it was from the enemy.

Two generations later, it is the F-16 that has fallen victim to a similar obsession with sophistication. Originally developed as a cheap, highly maneuverable air-superiority fighter capable of running rings around any and all opponents, it is being burdened with an ever-increasing array of additions and alterations that have rendered its dogfighting capabilities largely irrelevant. In a move chillingly reminiscent of the philosophy that inflicted the Mark XIV torpedo on the submarine fleet, the "new" F-16 will employ a fire-and-forget AMRAAM missile. As with the Mark XIV, few pilots will have had a chance to train with the new and very expensive weapons before they are called upon to use them in action since the Air Force has announced it will stage only ninety-four official tests. If the day ever comes when AMRAAMs are fired in anger, the results, like those for the torpedoes and acoustic range-finding system, may well prove less than anticipated. Early tests have shown that the missile's "active seeker" target selection computer often has trouble finding its quarry, particularly under the cloudy and foggy conditions that prevail over Western Europe for much of each year. During World War II, there was time to correct many of the weapons' various faults. The next war, if it comes, will be very different. If the much-touted conventional weaponry fails to live up to its designers' expectations, the temptation to "save" the situation by resorting to nuclear weapons may well prove irresistible.

One Electric Boat product that was allowed to live up to its potential was the patrol torpedo (PT) boat, arguably one of the world's most cost-effective weapons. In view of its success, it should come as no surprise that this innovative craft also suffered from intense opposition at the Navy's highest levels.

Even though the British had made extensive use of small, heavily armored motor launches during the First World War, the U.S. Navy allowed the idea to languish until 1938, when its General Board was prodded into action by no less an authority than the White House. During his short term as a Navy under secretary in the Wilson administration, Roosevelt had taken a keen interest in the first generation of boats, many of which were built in the U.S. for the Royal Navy. With the right tactics, they were versatile hunters for use against submarines, shore installations, and, when conditions were right, surface vessels many times their own size. They were also inexpensive. Built of twin layers of mahogany planking separated by a resin-impregnated sheet of canvas, the descendants of those first boats promised to provide a readily available addition to the Navy's short supply of capital ships.

By any reasonable expectation, Electric Boat's Elco division in Bayonne should have walked away with the contracts for all four of the experimental prototypes that the Navy ordered in response to White House pressure during 1938. The yard was an acknowledged leader in the field of large, fast pleasure craft, while its ability to fulfill accelerated naval orders was beyond dispute. During the final stages of World War I, Elco set a world record that has yet to be equaled, producing 550 eighty-foot motor launches for the Royal Navy in just 488 days. If the members of the U.S. Navy's selection group had bothered to seek additional confirmation of Elco's competence, they need only have cast their minds back to the days of Prohibition, when its boats were pressed into frequent service by a generation of rumrunners and smugglers. Incredibly, each of the prototype contracts went to rival firms. As subsequent tests would prove conclusively, there was not a good design among the lot.

The kind of boat favored by the White House combined massive horsepower with a planning hull and an assortment of torpedoes, depth charges, and heavy-caliber automatic deck guns. Whatever else they may have been, the winning entries were certainly not what the President had in mind. A pair of smaller boats were simply too small; at less than fifty-seven feet from bow to stern, they lacked the length and beam to handle the large, twenty-one-inch-diameter torpedoes that the Navy hoped to use. The two larger boats, on the other hand, were so big their underpowered engines could not meet the required top speed. Of them all, the most sadly deficient was the design in which the Navy invested the greatest pride and expense. Constructed with painstaking thoroughness in the Philadelphia Naval Dockyard, the all-metal boat was so heavy, so large, and so slow even the hard-pressed Royal Navy later rejected it when the administration offered to include it in a package of discontinued prototypes given to Great Britain as emergency war aid in 1942.

With the affair shaping up as another example of a promising weapon spoiled by an indifferent naval bureaucracy, Elco's Henry Sutphen was left with no choice but to pursue the prize through irregular channels. He was fortunate in being able to count among his allies one of the few men in a position to exert his authority over the members of the General Board, Assistant Navy Secretary Charles Edison. The two men met informally on a number of occasions throughout January 1939 to discuss Sutphen's charge that the Navy was about to bury a good idea beneath layers of administrative incompetence. Edison agreed with his friend's summation and promised to do all that the law allowed to secure Elco's involvement. Sutphen had only to come up with a more promising design.

For some months Sutphen had been following the British Admiralty's attempts to develop a motor torpedo boat of its own and had just learned that a sixty-foot craft designed by the Vosper company had just been selected to go into full production. It was, in his view, the wrong decision. The rejected competitor was the brainchild of the then well-known record-setter Hubert Scott-Paine, a member of that peculiarly British breed of tinkering gentleman mechanics who were to play such an important role in the development of so many wartime defense projects. Scott-

Paine's boat was big enough to suit the U.S. Navy's needs, priced to win White House support, and, unlike the lackluster American designs, capable of performing its mission without presenting an easy target or sinking beneath the weight of its own armaments. With nothing more than Edison's verbal assurance that he would be reimbursed for his trouble and expense, Sutphen set out for London and a meeting with Scott-Paine.

The Englishman was more than receptive, quickly agreeing to sell Elco the boat he had prepared for the British trials and volunteering to accompany it to America, where he would supervise the alterations required to bring it up to Navy standards. Sutphen returned home, and just as he had promised, Under Secretary Edison obtained the General Board's approval. The board's chairman, Admiral Thomas C. Hart, phrased his reply with a diplomat's concern for the dented egos of his own junior officers. The British boat was to be purchased not because it was superior to its American counterparts—something the comparative performance statistics confirmed beyond the shadow of a doubt—but because "said design is known to be the result of several years' development, [and] the General Board considers it highly advisable that such craft be obtained as a check on our own development."

The Scott-Paine boat arrived in New York as deck cargo in early September and was taken straight to Bayonne, where its designer was waiting to begin work. The boat was stripped, overhauled, and then rebuilt under Navy supervision, all within the space of two weeks. Edison, still Elco's greatest ally, was impressed by the preliminary sea trials along the coast of Long Island and dashed off a memo to the White House requesting that the President release $5 million, all that remained of an original $15 million set aside for the development of the four stalled prototypes. No doubt concerned at the possibility that this latest boat might prove to be another voracious consumer of research funds, Roosevelt scratched two vital questions in the margin of his under secretary's memo: "How many? How much?"

Heeding his master's demand for economy, Edison made sure that his friendship with Sutphen did not enter into the negotiations. Sutphen began by offering to build sixteen boats for the $5 million. Edison countered with a demand for twice that number. Much to Sutphen's annoyance, the final figure was set at twenty-three, enough to equip two experimental squadrons, with the addition of Scott-Paine's original boat, now known officially as PT-9. If Sutphen was vexed by the hard bargaining, he soon had reason to grow even more irate. When Scott-Paine's blueprints arrived from England, Sutphen and his chief designer, Irwin Chase, discovered that the Englishman had simply cleaned out his shelves and files, dumping thousands of pounds of documents and mechanical drawings into crates without making the slightest attempt to catalog or classify them. After several weeks of mounting frustration, Chase was forced to recommend starting from scratch. The PT-9 was "borrowed" back from the Navy and systematically measured, photographed, and copied. It was a laborious, annoying task that lent some small degree of credence to Sutphen's later assertion that the original contract involved Elco in a loss of more than $600,000. There was, however, much better news to come.

The following year saw Congress approve a further $50 million for PT boats, the vast bulk of the money going to Elco, which had made its own alterations to the original Scott-Paine design. The new models were slightly longer than the prototype, seventy-seven feet rather than sixty-nine, and they substituted a more powerful Packard 1,200-horsepower engine for the original Rolls-Royce power plant. By the end of the war, Elco's production stood at 398 boats, well over half the country's total production.

Elco came close to losing its premier position only once. In the summer of 1941, the Navy bowed to pressure from other shipyards and agreed to conduct a short series of competitive trials that reached their climax in the waters of Long Island Sound with an event that came to be known as the Plywood Derby. Four boats, each representing a different builder, were pitted against each other over a 190-mile course. Elco hardly needed the home-water advantage to come out on top, beating by almost an hour the only other boat to finish. The second-place competitor also reaped a reward. Built by Higgins Industries of New Orleans, it was the sole survivor of the four original prototypes commissioned back in 1938, and eventually became the U.S. Navy's second-string boat.

With the exception of the Reagan era, forty years later, World War II remains the most active and profitable chapter in General Dynamic's history, with close to one billion dollars' worth of contracts going to Elco, Electric Boat, and Electro Dynamic. After the long drought of the 1920s and the indignities of the mid-1930s, the company was finally back on the rails: out of debt, its reputation rehabilitated by the war, and flush with accumulated wartime profits. There was only one cause for concern: how to avoid falling victim to the treacheries of peace.

6. The Promised Land

There is always a temptation to view history as a picture postcard, the near and recent peaks dominating the landscape while those in the middle and far distance are shrouded by a haze of often selective hindsight. The forty years since World War II present just such a time-shrunken perspective. The furthermost pinnacles rising above the clouds mark the start of the long trail that runs from Pearl Harbor to Hiroshima and on to Korea, Vietnam, and parts even closer. As a rough guide to the topography of the recent past, the depiction is accurate enough, although it does suffer from one major deficiency. The valleys, those moments of relative calm that separate each great upheaval from the next, tend to be both overshadowed and overlooked. This is indeed a pity, because it was within one of those notches in the distant skyline that the modern world, and with it General Dynamics, assumed their present forms.

November 1945 was an anxious month, a moment caught between the lingering elation of V-J Day and the growing realization that the hard-won peace might not live up to the rosy promise of wartime propaganda. The world beyond America's shores, it was generally agreed, would be a very different place, but that was about as far as the consensus went. With the nation's attention focused on the twin novelties of peace and relative prosperity, a coincidence of good fortune not seen since the great Wall Street crash, there was little time and less inclination to dwell on the possibility of future conflicts. As if by common accord, the martial and patriotic motifs that had figured in every aspect of American life for half a decade began to disappear. Rosie the Riveter swapped her coveralls for Dior's voluminous New Look, while many of her sisters spent their evenings writing Washington to demand the early demobilization of boyfriends, husbands, and fathers. The stampede to peace was irresistible and soon swept President Truman along with it. In the face of opposition from his military and economic advisers, the President ordered an accelerated demobilization that saw the world's largest armed force shrink by 90 percent in less than a year.

Truman's acquiescence to the demands of the Bring Daddy Home clubs alarmed many in Washington, for a variety of reasons. Among the administration's economic advisers there were understandable doubts about the private sector's ability to absorb and sustain such a massive addition to the work force. Those whose attention was focused on the international stage were no less dismayed by the armed forces' sudden disintegration. It seemed to them a return to the pacifist isolationism that marked the end of the First World War, and they campaigned vigorously against it, demanding not only a slow and orderly demobilization but a permanent peacetime draft.

Main Street's preoccupations, when daddy did come home, reflected the fears of the administration's economists. Housing was almost unobtainable and the newspapers were rife with stories of decorated veterans who had taken up residence in garages, basements, and, in one of the most publicized cases, a renovated chicken coop. For those lucky enough to find a place to live, the problems associated with furnishing it were often insoluble. Throughout the last eighteen months of the war, the military had resisted all White House attempts to organize an early and orderly reconversion. The immediate result in late 1945 and early 1946 was that factories which might have been producing sewing machines and radios were only just finishing their last, backlogged orders for radar sets, boots, and gunsights that were no longer needed. The notable exception was the automobile industry, which had ignored the Pentagon's objections and made preparations for the transition to peace.

It was during the last week of that nervous and distracted November that one of the most anxious men in Washington left for New York, intent on injecting a note of sobriety into a victory celebration run long past its time. General George C. Marshall, the man who would soon lend his name to an ambitious plan to rebuild Europe on a foundation of U.S. loans, maintained that America was in danger of losing its national resolve. The high purposes of the war years, he said, were about to be subverted by "the pygmy confusion of peace." He had been pushing the same line in and around Washington since Roosevelt's death, and now, thanks to an invitation from the *New York Herald Tribune,* he was about to place his sentiments on the public record before a luncheon audience of some four hundred business leaders, political figures, and editorialists.

The speech amounted to a formal notification that the next war was already under way and a warning that the U.S. could not afford to wait on the sidelines as it had done during the 1930s. As he explained it, the postwar world was a prize that might fall either to the "operators"—a reference to the Kremlin that required no further explanation—or to the "cooperators," nations that were waiting to line up behind the United States to defend and expand the right of law. American leadership and financial support were essential if the "cooperators" were to survive and prosper; otherwise, they would be gobbled up with the same ease and certainty with which Hitler swallowed first the Sudetenland and then the rest of Europe. Marshall attributed part of the blame to the accelerated demobilization, which he said had prompted "a widespread demoralization of the American people . . . a disintegration, not only of the armed forces, but apparently of all conception of world responsibility and what it demands of us." As the first steps, Marshall called for an unqualified U.S. commitment to the United Nations and endorsed massive loan assistance to Europe, which he described as no mere act of charity but an essential investment in a stable political and economic future.

However, it was the final and, at the time, least likely proposal that proved to be the most prophetic element of Marshall's prescribed code of conduct for what was evidently going to be a dangerous new age. Returning once again to the lessons of the 1930s, he made a plea for a robust and innovative military: "As late as 1937 we might have convinced the Axis gangsters of the complete futility of their plans by simply matching our cigarette money with expenditures for national defense," he said. Now, while the country carped about the shortage of kitchen appliances, another generation of villains was rising out of Eastern Europe. It would be "monstrous," he said, if the United States chose once again to turn its back on the world and potentially suicidal if it "failed to maintain an adequate military posture."

Despite at least one editorial that described Marshall as the conscience of a nation, his warning had little immediate public impact. Even among the major wartime defense producers, a sector of the economy where he might have expected a sympathetic reaction, there was a curious disinclination to believe that the armed forces would be as important in the years to come as they had been in the war just won. History alone made it unlikely. Even during the heyday of Teddy Roosevelt's Great White Fleet, there had never been enough work to support more than a handful of long-suffering defense specialists like Electric Boat. It was true that the First World War had seen the creation of a massive munitions industry, but it returned to civilian pursuits just as quickly in the 1920s. Many wondered why the aftermath of this latest war should be any different. Marshall and Washington's other internationalists could talk all they liked about America's destiny as a benevolent global policeman, but the public mood made it seem most unlikely.

A large part of the problem was the atom bomb. Perceived as the ultimate arbiter of power, it was supposedly the one weapon that would do away with all others. Wars would be settled in hours and without recourse to the disruptive industrial mobilization that had made the last victory possible. Since it was generally believed that America's monopoly of nuclear weapons would continue at least until the mid-1950s, the outcome of such a war appeared equally certain.

The notion that tanks, capital ships, and even the common foot soldier had been rendered obsolete flourished in some of the most unlikely places. *Aviation News,* a trade publication whose editorial columns often served as an unofficial mouthpiece for the senior levels of the Air Force hierarchy, was quick to sound the death knell for the other services. In September 1945, for example, an unnamed writer claimed that the U.S. now required nothing more than a fleet of long-range intercontinental bombers capable of delivering an atomic punch to any corner of the world. And even the manned bombers' days were numbered, the writer enthused, predicting that the next generation of atomic weapons would be mounted in the nose cones of globe-girdling rockets.

Nowhere was big industry's blindness to the wider implications of the A-bomb more evident than in the pages of advertising that punctuated the December issue of *Fortune,* the same issue that reported and endorsed Marshall's New York speech. The major airframe companies, shipbuilders, and electronics companies expressed their uncertainties about the future by concentrating on the past, extolling their products' contributions to the war effort or, less frequently, announcing plans for civilian production. There was only one exception to the general rule. On the inside front cover, one of the magazine's most expensive display pages, there was an advertisement that indicated Marshall had at least one vocal disciple among the captains of industry. The illustration depicted a strange, futuristic vessel that looked like a cross between a submarine and an art deco apartment building. The caption explained it all in the best Madison Avenue prose: "The submarine of tomorrow may never match this atomic powered, jet propelled marvel. But, whatever science and invention may lead to, one thing is certain. We should maintain the development of the submarine as a key security weapon in proportion to our devotion to peace." The words were the work of a copywriter, but the sentiments were those of John Jay Hopkins at his aggressive best.

Hopkins was very nearly alone in recognizing the commercial possibilities presented by the emerging world order and unique in his willingness to wrap Electric Boat's postwar sales strategies in the cloak of national security. Today, of course, one need only browse through any major city newspaper or national magazine to find a slather of defense industry advertisements proclaiming a similar message. But in 1945, a matter of months after the war's final shots, the very idea that a munitions company would take its case to the court of public opinion was unprecedented. The arms companies were not averse to vigorous lobbying, but as the Nye hearings disclosed, it had always been done in a variety of circumspect ways. Congressmen and senators were leaned upon, admirals courted, and the merits of a particular kind of submarine, battleship, or plane telegraphed through the pervasive old boy network. If a manufacturer did go public in pursuit of an upcoming contract, it was done through some allegedly independent front organization, like the vociferous Navy League that had served as a mouthpiece for William Shearer and the arms companies during the twenties.

Hopkins' decision to dispense with the traditional pretenses was motivated in no small part by the realization that Electric Boat, more than any of the other major war suppliers, was extremely vulnerable to the vicissitudes of peace. The automakers, reveling in the booming postwar market, had no such problems. Even the airframe industry managed to subscribe to a buoyant, if misguided, optimism, which envisioned Americans taking to the air en masse, as both pilots of family planes and passengers on a new breed of sleek, fast, and luxurious airliners. Electric Boat, however, was in no position to seek assurances in fantasy. While it was just conceivable that Americans might acquire a sudden passion for recreational aviation, it would have required an altogether different form of madness to suggest that submarines would remain one of the country's most vital weapons systems. The Navy's inventories were swollen with boats, most destined to be scrapped, mothballed, or sold off to second-string nations. Moreover, now that the A-bomb had confirmed the Air Force as the backbone of national defense, it was painfully obvious that the other services would have to fight for whatever remained of the reduced Pentagon budget. Hopkins also realized the admirals would be hard-pressed to defend their battleships and aircraft carriers. Submarines, those small, slow, and dangerous relics, had no place in the strategists' visions of the type of weapons systems that would fight the next war. It was starting to look like the 1920s all over again.

The strings to the company's peacetime bow were few and dubious at best. The Elco division was about to switch from the PT boats that had sustained it since 1939 to a new range of weekend pleasure craft and larger, more expensive motor yachts. There was also a well-advanced plan to build and market an offset printing press, a scheme intended to retain the company's expertise in precision engineering and assembly procedures formerly employed in the fitting out of submarine weapons, buoyancy and navigation systems. Finally, there was a more distant hope that the submarine yard could find a niche as a major supplier of high-stress components to civil engineering projects like the national highway system for which the auto lobby was clamoring. Ironically, this has proved to be one of General Dynamics' chief contributions to America's postwar economic development. Throughout Connecticut and much of New England, hundreds of thousands of cars and trucks make daily use of bridges supported by prefabricated girders that carry the Electric Boat trademark.

At Groton itself, the main hope for the immediate future remained with the Navy, which was preparing to refurbish some of the more modern boats with innovations pioneered by its vanquished enemies. The Guppy—named for Greater Underwater Propulsive Power—proved to be a harbinger of things to come. The extensive modifications were touted originally as a relatively inexpensive way of extending the wartime submarine fleet's active life into the late 1950s and beyond; unfortunately the work proved so complex the average cost per boat often ran to more than $3 million, a figure that sometimes exceeded the original price. Electric Boat would receive the lion's share of the Guppy contracts, but in the first months of the peace, there were no guarantees that this would provide enough work to retain even a hard core of specialist technicians. The Guppy program was welcome for what it was worth, but it remained the last crumbs of the wartime feast rather than a foretaste of the next course.

Nothing expressed Electric Boat's uncertain future more graphically than its performance on Wall Street. Within ten months of the war's end the stock price had dropped thirty dollars to around thirteen dollars, roughly half the book value. It didn't require a financial expert to identify Electric Boat as the potential victim of a takeover bid and subsequent dismemberment. Its directors were novices in the ways of the civilian marketplace, while its technical expertise, with a few limited exceptions, could be applied only to a field of endeavor that the A-bomb appeared to have made redundant. To the predatory speculators already making their presence felt, Electric Boat was looking more and more like a sitting duck with every fresh month of peace. In the event of a successful takeover, the marginally profitable divisions like Elco and Electro Dynamic could be sold off, recouping a large part of the raiders' initial outlays. As for Groton, it could expect to go on servicing the Navy's diminished needs, perhaps collecting an order or two from overseas and generally biding its time until it, too, was placed on the auction block. By then, of course, the considerable reservoir of accumulated wartime profits that helped to make the idea of a takeover so enticing would be long gone, distributed as bonus dividends and increased fees for the new directors.

Given the situation, Hopkins mounted his defense with a surprising degree of confidence. Announcing a tax-free dividend of preferred stock that could be converted to common or sold off after a specified time, he pushed the price to an astonishing $35 in less than a month. The stratagem flushed out the speculators, but it did not solve the basic problems reflected in the latest balance sheets. Sales for all of 1946 amounted to a meager $15 million, the work force was down from its wartime peak of thirteen thousand in 1944 to around four thousand, while profits were scarcely sufficient to generate earnings of two dollars a share. If Electric Boat was to remain intact Hopkins knew that he would have to produce something more concrete than promises and patriotic advertisements in the financial press.

Electric Boat's dilemma was more than a little ironic. For the first time in its existence, the company was flush with cash; the problem was that there was nothing to buy. Speculators were correct in surmising that the management had no stomach for consumer goods, and while the newly installed president was known to be bullish about defense stocks in general, there was not one that met his specific requirements. The area in which he invested the most faith, the airframe industry, appeared to be out of the question. It was a closed club of existing producers, each clinging to a vanishing hope that the predicted boom in civil aviation would somehow guarantee its future. Still, Electric Boat had to do something. The inflated stock price owed much to the expectation that Hopkins was about to pull the proverbial rabbit from a hat. But as the jubilation of 1945 melded into the growing unease of 1946, the magician's props were still missing from the stage.

The solution arrived unexpectedly, and from a most unlikely quarter. At about the same time Hopkins was repelling the raiders, the cabinets of Canadian prime minister Mackenzie King was about to demonstrate its commitment to the principles of free enterprise by disposing of the diverse, government-owned munitions industry brought into being by the war. Peace had robbed the aircraft builders, ordnance factories, shipyards, and steel mills of their markets and now most lay idle, waiting for the entrepreneurial guidance that the Ottawa government was ill-disposed and ill-equipped to provide. Of them all, perhaps the most troubled was an aircraft assembly plant on the outskirts of Montreal, Canadair.

Like so many other war babies, Canadair was conceived in a moment of passion and disowned without remorse. Early in 1942, a team of Canadian Army engineers and civilian contractors began tearing up a meadow near the village of Cartierville, about twenty-four miles from Montreal. When they left, one year later, the site was covered by a forty-acre aircraft assembly plant so vast and so quickly constructed the awed locals concluded it had to be the world's largest. That was incorrect, but it was certainly the biggest commotion Cartierville had seen in quite some time. Less than fourteen months after the first sod was turned, over twelve thousand people were reporting for work every day, while the plant, now grown to include its own paved runway, continued to swallow the surrounding countryside. By the time the job was done and the first plane in the hands of the RCAF, the total cost amounted to around $23 million.

Canadair served with distinction throughout what remained of the war, building almost four hundred complete Catalina flying boats and the component parts for two hundred more, which were assembled in the United States. But then, in December 1944, all work came to a sudden stop when Ottawa canceled most major war production. While the work force was cut by two thirds, the private management company that oversaw day-to-day operations for the Ministry of Defense began searching for an American or a British plane that could be built under license. Canadair's stewards settled on the Douglas DC-4, a four-engine transport which Canadair hoped to build in two versions, an austere work horse for the armed forces and a second, more expensive model with additional soundproofing and a pressurized cabin, for use by Trans Canada Airlines. The plane was given a new name, the North Star, and subjected to what all but the members of Canadair's engineering staff considered minor alterations.

The deal with Douglas was the subject of much parliamentary praise from government ministers and backbenchers, who portrayed it as another example of the war cabinet's providential planning for peace. Conspicuously absent from the speeches, however, was any hint of the government's long-term plans for Canadair. In French Quebec, the news was received as a bitter disappointment after the tremendous exertions of the previous three years. Canadair's survival as one of the province's largest employers now rested upon firm orders for just forty-four North Stars and another, second-string program that aimed to transform war surplus DC-3s into passenger planes. Like so many of Canadair's postwar projects, the conversion program proved to be a tiresome, frustrating job that demanded more trouble, time, and expense than it returned. Only from Ottawa's distant perspective was it possible to describe Canadair as a success story in the making. From the more immediate vantage points of Montreal and Cartierville, the situation demonstrated every trait of an impending debacle.

It took less than a year for Canadair to live up to the nay-sayers' worst expectations. The first North Star was unveiled in June 1946, to an undeserved fanfare of speeches and public acclaim. While it was true that the prototype had been built on time and within the budgetary limits, those behind it on the production line were beset by a multitude of problems.

In taking on the North Star, Canadair set out to build a very different plane from the graceful but comparatively unsophisticated Catalina flying boats it had grown up with. The North Star was not one plane but two, each very different from the Douglas DC-4 on which both were based. The RCAF's twenty aircraft were spartan, general-purpose transports that could be set up to carry troops, medium-weight equipment, or even parachutists, while the civilian version was, by the standards of the day, the acme of airborne luxury. Passenger comfort demanded innumerable minor and major variations on the original theme, many beyond the means of the disorganized and dispirited work teams. The untried cabin pressurization system that TCA demanded proved particularly troublesome, since it had to be installed before work on the other internal fittings could even begin. When the system failed to live up to expectations, Canadair was forced into an irksome game of musical planes in order to meet the first of the delivery dates laid down in the sales agreements. Six of the RCAF's unpressurized North Stars were hauled off the assembly line, tricked out with seats, "silent" mufflers, additional soundproofing, a galley, and bathroom, and loaned to TCA as makeshift substitutes. By the time the first civilian North Stars were ready in late 1947, the unpressurized stand-ins returned to Cartierville, where they were stripped of all fittings, repainted in Air Force green, and handed back to their original owner.

The management company's many shortcomings were further magnified by the consequences of the government's disruptive intrusion into the design negotiations with Douglas. Ottawa's insistence on Commonwealth components, even when they were more readily obtainable in the United States, resulted in a slew of unplanned alterations. The DC-4's designers had built their original plane around four air-cooled engines that were light, compact, easy to service, and extremely durable. Since there were no Canadian-built radial engines on the market, the negotiating committee equipped the North Star with British-built versions of the famous Merlin engine that had powered Hurricanes and Spitfires in the Battle of Britain.

The proposal had a certain dubious merit. An American power plant would have required payment in U.S. dollars, which the government, husbanding its slim supplies of foreign currency, was unwilling to spend. The Merlins, on the other hand, were slightly cheaper and could be obtained without hard-currency troubles as part of Canada's direct trade with the mother country. Unfortunately, the Merlins were heavy and large, and in keeping with the military heritage, they demanded between two and three times the routine maintenance required by the engines they replaced. Since their greater weight increased the load on wing booms and spars while simultaneously changing the plane's center of gravity, Canadair's technical staff was forced to evolve its own compensatory changes to restore the original handling characteristics. The government's meddling was to cause it considerable embarrassment in 1949, when opposition leader George Drew alleged that the North Star was inferior to its foreign-built competitors. Drew spoke of engine failures leaving him stranded five times in as many months and praised the skill of Canadian pilots as the only factor that had prevented calamitous loss of life. When a government representative rose to the plane's defense, his words were drowned by a chorus of parliamentarians interjecting details of their own disrupted journeys.

To those members on the government benches who viewed the government's continued involvement in the aircraft industry with active disdain, news of Canadair's ills came as confirmation that business was best left to businessmen. Even as early as December 1945, some conservatives were speaking publicly of fulfilling all existing contracts and then closing the factory for good. Others saw it as an auto assembly plant and urged that efforts be made to find a U.S. car company prepared to take out a lease. The most pessimistic predicted that it would become a vast warehouse stuffed with forgotten public service files. What few expected was the speed with which Canadair was offered for sale when a prospective purchaser hove into view. The buyer was Electric Boat, and its prize, one of the greatest bargains in the history of aviation.

Among the many guests invited to inspect the prototype North Star, Hopkins was the one whose approval mattered most. Scarcely a month had passed since his first meeting with the Canadian minister for reconstruction and supply, C. D. Howe, but the deal was as good as done. Canadair was exactly what Hopkins had been looking for, and Howe's attitude made it clear that Ottawa would consider any offer, no matter how low. Thus did Canadair change hands. The details were announced, to a stormy response, in August, less than a month after the North Star's maiden flight, and ratified in Parliament three weeks later, on September 14, 1946. The first and most common observation among the government's critics was that Canadair had not been sold so much as given away.

Some years later, a General Dynamics official commented that Canada was a "real capitalist country," and the Canadair negotiations make it easy to understand his enthusiasm. In terms of Canadair's worth, the deal Hopkins obtained amounted to an investment of little more than small change. In return for an immediate cash deposit of $2 million and a pledge to make quarterly installments amounting to $8 million over the next two years, Electric Boat acquired everything Canadair had to offer. By the government's own estimate, the factory and equipment inside it were worth all of $22 million, and that figure ignored the value of the North Stars under construction, the spare parts inventory, and the licensing agreement with Douglas that guaranteed the management's right to build its own version of the DC-6 at no additional charge. And that was not all. As a final inducement, Howe agreed to provide a ready escape route in the event that the aviation industry failed to match Hopkins' expectations. Until such time as it found its feet, Electric Boat was free to postpone the final payments and continue leasing the facility for just $225,000 a year. Although many doubted the wisdom of a move into the airframe business at a time when the industry's veterans were becoming increasingly pessimistic, none could fault the manner, or the terms, on which it was done. The financial world's approval was reflected on Wall Street, where initial doubts about the submarine builder's ability to run a troubled foreign airframe company were mitigated by the knowledge that Hopkins had committed only a small part of Electric Boat's war years nest egg. The stock price rose, fell, and rose again, while the market waited to see what would happen.

The most strident criticism emerged in Canada, where the sale chafed one of the perennial sore points in that nation's relationship with its often overpowering neighbor to the south. That Canadair had been sold after cursory public debate and for what many considered a pittance was bad enough; that the new owner was an American corporation added salt to the wound. Some months before Electric Boat exercised its option to buy, in late 1947, an opposition member of Parliament took umbrage at the claim that the sale was in the best interests of the Canadian economy and described the agreement as a national disgrace: "Last year the minister reported in this house the gift of a $20 million aircraft factory. At the time it had 7,000 employees with huge, unfilled aircraft orders and it is now rented by a United States syndicate at a rate requiring one hundred and ten years to realize the Canadian capital included in the expenditures on that plant."

All political slants aside, it was certainly true that Canadair was not the only war baby put up for adoption. Not long after Hopkins installed his new team of managers, the government closed and sold an Ontario steel mill in Sault Sainte Marie on the Great Lakes because it was considered obsolete. The plant was dismantled, loaded on flatcars, and shipped to the U.S., where it was soon reassembled and back at work. The sale occurred, one vexed member of Parliament declared, "when we in Canada are crying out for every form of steel." In Canadair's case, the loss of a public enterprise that had been lionized by wartime propaganda was made even harder to stomach by the unthinking arrogance that characterized the coverage of the story in the U.S. press. *Time* magazine's account was typical. Distributed throughout Canada less than a week after the details were announced, the story asserted that Canadair had been sold because the Ottawa government believed its citizens lacked the "know-how" to run their own sophisticated industries.

Hopkins' response made diplomacy a key element of Canadair's speedy reorganization. There was, for example, no purge of the old regime's middle-level executives nor any confirmation of the fear that Canadair was about to be overrun by American technocrats imported to supervise the local, ignorant labor. For every newcomer like Ken Ebel, a Curtiss-Wright alumni who became the new vice-president in charge of engineering, or Bob Neale, a Boeing man who assumed total control of the shop floor, there was a greater number of native sons, many lured away from domestic rivals like de Havilland, Avro, and the branch operations of U.S. manufacturers. The new president, Canadian-born H. Oliver West, typified the breed. A tough, often stubborn specialist in mass production, he had been one of Boeing's most valuable men during the war years and accepted Hopkins' offer on the understanding that he would be given a free hand to do whatever he considered necessary. Major questions of finance and long-term planning would, of course, be settled only after consultation with Electric Boat's New York headquarters, but in all other areas, West was assured, his decisions would not be second-guessed. In truth, he had little reason to worry, since Hopkins always made a point of leaving technical matters in the hands of those who understood them best. With nothing more than an intelligent layman's grasp of the finer technical points, he believed in hiring the best talent and standing back to draft the grand design. His own energies were best employed in smoothing the ruffled feathers of Canadian nationalism. For the remainder of his life, Hopkins was an outspoken advocate for Electric Boat's second home. He served on the boards of numerous Canadian-U.S. trade and industrial organizations, praised the country's prospects and the caliber of its work force, and urged increased U.S. investment at dozens of speaking engagements, lunches, and receptions across America.

Hopkins' faith was not misplaced. Even allowing for a fortuitous rise in the airline industry's stocks, West's performance was remarkable. Throughout 1947, his new managers cut through the layers of ossified inefficiency that had accumulated under the previous administration and began a drastic overhaul of the production lines. One Canadair veteran likened the process to a discordant orchestra being brought into line by a new and demanding conductor. The inventory control system, the foundation stone of any successful production line, was reformed and new suppliers were found—some from the United States—who could guarantee many of the previously scarce parts. At the same time, another and equally intensive effort by the engineering and drafting departments, recently grown to include a contingent of Dutch aeronautical engineers who had come to Canada as war refugees, rectified most of the flaws that had slowed North Star production to a trickle. Finally, even the DC-3 conversion program came into its own. All but idled at the time West took charge, it went on to refurbish more than 250 aircraft and acquired a reputation as a world leader in the field of large-scale retrofitting.

Success bred success, just as Hopkins had hoped. The North Stars' reputation improved in step with their reliability, and orders began to roll in from customers previously wary of handing a substantial deposit to an aircraft company likely to end up in the hands of the receivers. Canada Pacific, the independent domestic and international carrier that had become TCA's chief rival, signed up for a dozen planes, while British Overseas Airways Corporation bought twenty-two long-range versions in a deal whose ironic consequences illustrate the strides Canadair was making under West's supervision. Having ordered the aircraft in the expectation that they would arrive at least six months late, the British air carrier's directors were dismayed when the first batch turned up more than eight months ahead of schedule. Not only was BOAC forced to make an unexpected payment for planes it was in no position to use, but it also had to honor a galling bonus which the sales agreement specified in cases of early delivery.

In themselves, neither West's drive nor Hopkins' willingness to gamble would have been enough to assure Canadair's future into the 1950s. That required the combined efforts of Washington, Moscow, and a specially convened panel of cold war strategists who manifested their views in a report to President Truman on the declining fortunes and dim prospects confronting every sector of the U.S. aviation community. Although it dealt solely with the U.S., Chairman Thomas K. Finletter's recommendation became Canadair's invitation to take its place in General Marshall's vision of a secure West united behind the alliance of American industry and an ever-vigilant military. Security, the report contended in words that might have been Marshall's own, lay in maintaining a peacetime force "greater than any self-governing people has ever kept."

The President's Temporary Air Policy Committee had been established in mid-1947, at a time when the airlines and airframe producers could not have been more despondent. Contrary to early expectations, business was terrible. The predicted boom in recreational aviation had failed to materialize, while sales of commercial multiengine aircraft were negligible, since the same airlines that had been predicting a decade of mounting passenger traffic and rising profits now faced a totally unexpected drop in ticket sales. Peering through the prism of strategic logic as it existed in those early days of the cold war, Chairman Finletter and his four fellow panelists viewed the situation with undisguised alarm. Just as General Marshall had warned in his New York speech, America was allowing the industrial and technological cornerstone of the national defense to crumble at the same time a new and intractable foe was looming over Europe.

At fifty thousand words plus innumerable graphs and charts, the report remains one of the most lavish meal tickets ever put on paper. It had something for everyone, and mostly in large amounts. As their first priority, the Finletter group demanded a standing peacetime force of seventy Air Force bomber and fighter groups, almost twice what the generals were gloomily anticipating. Next, the panel endorsed the accelerated development and deployment of long-range intercontinental bombers capable of delivering a nuclear punch to any corner of the world. The report concluded with a final, chilling observation that the U.S. might fall victim to a nuclear Pearl Harbor at any time after January 1953—the earliest date by which the authors believed the Soviets would have an A-bomb of their own. The military rationale that ran through the principal recommendations was underscored by countless minor themes. There was, for example, a demand that Washington sponsor the aviation industry's financial stability by using the airlines to carry first-class mail even "at the expense of railways and steamship companies." This was prompted not by a concern for efficiency but because the domestic air carriers represented "a potential military auxiliary that must be kept strong and healthy." Much the same thing was said of recreational pilots, on whose behalf the panel urged the creation of a nationwide network of air navigation aides. As for the manufacturers, Mr. Finletter noted that while the overabundance of producers made the industry inherently uneconomic, the interests of national security demanded that each be given enough government work to guarantee its survival. As a remedy for the producers' existing ills, he urged that $3.9 billion be distributed according to need over the next two years. *Time* magazine, somewhat awed by the panel's financial projections, calculated that the annual cost of all proposals would amount to $122.45 for every man, woman, and child in the country by 1952. And that figure did not include expenditures on the Army and Navy, which "would have to be appeased as well."

As far as Hopkins and Canadair were concerned, the Finletter panel's findings were a gift from heaven. True, the report seldom strayed far from the situation facing the U.S. industry, but its portents for foreign producers were easily read all the same. The wave of unprecedented spending was bound to spill over the border, first as orders for preassembled components and later perhaps in contracts for complete aircraft. Certainly, the Canadian government and the other members of the emerging NATO alliance could be expected to follow Washington's lead and modernize their own air forces. With much of its wartime equipment still in mothballs and vast tracts of floor space unused, Canadair was perfectly placed to meet the needs of its own government and maybe even those of Britain, Europe, and minor members of the Western camp in other parts of the world. The one cause for concern—the distinct possibility that the Finletter report might end up pigeonholed and forgotten like so many other presidential inquiries—was soon dispelled by three events on the other side of the world. The first, some four months after the report's release, was the Berlin blockade, in which, as if to confirm the panel's prophecies, the Allies' military and conscripted civilian aircraft were called upon to keep the city alive in the face of the Russian threat. Next, China was "lost" to the Communists, and finally, in the fall of 1949, there came the worst news of all. On September 9, the Atomic Energy Commission informed the White House that samples of radioactive fallout taken from the upper atmosphere indicated the Russians had just detonated their first A-bomb.

The surge in military work began in early 1949, when Ottawa ordered Canadair to begin production of the T-33 two-seat jet trainer for the RCAF. Six months later, Canadair landed its second military plum, beating Avro (Canada) and de Havilland for the rights to build an initial order of one hundred F-86 Sabre jet fighters. The Korean War soon saw the number increased, doubling it first to two hundred and then adding five hundred more. At that point, the backlogged orders from Canada, West Germany, and South Africa became a carte-blanche authorization to continue unrestrained production indefinitely. In all, Canadair manufactured 1,815 Sabres between 1950 and 1958, at an average purchase price of roughly $200,000 each. Although best known as the U.N.'s standard-bearer in the dogfights of the Korean War, the Canadair Sabres's made a more significant contribution to the colonial wars and boondocks' arms races that punctuated the 1960s. The nomadic final years of some two hundred Sabres originally purchased by the West German Bundesluftwaffe illustrate the way the West's outmoded weapons systems often end their days not as defenders of any one political ideology but as loose cannons careening about the poorly illuminated stage of the third world. The first fifty of the former German planes were sold to Portugal, which used them extensively against rebels in Angola and Mozambique. The second batch, seventy-four planes in all, went first to Interarms—the freelance arms dealership controlled by the notorious privateer Sam Cummings—before being passed on to Venezuela for $140,000 each, a markup of roughly 300 percent. In its turn, the Venezuelan government added a profit margin of its own and sold two dozen to Rhodesia, then the subject of an international arms embargo. The final group was purchased by the Shah's Iran, where they remained for only a few weeks before being reconsigned to Pakistan in a deal that drew an angry response from Washington. After several months of mounting diplomatic pressure, the Iranians agreed to repurchase the Sabres at the original price plus a premium of 30 percent, a sum many believed to have been discreetly funded by the United States.

By the end of 1950, military orders accounted for more than 80 percent of total production, and the figure was growing on a monthly basis. Against expectations, Canadair was selected to build the RCAF's fleet of long-range Argus reconnaissance transports as well as two military variations of the civilian Britannia passenger plane, which the company was also producing for several Canadian and overseas airlines. There were water-bombers for the government's parks and forests services, several twin-engine experimental models developed as possible successors to the DC-3s, and a constant flow of Canadian defense dollars for a variety of often ambitious research and development projects. The company even tried its hand in the most demanding field of aeronautical engineering and researched a bid to produce a wholly Canadian supersonic jet fighter. That effort, together with a rival project mounted by Avro, was abandoned only in 1959, when Ottawa decided Lockheed's controversial F-104 Starfighter represented better value for the money. As had been the practice since World War II, a large slice of that contract was also given to Canadair.

The company's luck was such that it sometimes received orders it had no right to expect. In 1950, for example, the U.S. Air Force broke with tradition and placed its first order with a foreign supplier for complete aircraft, a squadron of Canadair Sabres purchased when the U.S. manufacturer proved unable to provide enough planes for the war in Korea. Six months later, the U.S. Air Force turned once again to Canadair, ordering more than 250 jet trainers. Even past mistakes began to bear fruit. In the last months of the war, Canadair's former management had paid more than $1 million for a warehouseful of U.S. Army Air Corps spare parts for DC-3 Dakotas. The building and all its contents were dismantled and shipped north to Montreal, where they gathered dust for five years. In the early 1950s, when existing supplies of spare parts in the U.S. were all but exhausted, Canadair found that it was sitting on a gold mine. "Those spare parts made us an enormous amount of money over the next twenty years," former Canadair executive Harry Halton recalled. "It got to the point where people didn't bother looking anywhere else; if they wanted to keep their planes flying, they'd come to us."

There was really only one jarring note that disrupted Canadair's newfound harmony. Late in 1950, a festering disagreement with Hopkins about Canadair's long-term future came to a head, ending when West tendered his resignation and walked out. His motives were never stated officially, but Canadair veterans still recall the growing tension that existed prior to the announcement. West wanted Electric Boat's entire corporate structure turned on its head. His division was generating something like 80 percent of the group's total profits, while its thirteen-thousand-man work force was more than twice the size of Groton's. The tail was wagging the dog, as one contemporary U.S. commentator observed, and West regarded the situation as one fraught with potential disasters. Like Hopkins, he was delighted with the volume of military work, but he also recalled his own postwar years at Boeing, when the sudden halt to government-funded work left the airframe builder to flounder through a series of plant closures and retrenchments. Now riding the crest of a similar wave, Canadair would likely face the same fate unless it formulated its own plans to deal with the collapse of the military market that West believed was likely to occur at any point after 1955. Such far-reaching decisions could not be made in isolation, he maintained, nor could they be safely left to Electric Boat's executives, whose hands-on experience with the aviation industry was limited largely to expressions of joy at Canadair's escalating profits. West's solution was simple but drastic: Canadair should swallow its parent and become the nucleus of the group, relegating Electric Boat to the position of a profitable but minor subsidiary.

Hopkins would not entertain the idea. He could hardly dispute Canadair's importance, but his vision of the future differed markedly from West's. Far from being an end in itself, Canadair was to be Hopkins' stepping-stone to greater heights within the United States, where the military budgets were larger and the prospects almost unlimited. West lacked the energy and the inclination to pursue the confrontation any further. Recently diagnosed as having cancer, he handed in his resignation and went home to die. In his place Hopkins appointed J. Geoffrey Notman, a Canadian industrialist with a background in engineering, who was to remain in charge for more than a decade.

Years later, in 1976, when General Dynamics completed the cycle by returning a moribund Canadair to Ottawa's charge in return for $46 million, there were many critics who asserted the aircraft maker had never been given a fair deal. Ernie Regehr, a Canadian academic who lectures on economic history at Waterloo University, explained that the subsidiary soon became a poor relation in the General Dynamics family. The common criticism was that the association was a painless way for General Dynamics to fulfill its obligations under the U.S.-Canadian Joint Defense Production Agreement of 1963, an economic treaty intended to maintain a rough parity in defense sales between the two neighbors. "In the latter years there was little or no investment in technical capacity and no long-term industrial policy," Regehr said some years after the sale. "Canadair's primary importance to General Dynamics was that of a subcontractor assembling and fabricating parts for U.S. projects that could then be counted as part of total Canadian defense production. It was an exercise in bookkeeping." Harry Halton, a junior engineer when General Dynamics took over in 1946 and an executive vice-president when it relinquished control thirty years later, was more succinct in summing up Canadair's long decline through the 1960s and early 1970s. "With all its activities in the U.S. we tended to be overlooked—I suppose you could say we were about number 174 in General Dynamics' list of priorities."

Throughout the late summer and fall of 1951, the eyes of many senior members of the airframe and general aviation communities were fixed on San Diego and a building known as The Rock, the bluff concrete office building that housed the corporate headquarters of Convair Aircraft. The rumors doing the rounds of rival boardrooms and Pentagon project offices maintained that the investment company that controlled the giant airframe and missile manufacturer was about to sell off its holdings and get out of the aviation business for good. The news intrigued many but surprised few, because despite its solid nickname, there had been few moments during the previous decade when The Rock had not been convulsed by a major upheaval of one kind or another. But 1951 was different. For once, Convair was a reasonably happy realm. Thanks to the Korean War and a resurgence in the domestic airline industry, the numerous production lines were hard-pressed to meet the demand for bombers, military transports, and an increasingly popular range of twin-engine airlines. As would be the case with the M-1 tank more than thirty years later, the most arresting element of the various stories was not that Convair's management wanted to sell but the reputed identity of the prospective buyer, Electric Boat. As one magazine writer later remarked, Jonah was about to swallow another whale.

Jay Hopkins had demonstrated considerable pluck, and no little luck, when he removed an addled Canadair from the Canadian public payroll and watched it grow into a thriving enterprise. However, the idea that he was about to make Convair the target of his next foray into the aviation industry seemed almost laughable. Notable though Canadair's achievements may have been, they amounted to very little in comparison with the empire ruled from The Rock. Quite simply, Convair was huge, large enough to vie with Boeing and Douglas for the title of the world's largest aircraft company. Nationwide, Convair's work force ran to more than fifty thousand men and women, roughly half that number concentrated in a mile-long complex on the outskirts of Fort Worth, where the intercontinental B-36 bombers were produced. On the West Coast, its activities ran the gamut from orthodox jet fighters to a fantastic multimillion-dollar research and development project to build a nuclear-powered bomber. In another section of the San Diego plant, work had just begun on the Atlas rocket that would one day put the first American in orbit. And then there was Convair's chief claim to fame, the top-selling range of civilian transports that had carried its marque and reputation to every corner of the globe. Not surprisingly, the company's financial statistics conformed to the same enormous scale. In 1952, for example, the annual report put sales at close to $400 million and estimated the value of backlogged military and civilian orders at an additional $1 billion. Electric Boat, on the other hand, scarcely belonged in the same league. Its annual sales of only $110 million, its "mere" sixteen thousand workers, and a combined backlog for both Groton and Canadair that was not much greater than Convair's profits led many observers to believe that Electric Boat had fallen under the spell of its president's grandiose ambitions.

As it happened, the rumors were premature. Hopkins was interested, very interested, but after several months of tentative negotiations, both sides were forced to conclude that the time was not ripe. In any case, Hopkins had other things on his mind. Events were moving quickly in New York, where he was required to preside over the major details surrounding Electric Boat's transformation into the corporate entity he had chosen to call General Dynamics. It would not be until the next meeting of stockholders, on April 24, 1952, that the change became official, but there was much preliminary work to be done, including the orchestration of a public relations fanfare, before the proposal could be put to a vote. Hopkins was bent on depicting the event as Electric Boat's coming of age. Now that Canadair had emerged as the most active and profitable division, he considered the old name dated, almost anachronistic. The new name, on the other hand, would be vague enough to accommodate the widest and wildest of Hopkins' aspirations, while confirming, at least in the public imagination, the company's place as a standard-bearer on the cutting edge of applied technology. Still, while the change obliged the suspension of serious negotiations with Convair, there was no reason for regret. Not only had the early talks opened an official line of communication with Convair's parent, the Atlas Corporation; they had also earned Hopkins the right of first refusal when the aircraft maker came on the market once again. And that, Hopkins was assured, would probably be sooner than later.

Despite the company's recent return to profitability, Atlas' desire to be rid of Convair was easy to understand. It had joined Atlas' extensive portfolio in 1947, when the aviation industry's prospects and morale were clouded by the pall of postwar gloom. The purchase was followed by two years of heavy losses and the indignity of a minor Washington scandal, before the new owners' corrective measures began to produce results. By 1951, Atlas was ready to capitalize on the industry's resurgence by doing what it had done so often in the past—selling off a rehabilitated possession, pocketing part of the profits, and moving the rest into another, speculative investment.

Atlas itself was controlled by Floyd Bostwick Odlum, a man whose influence during the boisterous early years of General Dynamics' history was second only to Hopkins'. A corpulent, slow-talking dude rancher with a taste for self-deprecating jokes, loud ties, and ill-matched western work shirts, Odlum was one of the few people whom Hopkins might have counted as a close friend. Seldom prominent among the leading decision-makers—Hopkins' mercurial ego left little room for understudies—Odlum served often as fate's instrument. It was, first of all, his willingness to part with Convair that allowed Hopkins to acquire his greatest prize. Later, Odlum was the one man whose previous dealings with the capricious Howard Hughes might have stopped General Dynamics from sliding into a business deal that was to end in near-disaster. And finally, when Hopkins was only weeks from death, it was Odlum who provided the Palm Springs venue for Hopkins' dude ranch version of the last supper.

Like Hopkins, Odlum was a preacher's son who devoted his life to escaping the genteel poverty into which he was born. His father was an Irish-Canadian, a pious teetotal Presbyterian who drifted from one small town in rural Michigan to the next, always searching for a congregation large enough to support a consumptive wife and an ever-growing family. The youngest and smallest of five children, Floyd displayed an early talent for rapid-fire mental arithmetic and a willingness to try his hand at just about anything. Years later, he was to boast of once riding an ostrich at a county fair, an event that captured his lifelong readiness to risk ridicule in pursuit of a profit. Entrepreneurial instincts were so much a part of his personality, they even rated a snide reference in the University of Colorado Law School yearbook for 1919. The caption beneath his angular, bespectacled face declared that "Little Odd" had somehow managed to monopolize the leadership of every student club, guild, and society "that came with a stipend attached." That was only part of the story. On his graduation day he was not only the top student of his year but the proprietor of a thriving real estate business and a campus laundry service.

It was hardly surprising that Odlum moved straight into company law, pursuing the prospect of advancement from one job to the next until, less than five years after graduation, we find him in New York serving as the vice-chairman of Electric Bond and Share, an immense investment trust that operated power stations and public transport franchises throughout the U.S., Canada, South America, and Europe. He might have remained in charge of foreign operations almost indefinitely, had it not been for a particularly nasty ulcer which, early in 1923, forced him into semiretirement for almost four months. Late one evening toward the end of his convalescence, he happened to be dining with his wife and another couple, when he made a passing remark that those present might have some fun if they pooled a little of their spare cash into a small investment fund. With the stock market beginning its long climb toward the unforeseen disaster of 1929, the idea met with instant and unanimous approval. Three weeks later, Odlum formally took charge of the grandly named United States Company, incorporated with a total capital of forty thousand dollars for the stated purpose of speculation in utility stocks and general securities.

Odlum made a dashing debut, doubling and then tripling the initial investment in as many months. At the end of its first year, the company paid an impressive 65 percent dividend, and twelve months after that the bottom line indicated an annual growth rate of more than 1,000 percent. It was late in 1926, as Odlum later recalled, when he decided to get serious. Abandoning his position with Electric Bond, he spent the remaining years until Black Friday making merry, the scope and value of the portfolio exceeding $10 million for the first time in mid-1928.

Odlum's record during the late 1920s was impressive but hardly unique. To be sure, he had demonstrated an ability to play the market with a virtuoso's touch, but then so too had many other investors, whose bright lights were soon to be snuffed out by the Great Crash. In the final analysis, his great and most enduring gift was knowing when to get out.

In later years, the financial press took note of Odlum's preference for check suits, suspenders, and wide-brimmed hats, and dubbed him "Farmer Floyd." A more appropriate rural allusion might have been "the Horse Trader," because when it came to judging the market's mood, few could match his ability to get in and out at the most opportune moment. At the start of 1929, Odlum took a close look at what was happening on Wall Street and, possibly recalling his father's sermons about the Tower of Babel and the false arrogance of man, decided to sell everything that was not tied down. When the grim day finally arrived, the recently rechristened Atlas Investments Company was sitting pretty atop a $14 million pile of cash and short-term notes. Odlum proceeded to have the time of his life.

His first targets were those many small and less providential investment companies whose remains littered Wall Street like the bodies of so many defenestrated bankers. Whenever he noticed a situation where issued shares were worth less than the physical assets, Odlum bought a controlling interest and forced the sale of all liquid items. Minority shareholders were offered the choice of taking what was owed or accepting a block of Atlas stock, a course many accepted. Odlum became a household word and something of a legend as his successes mounted throughout the early Depression years. One chronicler of the rich and powerful listed him high among "the 30 men who run America," while Chicago congressman Adolph J. Sabath commented that the day was not far off when J. P. Morgan would control one half of America and F. B. Odlum the other. *Fortune* magazine began calling him the "Depression phenomenon."

An avid and early supporter of the Roosevelt administration and its economic policies, Odlum switched tactics not long after the new President moved into the White House; having already accumulated an array of properties whose value could only improve, he began playing doctor to companies gone to seed for want of a firm hand or guiding philosophy. The Greyhound Bus Company and Madison Square Garden both responded to his touch, as did the ritzy Fifth Avenue department store Bonwit Teller, which passed to his wife in 1935 when the couple were divorced.

The next year was even more active, beginning with Atlas' acquisition of the RKO cinema studios in Hollywood and ending in Odlum's marriage to one of the most remarkable women of her time, the record-setting aviatrix Jacqueline Cochran. Like her husband, the new Mrs. Odlum was destined to play her own, albeit smaller, part in General Dynamics' rise. Seventeen years later, at the controls of a Sabre jet fighter borrowed from Canadair, she capped an illustrious career by becoming the first woman to fly faster than the speed of sound. The attempt was organized by Jay Hopkins, by then a close family friend, on the theory that a woman breaking the sound barrier "will give us prestige and publicity we couldn't buy."

Cochran and the members of her circle like Wiley Post and Amelia Earhart, who spent several weeks at the couple's mountain ranch near Indio, California, before leaving on her last, fatal flight, awakened Odlum's interest in aviation and led directly to his first investments in the field during the late '30s. An admirer of Roosevelt since his days in New York and a generous contributor to three presidential election campaigns, Odlum was a certain candidate for a senior position in the wartime administration. While his wife went north to Newfoundland to take command of forty women pilots ferrying bombers across the North Atlantic to England, Floyd left for Washington and an appointment in the Oval Office. He was asked to accept the directorship of the Office of Production Management's contract division, and took up the offer with alacrity, later informing the press that while he could have made several additional fortunes by remaining at his post with Atlas, his conscience left him with no option but to attend to a "matter of duty." Columnist I. F. Stone was moved to admiration by the tycoon's patriotism, commenting that his "resourcefulness, ingenuity and hospitality to new ideas in the operation of his own diverse business promise success in his new assignment." Alas, the promise was not fulfilled.

The new job was different from anything Odlum had tackled in the past, not least because it proved entirely beyond him. The executive order that laid out his responsibilities was such a sweeping charter, one contemporary observer remarked that Odlum could have made himself "a Minister of Supply with complete authority over the Army and Navy procurement systems." The system certainly needed an overhaul. Congress had heard innumerable complaints about inefficiency and inequity in the distribution of war work, the most frequent being that the major industrial corporations were getting everything while the smaller, regional producers were being starved of manpower, materials, and government funds. This was no paranoid fantasy of New Dealers dismayed at the appearance of "a financial and military oligarchy," as one critic later described it, but a reflection of the official statistics. Six months prior to Pearl Harbor, 75 percent of defense contracts were held by fifty-six firms, and the situation altered only slightly when hostilities began in earnest. The future Pulitzer Prize–winning author Bruce Catton, working as a newspaper reporter, was so incensed by what he saw that he wrote a book about it, *The Warlords of Washington.* Catton accused the major industries of hanging back, of encouraging inefficiency in order to make the most of the crisis. "The giants themselves had all manner of machinery that was not turning out a dime's worth of production for the Army or Navy," he wrote. "Heavy contracts had, for instance, gone to the automobile industry; but most of these involved the building of new plants and the purchase of new tools. The great bulk of the plant and machinery already existing in the industry was doing nothing whatever for defense production." Odlum's first task was to remedy the procurement program's ills, in part by organizing small factories and regional industries into competitors for the corporate monoliths that had shown themselves unequal to the task. In this, as Catton and other observers have noted, he failed miserably.

The organizational genius that had marked his business career deserted him, leaving only a flair for public relations. Odlum made a lot of speeches, often from the observation platforms of three red, white, and blue trains packed with examples of the kind of hardware the armed forces were purchasing. Local manufacturers were invited aboard at every stop and asked if they could duplicate anything on display. Many could and eventually did, but many more were forced to decline when Odlum proved unable to guarantee a regular supply of raw materials and labor. In fact, his greatest success was a very dubious achievement indeed. Whenever he spoke of the various material shortages, local collection depots were inundated with old tires, aluminum saucepans, and countless varieties of other dusty junk that, patriotic myths to the contrary, contributed little to the war effort other than a sense of achievement among those who donated and collected them.

Odlum's failure to rationalize national defense production did have one compensation—a splendid education in the ways of the military procurement process, which would stand him in a good position to make the most of the peace. America's response to the challenge of Pearl Harbor, although lopsided, was impressive all the same. In 1939, the term "defense industry" was a misnomer, a hollow euphemism for the motley collection of small-arms companies, government arsenals, and long-suffering stalwarts like Electric Boat. The industry's horizons were as limited as its clients, which were very limited indeed. In 1941, for example, Major General H. H. (Hap) Arnold, the head of the Army Air Corps, asked his senior staffers to estimate the total number of fighters, bombers, and transports required to win a global war against two determined and well-equipped enemies. Their initial estimate that 7,000 planes would suffice illustrated Arnold's point that the war could not be won by "thinking in small terms instead of looking at the big picture." The actual number of aircraft built in U.S. plants during the war years amounted to almost 230,000, a tally that far exceeded even Arnold's private predictions.

When the war was over, it was obvious that the immense industrial muscle so recently summoned into being would never be allowed to return entirely to fat. And of all the wartime industries, none appeared better able to weather a severe, short-term contraction than the airframe industry. The Allied strategy in World War II had revolved around air power; so, too, would all future conflicts. The A-bomb's ability to flatten vast tracts of enemy terrain made that conclusion inescapable; it was the missing ingredient needed to vindicate General Billy Mitchell and the other advocates of air supremacy, whose theories had taken something of a drubbing when first the Nazi bombardment of Britain and later the Allies' round-the-clock bombing of Germany had failed to make an appreciable dent in either side's war-making capacity. Moving after the war with his customary blend of speed and prudence, Odlum returned to his Utah ranch, where he became an avid student of the general panic that wracked the airframe industry throughout the last months of 1946 and all of 1947.

Less than an hour's flight from the Indio ranch, another and only slightly less famous millionaire was casting about for a deal that would get his Aviation Corporation (AVCO) down from the troubled skies and back to earth. Victor Emanuel had taken charge of the company he was to rechristen Convair only a few weeks before Pearl Harbor, when one of his subsidiaries, Vultee Aircraft, merged with the much larger Consolidated Aircraft. Convair, as the new company was named, prospered mightily throughout the war, producing almost 33,000 bombers, flying boats, observation planes, and naval attack planes. Business proved so lucrative that V-J Day found its management fingering an accumulated nest egg of no less than $60 million, which had been set aside to facilitate the production lines' smooth transition from bombers and transports to civilian airliners, cargo planes, recreational aircraft, and, of all things, kitchen stoves.

The massive outbreak of self-delusion that gripped the aircraft companies in the final days of World War II cast a particularly heavy spell over Convair's management. Almost every industry authority acknowledged that the flood of military orders would soon diminish to a trickle, but according to many of those otherwise sober minds, this represented not the slightest cause for alarm, since American families would soon be taking to the air en masse. Against all reason, the industry's Sibyls predicted that the day was not far off when there would be a pilot in every family and a plane on every block. Americans would fly to work at the controls of their own helicopters and spend weekends scouting the countryside for idyllic locations, where they could set down to unpack their picnic baskets beneath the sheltering wings of faithful family monoplanes. The delusion proved so enticing, no serious objection was allowed to stand in its way. Unlikely prototypes proliferated, one long-vanished automaker even unveiling something called an aerocar, which was billed as the ultimate solution to freeway snarls. When a traffic jam threatened, Mom and the kids could fasten their seat belts while Dad flicked a few switches, engaged a propeller mounted above the rear bumper, and took the family soaring into the blue.

At Convair, the postwar madness manifested itself as something called the Stinson, a "family" monoplane that first emerged from an expensively refurbished Detroit factory late in 1946 and went on to consume almost half the $60 million nest egg. The situation at the San Diego and Fort Worth plants was little brighter. An ambitious plan to produce a new line of commercial airliners and transports suffered not only from grave production problems but also from a grand indifference on the part of potential customers, who preferred to buy DC-3s and other war-surplus aircraft from conversion specialists like Canadair. The one project that appeared to have any chance of surviving in the civilian market was the Convair 240 and its future was blighted by a severe and unexpected recession in the airline industry that generated a string of canceled orders. Evicted from dreamland, AVCO decided to get out of aviation before things became any worse. Fortunately, there was an alternative, since the company also numbered among its possessions a group of civilian plants that had been working around the clock to meet the booming postwar demand for kitchen appliances and other consumer hard goods. If someone could be found to take charge of Convair, AVCO would be free to concentrate its energies on filling the kitchens of newly built Levittowns with appliances, and its balance sheets with black ink.

It was at that moment Odlum decided to make the move for which he been preparing over the previous twelve months. Beginning in late

1946, large and small parcels of AVCO began changing hands, most of them ending up in the Atlas portfolio. By May, Odlum controlled the largest block of stock, some 120,000 shares, and announced that he was ready to talk. The brief round of negotiations that followed resulted in a direct swap. Odlum walked away with Convair's plants in Fort Worth, San Diego and the Stinson factory in Detroit while AVCO retained exclusive control of the consumer division in Nashville.

Once again, Odlum had demonstrated a sense of timing that verged on the clairvoyant. Less than sixty days after the AVCO negotiations were finalized, President Truman announced the establishment of Thomas Finletter's wide-ranging inquiry into the aviation industry. Now certain of government support, the industry returned to work assured that its labors were not only in the best interests of shareholders but also vital to the noble cause of national security, a theme grimly captured by the official subtitle of the Finletter panel's report, "Survival in the Air Age."

The Finletter report was encouraging news, but its recommendations fell far short of the universal elixir required to remedy all of Convair's ills, many of which were self-inflicted. It took Odlum only a few days to discover that things were even worse than he had imagined. The San Diego production lines, for example, were fouled by rampant disorganization and a chronic shortage of parts that saw many of the 240s—already underpriced to compete with a similar aircraft produced by the Martin Company—rolling out of the assembly hangar with vital components missing. Eight days before Finletter's findings became public property, the new management was forced to announce the loss of a further $17 million. Even worse, there were strong indications the figures for 1948 would range even further into the red.

Odlum's prescription was painful and rather embarrassing for one of the most respected names in aviation. He first paid American Airlines to cancel twenty-five of the one hundred 240s it had ordered. Next, he limited the production run to two hundred aircraft, the total number for which a full complement of parts was available. And finally, he ordered the closure of the Stinson plant and two minor facilities in New Orleans and Miami. At the conclusion of his first full year in charge, losses had been reduced to $11.7 million, hardly small change but a good deal less than generally anticipated. The turning point came one year later, in 1949. A resurgence in domestic airline traffic and a substantial increase in military orders, prompted by the first Russian A-bomb test and the Berlin blockade, pushed Convair to its first profit since V-J Day. It was not a large amount, a paltry $3.7 million, but it was the first of the many better things to come. The two-hundred-unit production ceiling on the 240 series was lifted and the plane began to make money, as did a new and larger aircraft, the Convair 340. There were new contracts for alterations to the B-36 bombers, government research funds for rocket and guided missile experiments, as well as a standing Air Force order for hundreds of F-102 fighter-interceptors, the world's first delta wing combat plane. All in all, it looked like another example of the Odlum magic.

Any investor casting a superficial eye over Convair might have concluded that everything was bliss. By the start of 1952, the work force was back almost to wartime levels, there was a huge, funded backlog of military work, the 240s, 340s, and even more modern 440s were selling well, and the division was making a handsome contribution to Atlas earnings. But if the same investor had taken a closer look, he might have noticed a number of perennial problems hidden by the gloss of a booming economy. The most vexatious was the complicated question of turf.

Fort Worth was the most convincing evidence of the war's good fortune. Commenced in April 1941, the main construction hangar was finished within twelve months and delivered its first B-24 by May 1942. Unlike San Diego where components were fabricated, bolted together, and flown away, the Texas site was, at least in the beginning, purely an assembly plant. Wing sections, fuselage, and tail assemblies were put together at Ford's Willow Run factory in Georgia and then shipped south to Texas, where they were mated with the gun turrets, electronics, and the myriad systems and subsystems. The success of the operation depended more on inventory control than creative engineering and for a while the new plant represented little more than a distant annex of San Diego's head office. That cozy relationship changed rapidly as Fort Worth came into its own. Divided both by geography and the entirely natural desire of local management to be master of its own domain, the child began to demonstrate a definite personality of its own.

By V-J Day the rifts, while scarcely noticeable to an outsider, were beginning to exact their toll. Neither San Diego's nor Fort Worth's technocrats welcomed the intrusion of AVCO's lawyers and head office financiers, who were their nominal superiors, while the divisions each continued to pursue their individual and divergent interests. Texas dealt almost exclusively with Curtis LeMay's Strategic Air Command while San Diego concentrated on airliners and military projects for other branches of the armed forces. It was not a case of the left hand not knowing what the right was doing—although that was certainly true at times—so much as both hands completely ignoring the brain. The situation changed slightly when AVCO passed the scepter to Odlum, but despite his attempts at reform, it was still far from perfect. Some time later, not long after Convair changed hands for the third time in less than twelve years, this lack of corporate motor skills would cost General Dynamics dearly.

The difficulties in dealing with his intractable minions must have been prominent in Odlum's mind during those first talks with Hopkins. And he must also have been contemplating a related but much larger question: What to do with the company itself.

During the course of his spectacular career, the tycoon had built his coups, and his reputation, around what he called "special situations," ailing companies whose assets, technical expertise, or position in a depressed market made them likely candidates for rehabilitation. Never had he mounted a takeover bid with the intention of acquiring a permanent possession. The style was cut and thrust—buy, repair, and sell at the most opportune moment. The process was not always quick, but the result was usually the same. The RKO movie studios, for example, entered Odlum's collection in 1936 and departed eleven years later at the conclusion of an interminable and increasingly bizarre round of negotiations with Howard Hughes. Now it was Convair's turn to find a new home. Odlum realized the aviation boom was leveling off and the period of spectacular growth associated with the early cold war and Korea almost over. If Convair's position was to improve further, it would not be as the sole result of a growing economy and an escalating defense budget, but as the goal of a much-needed review of its internal affairs, strategies, and forward planning. An agile speculator rather than the founder of an industrial dynasty, Floyd Odlum lacked both the inclination and the shop floor experience to see such a potentially divisive exercise through to its conclusion. It was simpler, and easier, to take the money and hunt about for another "special situation."

If Odlum was a shifting cultivator, Hopkins was a pioneer determined to claim a large slice of the new frontier for Electric Boat, which was already homesteading the virgin territory opened up by the technological advances of World War II. Thanks to Admiral Hyman Rickover's decision to award the contracts for the first two atom-powered submarines to Groton, the company held a near-monopoly on the field of applied nuclear propulsion technology. At Canadair, the military thrust was diluted by civilian work, but in terms of growth, its performance had been even more impressive. Jet fighters for NATO and Korea, trainers and transports, all helped to boost the work force from less than seven thousand in 1946 to more than fifteen thousand in 1952. Canadair might have done even better had it not been for one major problem: its nationality. As a Canadian business, it was effectively precluded from competing for large U.S. Air Force contracts. True, it had contributed parts to U.S.-built planes and even sold a few fully assembled trainers and Sabres, but they were the exceptions to the rule. Convair, four thousand miles south of the Great Lakes, had no such worries. It was American to the core, and its well-established contacts with the Pentagon were legion.

Convair meshed perfectly with Hopkins' intention to establish what he once boasted would be "the General Motors of defense." As *Business Week* commented several months before the eventual takeover, "The single weapons approach is not enough for Convair. Its thinkers are scanning the whole range of future war, trying to conjure up the entire arsenal of atomic conflict." In this they had been entirely successful. The list began with the blue-chip contracts for the B-36, the Atlas, Tartar, and Centaur rockets, and the F-102 family of fighters, recently grown to include a faster, better-armed, and more expensive member, the F-106. Beneath these mainstream projects there existed an assortment of prototypes and experimental designs that dotted the sometimes bizarre leading edge of military aviation. At San Diego, considerable time and effort was being invested in an attempt to build a supersonic jet fighter that could take off and land on water. Despite generous Navy backing, the attempt succeeded only in killing one of Convair's most experienced test pilots before it was scrapped. At Fort Worth, another design group was tinkering with a proposal to turn a squadron of B-36 bombers into flying aircraft carriers by implanting pocket-sized fighter planes in their cavernous bomb bays. Later models, it was hoped, would go into action alongside flights of the large, slow, and highly vulnerable bombers, releasing swarms of tiny defenders at the first hint of enemy interceptors. And then there was the Pogo, the world's first and least successful vertical takeoff fighter. Sitting on its four-pronged tail, the Pogo clambered into the sky with the aid of an enormous turboprop engine turning two counter-rotating propellers. The Navy had dreams of putting one Pogo aboard every major surface ship, but this idea never came close to fruition. And perhaps that was just as well, since the combination of raw horsepower and small control surfaces made it almost impossible to fly, even for Convair's test pilots. So bad was the plane's reputation, one pilot won an award for simply getting it up to two hundred feet and then setting it down without mishap.

It was, however, the strangest of all Convair's projects that most captivated Hopkins' imagination, the nuclear-powered bomber. The NX-2, as it was known in official circles, represented a joint Air Force–Atomic Energy Commission attempt to duplicate what Electric Boat's Groton shipyard was about to do for submarines. A "small" nuclear reactor was to be hooked up to a series of modified jet engines, providing the NX-2 with the theoretical ability to zoom about the stratosphere for weeks on end. The second generation of nuclear-powered aircraft, so it was hoped, would remain aloft for even longer periods of time, each carrying at least one air-launched intercontinental ballistic missile.

Today, of course, the entire concept seems absurd, but in the early 1950s it was deemed no more improbable than the other nuclear-powered oddities that the popular press maintained were just around the corner. The Age of the Atom was going to be paradise on earth. Every home would have its own reactor in the basement, the seas would buzz with nuclear-powered ships, and electricity would be so cheap the utilities might just give it away. Nucleonics, as Hopkins liked to call it, was going to change the social fabric of the globe. Why should the skies remain immune to the atom's influence?

Among the legion of nuclear prophets, Hopkins was one of the most vocal advocates, a tireless crusader *Time* magazine dubbed "Mr. Atom." Not long after Groton began work on the Nautilus and the Sea Wolf, he established a subsidiary that was given the grand name of General Atomic Corporation. It employed no people, possessed no facilities, and existed only as an expression of Hopkins' hope that General Dynamics would someday emerge as a commercial leader of the nuclear movement. As with many visionaries, his faith sometimes ran away with his imagination, although seldom as far as it did one day in 1954 when he told a meeting of world business leaders in Tokyo that the developed world should donate "simple" reactors to the emerging nations. They were to be lifted by helicopter into remote villages across the third world, where the grateful natives, luxuriating in the bounty of unlimited nuclear power, would renounce their primitive ways, establish modern industries, and become devoted citizens of the West. In the months leading up to Hopkins' second round of talks with Odlum, the scheme, daffy though it may have been, added much to Convair's appeal.

Electric Boat's dominance of the atomic submarine construction program made it the leading authority in the field of applied nuclear propulsion. If Convair, together with its contract for the NX-2, could be brought into the family, General Dynamics would be in a prime position to lock up its early lead over any and all potential rivals. Hopkins was well aware that problems of one sort or another were inevitable, but like most technical considerations, the prospect bothered him not at all. The most important thing about the A-bomber was not whether it could be made to work —and many authorities were already saying it was impossible—but its seemingly endless potential for growth. The Air Force had been toying with the idea since the end of the war, and by 1951 it had succeeded in winning the funds to turn Convair's preliminary designs into a flying test bed for several different kinds of nuclear-powered jet engines. The Atomic Energy Commission, another active supporter of nuclear aviation, was investigating a number of other, even more farfetched proposals, while the Navy, not wanting to be left out, was considering a nuclear-powered seaplane. If the NX-2 proved to be even half the success its supporters predicted, it was not unreasonable to imagine that Convair might one day become a prime contractor for as many as five different planes, missiles, and, eventually, space vehicles. As the technology advanced, General Atomic would be waiting, poised to investigate and exploit any civilian spinoffs.

Of course, it did not happen that way. A nuclear plane never flew, nor was it ever proved that one could. Through the late 1950s and early '60s, the annual budgetary allocations grew ever smaller, until even the Air Force lost the final vestige of its early enthusiasm. By the time Defense Secretary McNamara put an end to the saga in 1962, all but diehard devotees had been forced to admit that it was a pipe dream. Death spared Hopkins the pain of watching a favorite program wither away, but as a businessman, he would not have been too disappointed. For fifteen years Convair received a steady flow of cash and contracts, in exchange for which it produced partial mock-ups, computer models, and assorted theories, but nothing that could have been put on display at a Veterans Day airshow.

It had begun in 1942, when the physicist Enrico Fermi wrote a speculative report for the Manhattan Project that mentioned nuclear fission as a promising source of power for ships and perhaps even planes. The paper finished, Fermi returned to his work on the A-bomb and forgot all about it. The Air Force did not, and in 1946 the idea was dusted off and handed over to an ad hoc group of scientists recruited from places like Pine Ridge and Los Alamos. Two years later, they were joined by a second panel, sponsored by the Atomic Energy Commission and drawn largely from the Massachusetts Institute of Technology. It was, to use an analogy drawn from nuclear physics, a case of two academic bodies becoming one and going critical. Individually, neither group was prepared to predict anything more than limited success; together, the combined weight of their guarded optimism temporarily obscured the very real objections. While it was generally agreed that a nuclear-powered plane could be made to fly, there was considerable doubt that one could ever be made to fly well. Entirely absent were all realistic assessments of an N-plane's military usefulness. That question was left to the Air Force, where those in charge gave it scant attention in their rush to expand the service's atomic horizons. In this respect, the N-bomber was the first of the postwar weapons developed not because there was an explicit need but because the concept provided a challenge too exciting to resist. For this reason alone, it is worth leaving Odlum and Hopkins for a moment in order to examine how a military procurement program that was sold to Congress as vital to national security was, in reality, an expensive practical joke on the American taxpayer.

The experts agreed that it would not be easy, or cheap, putting the cost at around one billion dollars over the next fifteen years to perfect a new way of performing an utterly simple task. A jet engine works because it sucks in air at the front, mixes it with fuel, ignites the mixture, and shoots the resulting gases, now superheated and much expanded, out of the tailpipe. Since nuclear reactors also produce heat, the atom appeared to present an alternative to aviation fuel. All that was needed was a way of piping the radiant energy of the atomic pile into the bowels of a jet engine, where it could heat incoming air and spin the turbines. The military advantages of such an engine were obvious even if the solutions to many of the finer technical points were not. A nuclear plane would have almost unlimited range and be capable of flying higher and farther with a heavier payload than was ever thought possible. Permanently on patrol in the upper atmosphere, their crews lounging about in commodious living quarters while robot pilots did the flying, the N-bombers were to fulfill a role now played by the Navy's ballistic missile submarines. At the first sign of trouble, the crew would swoop down on America's enemies like H. G. Wells' invincible airmen. Some speculated that nuclear technology might one day render even men obsolete. The most farfetched scheme poured some millions into an active investigation of a plan to build an enormous cruise missile powered by a nuclear rocket engine. Instead of delivering a warhead, the missile was supposed to roam at treetop height over enemy soil, setting fire to the countryside and flattening targets with its superheated shock waves.

The optimistic gloss did little to solve the formidable obstacles. The first was a matter of sheer bulk. If an N-plane was to hold all of the essential nuclear hardware, radiation shielding, crew quarters, engines, and still be able to carry a payload of weapons, it would be immense, possibly weighing one million pounds or more. Radiation leakage presented another problem, or rather a long list of problems. Goodyear, for example, received a $1-million contract to develop a range of radiation-resistant tires for the landing gear, while several other firms worked to perfect the protective suits that all maintenance personnel would be required to wear whenever an N-bomber returned from the stratosphere for "routine" servicing. In the air, excessive radiation from the straining reactors forced Convair's designers to turn the cockpit into something resembling a flying bank vault. They did away with windows and sealed the crew behind layers of lead shielding, obliging the pilot to view the world through a series of stereoscopic periscopes. The consequences of a crash—an entirely likely event in view of the plane's bulk and its pilot's limited visibility—were noted but not dwelt upon. Like all the other safety-related issues, it was put aside to be solved at some unspecified date in the future.

Despite the many unresolved problems, the N-plane rolled relentlessly ahead. The Air Force and the AEC research efforts were united under one roof, the Aircraft Nuclear Propulsion Group (ANP), under the chairmanship of Major General Donald J. Keirn, the Air Force representative with the AEC and the head of the service's own nuclear aviation office. Under the terms of the congressionally sponsored merger, the AEC agreed to finance the research and construction of several different reactor-engine combinations, while the Air Force met the cost of all other systems, including Convair's work on the airframe.

With its bureaucratic structure in place, the nuclear aviation program became one of those perennial defense phenomena, a project whose budget and prestige grow in inverse proportion to its success. The original predictions that nuclear aviation was within the bounds of existing technology rested upon the assumption that it would be a relatively easy task to "shrink" a reactor, minimizing weight and bulk without sacrificing the power output. This rapidly proved to be a false hope, something evident in the General Electric Tory II reactor–jet engine combination that was subjected to a series of static tests throughout the mid-1950s. Almost forty feet long, fourteen feet high, and weighing more than a small steam locomotive, the Tory II test unit was so large that AEC engineers were forced to mount it on a modified railway flatcar. Yet despite its immense size, the tests revealed that it lacked the power to lift even a medium-sized aircraft off the ground without the assistance of jet or rocket boosters.

Even though some of the most respected names in nuclear physics dismissed the entire affair as a waste of time and money, the research drive continued, propelled by its own growing momentum and the fear, adroitly fanned by ominous Air Force warnings, that the Russians were about to fly a nuclear plane of their own. Perhaps the most notable skeptic was Robert Oppenheimer, the father of the A-bomb and a man whose active career at the inner circle of atomic policymakers was cut short partly because he dared to criticize the NX-2. In January 1951, Oppenheimer informed Major General Roscoe Wilson, the Air Force delegate to an AEC panel investigating long-term weapons research proposals, that the nuclear plane would never fly. Roscoe, a self-proclaimed "dedicated airman," took an inquisitor's view of such heresy. When the meeting adjourned, he went straight to the office of the head of Air Force Intelligence and reported Oppenheimer's comments as another example of "a pattern of action not helpful to the national defense." Realism and scientific objectivity had become grounds for suspicion; unquestioning faith in a technological phantom, a prerequisite for patriotism.

And so the N-plane stumbled on, always appearing to be on the verge of the major breakthrough that would make a faulty theory tenable. Goaded by growing but far from universal skepticism in Washington, the Air Force orchestrated a series of tests intended to prove that progress was indeed being made. In 1953, for example, the project office even managed to get a reactor airborne, another General Electric prototype that was fitted into a much-modified B-36 but, luckily, never hooked up to a jet turbine. The flying laboratory rumbled about the skies of Texas and New Mexico for several months, garnishing a good deal of attention in the technical press but producing little in the way of tangible results. It was the beginning of the end and the start of the nuclear plane's long, expensive decline.

The following year, Defense Secretary Charles F. Wilson expressed his growing impatience by eliminating all flight test funds from the Air Force budget, a move that was meant to put an end to the entire project. The Air Force response, although unexpected at the time, now seems entirely typical. Keeping the effort alive with funds diverted from other projects and the service's unallocated revenues, the ANP team began work on a proposal for a second N-plane that was even more ambitious than the first. Unfortunately, all of the original defects were still in evidence and the second plane, like the first, slipped further and further into disrepute. When that plane was also canceled, the Air Force launched one last desperate bid to stave off the end. This time the ANP team came up with something called the CAMAL, an anagrammatic description of the plane's predicted ability to perform continuous airborne alert duties, air-launch ICBMs, and perform low-altitude penetration of an enemy's air defenses. For the third time, Convair won the right to design and build an experimental airframe, this time in conjunction with Lockheed, and cemented its claim to roughly one third of the $150 million allocation included in the 1958 defense budget. It was the final, grand moment for a doomed idea. From that time on, the annual appropriations began to wither away until, by 1962, the Air Force was obliged to withdraw completely and surrender what little was left of the program to the AEC.

There were many factors that helped the N-plane to survive as long as it did, but the most important was an enduring dose of cold war paranoia. Throughout the history of the various research efforts, the threat of a Russian nuclear-powered bomber remained a constant but unknowable specter able to loosen congressional purse strings with astonishing ease. One of the main reasons the CAMAL project was funded when all available evidence indicated it was a waste of money lay in a slim Soviet pamphlet, "Applications of Atomic Energy in Aviation." Although it overflowed with fanciful sketches of nuclear planes, helicopters, and even trucks, the underlying element of whimsy was entirely ignored by the Air Force translators attached to an intelligence unit at Wright-Patterson Air Force Base. This was an astonishing oversight, because the three Russian authors appear to have found their inspiration in Oz rather than Omsk. The big joke, as we now know, was that the Russians abandoned their practical research not long after it started and limited subsequent efforts to the production of misleading pamphlets.

In Washington the Soviet conceit was taken in deadly earnest, so much so that *Aviation Week* published one of the silliest, not to mention most ill-informed, account of a foreign weapon ever to figure in an appropriation campaign on Capitol Hill. Illustrated with "exclusive" drawings of the Soviet warplane, the article asserted that the first of several prototypes had just completed its maiden flight, a spectacle that the magazine maintained had been witnessed by several awestruck but unnamed western diplomats. A passionate supporter of the various American N-plane research programs since their inception, the magazine directed its anger at the naysayers in the Eisenhower cabinet who had refused to invest sufficient faith and money in the NX-2. The news from Russia was, the magazine asserted, "a sickening shock to the many U.S. Air Force and Naval aviation officers who have been working doggedly on our own nuclear aircraft propulsion program despite financial starvations, scientific scoffing, and top level indifference, for once again the Soviets have beaten us needlessly to a significant technical punch."

Eisenhower took the trouble to discount the report some two weeks later but by then it was too late. Quoting *Aviation Week* as their authoritative source, various Congressmen cited the fictional Soviet warplane as a further example of the enemy's sophistication. Coming as they did only a few months after the first flight of a Russian sputnik—a feat which the U.S. was still a long way from duplicating—the speeches won the N-bomber a reprieve that was to last for the next five years.

All this belonged to the future when Hopkins picked up the phone at his Washington home on the morning of March 31, 1953, and heard Floyd Odlum's baritone booming down the line from California. The call was unexpected, but Odlum wasted no time with pleasantries, asking only if Hopkins still wanted to buy an aircraft company. When Hopkins replied in the affirmative, Odlum began to describe the kind of deal he was after.

Odlum loved the telephone with a true twentieth-century passion and had once held the record for the longest, and most expensive, transatlantic telephone call. Throughout that weekend he was at his best form. Often floating in a heated swimming pool, where he spent much of every day seeking relief from chronic rheumatoid arthritis, he barked his demands into a poolside phone while Hopkins took notes at the other end. The negotiations progressed in staccato bursts, the principals discussing the major details with each other and then breaking away to clarify minor points with their subordinates. The deal was struck late on Sunday night amid much mutual congratulation.

In return for $8.7 million in cash—money Hopkins proposed to raise through a series of bank loans—and an additional $1 million worth of Electric Boat stock, Odlum agreed to hand over 17 percent of Convair's issued stock, roughly 400,000 shares. It was not a majority interest, but it was by far the largest holding, and more than enough to guarantee Hopkins' right to appoint eleven of the nineteen directors and replace Odlum as the new chief executive officer. He would also be able to count on a degree of support and advice from Odlum, who retained 30,000 shares and the control of one board seat.

There was only one cautionary voice amid the chorus of congratulatory editorials and commentaries in the financial and general press, a short observation in a *Fortune* magazine column devoted to corporate intelligence. After speculating whether General Dynamics would choose to absorb Convair in toto or content itself with the 17 percent interest acquired from Atlas, the anonymous writer noted that there was ample reason to proceed with caution: "Convair has its problems and Hopkins might find it unwise to involve General Dynamics too deeply in them."

As subsequent events were to prove, it was a prophetic observation—one that went entirely unheeded.

7. The Flying Edsel

At the small Tri-County Airport, not far from Boulder, Colorado, there is a restaurant where you can get a coffee, a doughnut, and if proprietor Rick Rush happens to be on the premises, a little background to one of the most bizarre episodes in the history of the U.S. aviation industry. "We paid four thousand dollars for it a few years ago," Rush once said about the Convair 990 jetliner that now sits atop his establishment like an angel on an eccentric's tombstone. "From what I can tell, this one started off with an airline somewhere and then a sheik had it, and then it went to a charter operator before we finally picked it up," he said. "I hear they were great planes in their day, but after the disaster in the early sixties, none of the big airlines would touch them."

The disaster to which Rush referred was not a jetliner crash but the near-total collapse of General Dynamics, for which the 990 and its smaller but no less troublesome sister, the Convair 880, were directly responsible.

The comedy of errors that produced both planes would have broken John Jay Hopkins' heart had he lived long enough to hear the jokes about "Convair's Edsel." Death spared him the embarrassment and left his successors to cope as best they could with the consequences of their monumental follies. By the time the last plane had been delivered and the last loss written off, Convair's long and illustrious career as a builder of passenger planes was over forever. With a new management, a new chief stockholder, and a gaggle of nervous bankers waiting anxiously to get their money back, Convair fled the risky business of commercial airframe production to concentrate on a safer, more lucrative market. Although it would still handle a considerable volume of subcontracted piecework for its former rivals, Convair turned almost exclusively to military pursuits.

The story really begins with Floyd Odlum, whose talent for sensing the best time to move on to something new played a large part in the timing of his negotiations with Hopkins. On the face of things, Odlum had no reason to complain. Convair had done nicely from the aviation boom that followed the release of the Finletter report, and in 1952 it was a picture of prosperity. San Diego had its oddball fighters, most of the nuclear bomber program, and a line of twin-engine passenger planes that were some of the most popular of their kind in the world. To cap it off, there were plentiful U.S. Air Force orders for the delta wing F-102 and its much faster successor, the F-106.

The Fort Worth operation was no less a picture of industry, with its 24,000 workers and its reputation as the builder of the monstrous B-36, by far the largest and most ungainly bomber ever seen. Fort Worth, too, had a slice of the nuclear plane program and there were enough retrofitting and redesign projects to occupy fully one third of the work force. The problem as Odlum saw it was not the present but the near and long-term future. The B-36, for example, had generated enormous amounts of money since its inception six months before Pearl Harbor as an experimental prototype capable of reaching Berlin from fields as far away as Nova Scotia. It had been denounced before Congress as a "billion-dollar blunder" and had come close to prompting open warfare in the Pentagon in 1947, when the Navy mutinied in the so-called Revolt of the Admirals, attempting to have the Strategic Air Command's escalating slice of the defense budget diverted to its own aircraft carrier battle groups. SAC fought back and won the right to continue funding production, not because the B-36 was a worthwhile aircraft—even with ten engines it was a lumbering, grossly underpowered sitting duck—but because it was the only way to freeze out the Navy until something better, like Boeing's B-52, could be swung into the breach.* No doubt Odlum also had a few good things to say whenever Air Secretary Stuart Symington dropped in at his Indio, California, home. "Congressman Van Zandt pointed out that . . . Symington is reported to be a frequent house guest at the California ranch of Floyd Odlum, the financier who now controls Consolidated Vultee," *Life* reported in 1949. "Odlum, according to gossip Van Zandt had heard, had helped raise anywhere from $1.5 to $6.5 million for the Democratic campaign chest."

By 1952, the arrival of the first large jet bombers was drawing very close and Fort Worth had only a few dozen B-36s left to deliver before the 385-plane contract was complete. Odlum concluded he would be

*An indication of the B-36's inadequacies can be seen in the fact that the Air Force wanted to base its next big bomber on Northrop's famous flying wing design. The plan fell through, so defense veterans maintain, because Jack Northrop refused to allow his design to be handled by Fort Worth. With no available alternative, the Air Force returned to the B-36, which, lacking any military virtues to speak of, was praised for its sheer size. "A B-36's de-icing system could heat a 600 room hotel," *Air Force* magazine was still boasting as late as 1959.

better off allowing someone new to run the risk and expense of developing a successor.

Less obviously, San Diego was also facing a new era and a different, though similar, set of problems. Government test programs had breached the sound barrier time and time again, and now, as the inevitable next step, people like Odlum realized that it would only be a matter of time before jets dominated commercial routes as thoroughly as they did military aviation. Cashing in on the impending jet age would be a race, with the richest prizes going to the swift. The first airline to introduce jetliners would have no trouble finding customers eager to buy seats on these faster, more comfortable, and infinitely more glamorous planes. The first airframe company to produce a practical and economical jet would have the airlines lined up and eager to buy.

This was to be a high stakes game reserved only for the biggest players. Any airframe company could build a jet fighter or a medium-sized military warplane. Aircraft like the famous Sabre were relatively small and simple, even with their jet engines and supersonic speeds. And of course, if anything went wrong, the Pentagon was always on hand to finance revisions and take the sting out of failure. Commercial jetliners, on the other hand, were an entirely different matter since they presented a new and uncharted set of economic and engineering trade-offs. They were to be larger than most previous piston-engined passenger craft, more complex yet requiring much less routine maintenance than their military counterparts. Most daunting of all to the smaller companies faced with finding the enormous amounts of money required for research and development, they had to be sufficiently cheap to make the airlines' transition to jets worthwhile.

Convair was certainly large and prosperous enough to take on Boeing and Douglas in the coming battle for the jetliner market. What it lacked was Odlum's faith in its ability to succeed. As with Fort Worth, a prolonged period of massive investment in research and development would be required, which he was simply not prepared to make. Just as he had done throughout his business career, Farmer Floyd decided to sell out and take the proceeds, in search of another of his famous "special situations."

That was more or less what happened. Convair changed hands and was securely in Hopkins' care when Boeing officially unveiled its contender for the new market. At the start of 1954, in what remains one of the U.S. aviation community's most impressive engineering achievements, the Seattle manufacturer began work on a long-range passenger jetliner. Eleven months later, the prototype 707 rolled out of the factory, taking to the air for the first time two months after that. It was not the world's first jetliner—Britain's de Havilland Comet had already taken that honor—but it was by far the better of the two. Moreover, being the first American jet on the market placed it in a position to seize the lion's share of fleet sales when the airlines began their inevitable rush to replace their now outdated piston-engine planes. The Douglas company, much later to become McDonnell Douglas, also saw the writing on the wall and quickly began work on its own contender, the plane that would eventually become the DC-8.

Convair executives noted these developments with mounting alarm. Unless the move into jets was made quickly, the company would be left high and dry while its competitors walked away with the show. Boeing, for example, faced almost no chance of losing money, since it had already signed an initial contract with the U.S. Air Force to supply eighty-eight models of a military variant that would be put to work first as the KC-135 air-to-air tanker and later, in a much altered form, as the basis for the AWACS reconnaissance-surveillance aircraft that was to figure so prominently in the United States' arms diplomacy with Europe and the Middle East. Douglas, the second contestant to enter the race, lacked the direct financial support of the Pentagon, but it, too, was off to a good start. A bare four months after the project was announced, in June 1955, Pan Am placed an order for twenty of the new jets. Like penguins jumping off an ice floe after the first of their number has shown it is safe to enter the water, six other major airlines, including two of the biggest, United and Eastern, added their names to the list before the end of the year. The last of the big three manufacturers, Convair was still languishing at the starting blocks. The industry's observers wondered what on earth could be holding it back.

There was a two-word answer to that question: Howard Hughes. A short drive up the coast from The Rock, in any one of several shifting domiciles scattered throughout the greater Los Angeles area, the millionaire was pondering the future of his "personal" airline, Trans World Airlines. Like all its competitors, TWA had been considering the move into jets for quite some time—so long, in fact, that Hughes had lost sight of the reasons that made a speedy decision imperative. As early as 1951, TWA's chief engineer, Robert W. Rummel, had been pressing for an exploratory test program with turbojet aircraft like the Comet. Rummel soon changed his mind about the British plane when it began to encounter unexpected technical problems, but he remained insistent about the need to prepare for a rapid, and inevitably expensive, transition to jets. Hughes listened, often calling Rummel in the wee hours to discuss esoteric technical points, sometimes keeping the unfortunate engineer away from his bed until long after dawn.

But despite the mounting urgency of the situation, the closest Hughes could come to a decision was his selection of Convair as potential sole-source supplier. Rummel and others tried to point out that even if he was determined to buy from the San Diego manufacturer, keeping a second, competing company in the background would place him in a much better bargaining position. As usual, Hughes listened but refused to act. So long as he dealt with only one firm he could exert a major influence on the design process and that, much more than the airline's commercial success, was what the millionaire enjoyed most of all.

By the first months of 1955, Convair's bosses, the independent-minded airframe veteran John V. Naish and former U.S. Army general Joseph T. McNarney, were sending couriers up to Hughes' Romaine Street headquarters in Hollywood with progress sketches for two long-range passenger jets. The first would have been the marvel of the day, comparable to a modern jumbo jet, with six underslung engines and room for three hundred passengers. The second, a smaller, four-engine model similar to those soon to be announced by Boeing and Douglas, was less adventuresome but entirely equal to the demands of TWA's long-haul routes. Hughes seized upon the drawings and called eagerly for more, poring over the technical data, correcting minor computational errors, and firing back demands for further information on hundreds of tiny points. So engrossed did he become in putting his individual stamp on TWA's new fleet, Boeing and Douglas had time to begin work on their own long-range aircraft and soon snapped up most of the potential customers. By the time Hughes felt finally free to make a decision, it was far too late for both parties. Even with a large initial order from TWA, Convair could not have covered its costs on what little was left of the long-range market. Reluctantly, Convair's liaison people packed up their plans and went back to San Diego to reconsider the situation.

Bob Rummel and TWA's long-suffering president, Ralph Damon, were both dismayed by their chief stockholder's latest infuriating and irrational display. Late in 1955, with Boeing and Douglas now holding backlogged orders for at least 120 planes destined for TWA's rivals, Damon drew on Rummel's research in a three-thousand-word confidential report to the board of directors that warned of "disastrous consequences" unless the airline joined the rush to jets. Eastern Airlines would be ordering its first new planes within the month, he warned Hughes, adding that this would leave TWA "in a position of splendid isolation." Any further delay might see the airline forced from the international market altogether. "All our international routes are on a temporary basis expiring in 1959," he said in reference to the Civil Aeronautics Board's licensing and review system. "Hearings to extend the temporary routes will probably begin in late 1958 and extend through the early part of 1959, by which time the full blast of jet propaganda of our competitors will be sweeping the country and will be an important psychological factor in the decision of the CAB as to our 'fitness, willingness and ability' to compete in international routes." The situation was loaded with lethal possibilities, as Damon, no doubt hoping that repetition would succeed with Hughes where logic had failed, asserted over and over again. "Not only do we have no chance of extending our international routes around the world," he concluded, "we may even be denied further operations to many points we now serve under temporary certificate, including London and Frankfurt, unless our plans are competitive equipment-wise."

For once, Hughes responded, albeit after a further three-month delay, and turned reluctantly to Boeing's 707. Rummel and Damon were relieved but far from happy, since all the early spots on the Seattle manufacturer's production roster had long since been appropriated by TWA's competitors. Once again, Hughes had shown himself to be both the airline's undisputed ruler and its chief liability.

This one encounter with Hughes might have persuaded another company to stay away the next time, but not Convair. The prevailing mood at The Rock can only be described as one of "jet mania," a passionate, almost intoxicated rush to get into the jet market. This ran directly counter to the official policy of only a few years earlier, when Thomas G. Lanphier, the company's chief planner, told *Fortune* that it would be foolhardy for Convair to abandon its traditional emphasis on medium-size, twin-engine planes designed specifically for short and medium-haul routes. Plans for a jetliner were "a self-imposed limitation," he explained, adding that he was optimistic about the prospects for a slightly larger version of the existing 440 model equipped with more powerful turbo-prop engines.

At some point between Lanphier's comments and the first jetliner discussions with Hughes, the mood underwent a drastic change. Was Jay Hopkins responsible for the about-face, as some would later allege? The answer would appear to be "yes, but not entirely."

Hopkins exuded a unique brand of aggressive, overly optimistic ambition that could, given the right circumstances, act as a potent catalyst on his underlings. Convair was, however, the wrong place to encourage grandiose ambitions. The division's attitude was one of fierce independence, and as Odlum had learned, a strong and ever-watchful master was required to lay down the law. Later, when General Dynamics had been pushed to the edge of bankruptcy by Convair's excessive secrecy, chairman Frank Pace summed up the one problem that the company has never been entirely able to solve: "When you have a company employing 106,000 people, made up of nine different divisions, each really a corporation in its own right, most of which were separate enterprises before they joined the organization, and each headed by men who were presidents of corporations, with their own separate legal staffs, financial staffs, etc., all of these highly competent men . . . your capacity to know specifically what is happening in each division just cannot exist. If you did try to know everything that was happening and controlled your men that tightly, your men would leave you or would lose the initiative that made them effective." Convair's problem was that there was much too much initiative, just waiting for an opportunity to do its worst.

With scant regard for what the head office might think and little advance notice, division president McNarney was soon directing his engineers to take a second look at the jet question. The company had missed the bus as far as the long-range market was concerned, but there remained other potentially lucrative possibilities to be explored. If the airlines liked jets on their long-haul routes, it was only reasonable to assume they would eventually start looking for slightly smaller jet transports to handle middle-distance traffic between points like New York and Florida, or Colorado and the West Coast. With the exception of a new and as yet untested French plane, the Caravelle, the short to mid-range market was wide open. Convair's engineers were told to dust off the plans for the four-engine transport that had held Hughes' attention for so long and perform whatever alterations were necessary to make it fit the new task. That decision was sound, even inspired. The next certainly was not. Even though they had the entire market before them and no domestic competitors in sight, McNarney and Naish took the idea straight back to Howard Hughes.

The ever more bizarre episodes in Howard Hughes' later life read now like a casebook history of psychotic contraction. Once, back in the thirties, he had reigned as one of the heroes of the day. Rich beyond imagination, adorned with starlets, a successful Hollywood producer and a record-setting aviator, he was as much a subject of popular fantasy as the actors and actresses he hired for his movies. But then, slowly and imperceptibly at first, his talents and his world began to shrink about him. The success and the adulation of the thirties turned into the scandal-charged congressional hearings of the late forties, when he was accused of plundering the nation's war chest to build his infamous, and hangar-bound, "Spruce Goose," the largest and costliest airplane in the world to date. The Hollywood touch also deserted him. Vast and successful epics like *Hell's Angels* gave way to the $23 million in losses he inflicted over the course of just two years on RKO, the studio he acquired from Floyd Odlum in 1947. Increasingly, as he began to retreat further and further from the world, he fixed his interest on aviation, the first and most consuming passion of his life. A King Midas in reverse, TWA would follow RKO in being driven to the verge of extinction.

The fact that Convair was prepared to deal with Hughes at all remains the most telling indication of the company's misdirected gusto for the jet game, since it would have been difficult to find anyone at San Diego who did not know how troublesome the man could be. Long before he forced Convair to abandon its plans for a long-range jet, he had entered into protracted negotiations for a new fleet of propeller-driven transports, only to walk out on the deal at the last moment. Convair vice-president Jack Zevely almost cried when he heard the news. As chief negotiator, he had become a regular midnight visitor to the Palm Springs city dump where Hughes insisted the meetings be held, the blueprints illuminated by the beam of a hand-held flashlight and the odorous glow of burning trash. Hughes explained that the odd venue was essential for security reasons, although, as with most of the millionaire's pronouncements, this made no sense at all since TWA's need for a new medium-range airplane was common knowledge throughout the industry. That had happened in 1952, and the memories were still fresh enough to make those facing the prospect of yet another round of negotiations dread the mere mention of the millionaire's name. Still, then as later, Convair did everything it could to humor its potential customer. Hughes might be mad, but a crazy man's money was as good as any, particularly when the amount might run to $160 million or more.

The interminable conferences were as bad as ever before, although by now Hughes had abandoned public dump sites in favor of more congenial surroundings. One of Convair's engineers would be sent to a Las Vegas hotel, where he would be told to sit in his room and wait for the phone to ring. Sometimes the call came, sometimes it didn't. They were the petty aggravations; Hughes was capable of much worse. On another occasion, Jack Zevely and an executive were ordered to a secluded apartment-hotel, this time in Los Angeles. As usual, Hughes insisted on debating some very minor points and pursued this fancy until almost 4 A.M., by which time Convair's people were falling asleep in their seats. Deciding that a movie was just the thing to give everyone a new lease on life, he rang for a limousine and had the party shipped across town to RKO. With Hughes leading the way like a demented tour guide, they were shown into a private screening room and forced to sit through a screening of *Jet Pilot,* a B-grade clunker of the worst kind. For good measure and a dash of color, Hughes had actress Janet Leigh hauled out of bed to provide the sleepy audience with personal recollections of life on the set. Not for the first time, Zevely left for home well after dawn.

Perhaps it was the regular disruption of the negotiators' sleep patterns that allowed Convair to slide into the deal that would bring about its downfall. From start to finish, the contract was a compilation of pitfalls. TWA had ordered thirty planes and Delta had taken out an option on another ten, while the Dutch carrier, KLM, was also expressing interest. That was, however, the sum total of the advance orders Convair would be free to pursue for the next twelve months, since Hughes had extracted a promise from McNarney that stopped the company from doing business with any other airline until late in 1957. Incredibly, Convair agreed and committed one of the principal blunders of the entire sad affair. When Boeing and Douglas recognized the potential of short-medium-range jetliners, each reworked its own long-range plane to produce a version better suited for work on mid-distance routes. By the time the Hughes moratorium came to an end, Convair found that it had missed its opportunity yet again. Nor did Convair's engineering department help matters. Instead of the short-medium plane they had been told to design, the engineers came up with a maximum range of 3,400 miles. This was only a little shorter than Boeing's 707 and very close to Douglas' DC-8. Neither a long-range nor a medium-range aircraft, the Convair 880, as it would soon be called, was just big enough to demand too much turnaround time at every landing and a little too small to tackle the longer, more profitable intercontinental routes.

The next and ultimately the most costly mistake saw Convair allow Hughes to deposit a down payment of just $15 million, or a meager 10 percent of the total projected cost. For once in his life, Hughes had managed to strike a good deal. The next payment was not due until late in 1959, when Convair promised to have the first aircraft ready for service.

McNarney took a précis of the deal to the next General Dynamics board meeting in New York, in May 1956, where he convinced Hopkins and the executive board to vote unanimously in its favor. The former U.S. military governor of Berlin laid out a forceful and persuasive case. He put potential sales at 257 aircraft and maintained that the project would break even after the sixty-eighth plane had been delivered. The profits might easily amount to $300 million, against a total, worst-possible-case loss of no more than $45 million.

In presenting the facts, McNarney omitted all details of what amounted to a considerable body of opposition in San Diego. Thomas Lanphier, the division's chief planner, who had warned against a jet program three years earlier, was still adamant, arguing that the floor space, manpower, and start-up costs could be more profitably invested in military projects, particularly missiles, which he saw as the wave of the future. Perhaps because of his notable war record—it was Lanphier who led the flight of Navy aviators that shot down Admiral Yamamoto—McNarney dismissed his opinion as the emotional preference of an old hero. Naish's argument was harder to contest. Like Lanphier, he saw greater profits in military work and modest, turboprop passenger planes, and he also shared the younger man's skepticism for the cost projections McNarney had "sold" to New York. Even if the plane did a little better than expected, it would still contribute just 10 percent to the division's gross sales receipts.

McNarney surged ahead regardless, supported by Zevely and several other members of the division's board. He was an old soldier himself and accustomed to being obeyed. McNarney once faced down George Patton when the tank commander took his time disarming several crack German mountain battalions, apparently on the premise that they could be put to good use in the coming war against the Russians. After encounters with Patton and Soviet General Georgi Zhukov, the commander of the Russian occupation forces in Germany, any slight differences of opinion with his own subordinates in San Diego must have been easily settled.

Much less easy to understand was the failure of Jay Hopkins' previously infallible nose for a bad deal. This one reeked long before the first papers were signed, but Hopkins simply let the question pass, insisting only that McNarney run a credit check on each of the prospective customers before taking their orders.

Hopkins appeared to be his usual buoyant self throughout the meetings leading up to the May decision, but the truth of the matter was that he did have a lot on his mind. By rights, this should have been a time to enjoy the fruits of his spectacular success, his "second golden youth," as he told one interviewer. And to all outward appearances he was doing precisely that. After eighteen months of sea duty and a string of broken records, the *Nautilus'* success had swamped Electric Boat with orders, each larger and more expensive than the one before. Of all his gems, the submarine yard was the possession in which Hopkins took the greatest pride. It had been just another shipbuilder when he took charge back in 1947; now it was marching bravely in the vanguard of the atomic revolution. On the cover of *Time* one month, in Tokyo for a nucleonics conference the next, or preaching the gospel of containment at countless speaking engagements across the United States, Hopkins had become something of a celebrity and he owed much of his newfound prominence to Electric Boat.

Canadair had also turned a recent technological advance to its financial advantage, adding further to Hopkins' reputation as the high priest of high-tech. In the early fifties, U.S. government aeronautical engineers pinned down one of the immovable constants in the physics of transonic airplane design by defining the ratio between surface area and air resistance that determines the ease with which a jet can knife through the fierce turbulence encountered just short of the speed of sound. San Diego's delta wing F-102, for example, could exceed the sound barrier only by using so much fuel that the exercise was essentially meaningless. Then an engineer at the Air Force's Langley research facility, John Stack, perfected a supersonic wind tunnel which another researcher, Ken Whitcomb, used to test aircraft models with nipped, "wasp" waists. These tests produced what came to be known as the Area Rule, a table of coefficients that promised to reduce transonic drag by 25 percent or more. Canadair applied the lesson to its venerable Sabres, added a new and more powerful jet engine, clipped the wing angle, and discovered that it had developed a vastly improved warplane at minimal expense. Super Sabres—the only logical name for the new planes—were soon being shipped off to second- and third-string nations all over the world.

And there were also the small, though hardly minor acquisitions in which Hopkins was inclined to take great pride. Back in 1954, he had acquired the Stromberg-Carlson company, one of the pioneers of the U.S. telephone system and still a leader in what was just coming to be called telecommunication electronics. The new holding deviated slightly from the usual emphasis on war work, but Hopkins was certain it could be made to mesh with the rest of his empire. He had always depicted General Dynamics as "the future General Motors of defense," and a company that might soon turn its hand to guidance systems for missiles, jets, and submarines was not at odds with the grand design. With General Atomic, the other subsidiary for which he held such high hopes, it was possible to imagine the day when General Dynamics would handle every major contracted system that went into one of its own nuclear submarines or, later, the predicted nuclear-powered bomber.

Now there was the final glory of Convair's ambitious challenge to Boeing and Douglas, an impressive cap to a remarkable career. Hopkins had only one problem during that spring of 1956: He had just learned that he would be dead within the year.

Not much more than a year earlier and not long after the *Nautilus* first put to sea under its own power, doctors diagnosed cancer and removed a large section of his stomach. He was back at work and on the lecture circuit almost straightaway, a little thinner but with no less energy than before. "They got it all," he assured friends and well-wishers wherever his plane touched down. Now it had come back, and this time his doctors could see little point in surgery.

The son of a small-town minister, Hopkins must have heard all the standard eulogists' clichés many times over the course of his childhood and early youth. In his own case, it was one of the most common that best summed up his attitude: "You can tell a lot about a man's life by the way he prepares for death." Before death dragged him from the game, General Dynamics' chief architect resolved to set his company's house in order. The loose ends would be tied up and the lines of succession established beyond dispute. Set firmly on the path he had ordained, the company would be both a monolith and a monument to its founder's genius.

All in all, it had been a remarkable career, the kind of endless success story more common to fiction than to real life. From the genteel poverty of his father's California rectory and the proverbial newspaper delivery rounds, Hopkins had moved through college and a succession of profitable small-business ventures to a law scholarship at Harvard, breaking step only for a short stint in the Navy during World War I. The war left him with a commission in the naval reserve but no particular inspiration to make a career in armaments. That happened almost by accident, at least according to some of the later legends that came to surround him at General Dynamics.

Hopkins was a well-established lawyer and investment counselor, with offices on both coasts, when Henry Carse approached him about a job at Electric Boat. The submarine builder had taken a severe drubbing during the recent Nye hearings and it needed a capable man to repair the damage, particularly in Washington, where the company's reputation had been hurt the most. Any reluctance to come aboard would have been entirely understandable. Hopkins already had his law practices and a private fortune that had remained untouched by the Depression. Nor was he entirely in step with Roosevelt's New Dealers, having served as a special assistant to two Treasury secretaries during the declining years of the Hoover administration. He had remained a keen supporter of the President after returning to private life in 1933, although not keen enough to join the often rabid opposition to Hoover's Democratic successor. All the same, Hopkins took the job and, as usual, handled it effortlessly.

As with so many millions of others, it was the war that changed his life. A month or so after Pearl Harbor, he joined in the war effort by closing down his law firm and taking full charge of Electric Boat's finances and contract negotiations. Carse and the others were getting on in years and Hopkins' responsibilities soon grew to the point where he was pretty much running the company while his older colleagues coped as best they could with the pressures of around-the-clock production.

So much for the career; the man behind it was something else again. "Mercurial," "boundless energy," "indefatigable," were words that would feature in the later press stories about Hopkins and his transformation of Electric Boat. The words were certainly accurate, but they reflected only a part of the truth. Hopkins was a man with an inability to approach any problem at half throttle; be it golf, business, or the consumption of alcohol, he attacked everything with the same unrestrained zeal. There were few people who could keep up with him under normal circumstances, and fewer, it is said, who came close enough to enjoy anything resembling close friendship. He had his golf buddies, drinking partners, and countless business associates, but always, beneath the public Hopkins, there was that formidable intensity so much in evidence during his last months.

The defense industry on which he would make such a lasting impression appealed to both his Dullesian view of the world and his best business instincts. "If anything will call a halt to the Communist plan for world domination . . . it is the picture confronting history's most powerful Czar of a forbidding industrial strength standing behind the armed forces of the United Nations," he had told a Canadian audience in 1951. It was not merely good sense to contain the Soviets behind a skirmish line of modern weaponry fielded by the U.S. and its chosen allies; it was also inspired economic management. Hopkins never went quite so far as his successor at General Dynamics, Frank Pace, who once asserted that defense spending actually stiffened the nation's moral fiber, but he was always prepared to praise its potency as an economic stimulant and general antidote to the vagaries of the business cycle. Of course, General Dynamics' prosperity was his favorite example of the honorable profits produced by a state of constant siege. "Wars will be fought under the sea and in the air. Either way we are ready and we can also hit the civilian market," he boasted to *Business Week.*

Hopkins' detractors were few and not given to public utterances, but they existed all the same. They were, in fact, the chief reason Frank Pace had joined the firm back in 1953. To the public at large, the defense tycoon appeared as a symbol of American corporate patriotism. With golf clubs stowed beside his desk or, more commonly, in transit aboard the company plane between pro-am tournaments, Hopkins was the quintessential Eisenhower-era businessman, even down to his preference for martinis. Washington, however, was unimpressed, as *Fortune* reporter Richard Austin Smith wrote in an incisive postmortem on the eventual Convair 880-990 fiasco. "The truth was Jay was drinking too much and Washington had lost confidence in him," Smith quoted another member of General Dynamics' board. "We had to get someone in there who could restore that confidence." Frank Pace was the man chosen for the job. Unlike Hopkins, he was sober and predictable, and while he was also a low-handicap golfer, he never brought his clubs to a board meeting or flew off to Canada for an event of the International Golf Association, one of several golfing institutions and tournaments Hopkins founded or sponsored.

Pace came to the firm with impeccable credentials, having only just left government office with the Truman administration, where he had been first budget director and later secretary of the Army. He was still young by the time he arrived at General Dynamics, and there were some who considered him to be merely punching a ticket, filling out his résumé with a short and spectacular stint in the business world before returning to Washington for a stab at a Senate seat or perhaps some even loftier elected office. Pace would maintain that he won the job on his merits, although *Fortune* found an unnamed member of the board who disagreed. "Well, I just bought myself a show window," Hopkins was reported to have said. "It cost me $75,000 with a secretary. They give me hell because I'm never in the office, and I've got to have someone to answer the phone."

All the same, there was a great need for a man of Pace's abilities. During his time as budget director, he had straightened out the indecipherable fiscal bills presented for congressional consideration, distilling the information into a readily comprehensible, itemized presentation very similar to the one that remains in use today. Transferred to the post of Army secretary, Pace confronted the numerous problems of mobilizing an army and its suppliers for an unexpected war, and handled the exercise with competent aplomb. General Dynamics required a similar gift for organization, which Hopkins, for all his other talents, was unable to provide. Hopkins had assembled the ingredients and had stopped the resulting mixture from turning sour only through the strength of his domineering personality. The divisions seldom knew when to expect him. He would fly in, spend a few hours quizzing the local bosses, and then depart for the local golf course to mull over what he had learned. Should there be anything of which he disapproved, his observations and criticisms would be recorded in a short but explicit note, usually dashed off in the blue pencil he reserved for his exclusive use, and dispatched from the next port of call on his endless travels across North America. There was little need for the divisions to communicate among themselves, since he was both their master and their corporate switchboard.

Hopkins recognized the potential pitfall and was working on solutions right up until the end. Someone like Robert S. McNamara, the future defense secretary, who was then revamping Ford, might have made a suitable deputy in the campaign, a man with pretensions to understanding the often subtle human emotions and bureaucratic imperatives at work in vast industrial organizations. Hopkins, however, had only Pace and his own flagging reserves of energy. Still, even as Convair plunged deeper into the jetliner disaster, Hopkins did his best to pin down solutions to the problems posed by his creation's sheer sprawling size.

Late in 1956, he moved to a favorite haunt near Palm Springs, close to some of his best-loved golf courses and not far from Floyd Odlum's Indio ranch. The golf games and the flying visits to the divisions came less often and pretty soon stopped altogether. Early in the new year, Hopkins learned that the French government wished to present him with a medal honoring his contributions to aviation and science. Under normal circumstances he would have leaped at the invitation to visit Paris, eager to promote his company's name in Europe and preach a united Western defense against the common foe. Instead, he stayed at home, husbanding his energy for the meetings with Odlum and one or two other close allies from the board, like fellow lawyer Ellsworth C. Alvord.

The trio agreed that General Dynamics' management was simply too large. The board had grown with every fresh acquisition and now numbered a cumbersome thirty-two, most owing their first loyalties to the divisions they represented and supervised. By Hopkins' estimation, the number was twice the size it should have been and he decided to wield an ax at the next available opportunity. In order to spare his friend the agony of the long trip east, Odlum provided the venue for what was to be General Dynamics' version of the Last Supper. The difference this time was that there was not one Judas in attendance but many.

The vital meeting was held at Odlum's ranch in mid-February 1957, just as things were starting to go seriously wrong in San Diego. The West Coast delegation made no mention of KLM's recent decision to let its option drop, or of the latest cost projections, which indicated that the new plane's $4.4 million price tag would be far from sufficient to cover expenses. It was San Diego's project, good or bad, and the division was determined to see it through to the end, even if this meant keeping the head office in virtual ignorance of the facts. If Hopkins was curious or perplexed about the state of affairs, he made no mention of his qualms, concentrating what little energy he retained to push through the matter at hand. He had been a big man in his prime, heavy-jowled, with large, beefy limbs and hands. Now he was drawn, pale, and in obvious pain as he asserted his will for the last time.

His audience was shocked by what it heard. Half their number were to be voted out of office on the spot. It was nothing personal, Hopkins explained, simply a response to the demands of business.

Those he had allowed to survive were hand-picked men, and Hopkins must have gone home—he lacked the energy to stay and share a steak and drinks with his colleagues—confident that there would be no further problems. It was one of the few misjudgments of his career.

Back at their home bases, some of the survivors began talking and a consensus soon emerged: Hopkins was too ill to continue as the company's chief executive officer and should be replaced as soon as possible. The common choice, indeed the only choice, was Frank Pace, the man Alvord would later claim Hopkins had already rejected for the job. The mutineers kept their intentions to themselves at first, giving no indication of their plans until the day after the annual stockholders' meeting in Dover, Delaware, on April 25. In a final demonstration of his prodigious will, Hopkins had set out from Palm Springs intent on facing the stockholders for what even he must have realized would be the last time. It was an epic journey for a man in his condition, and by the time his plane touched down in Washington he was forced to retreat to his old apartment for what he insisted would be just a day or two of rest and recuperation. It was then that the mutineers declared themselves.

First in small delegations and then in larger numbers, the directors trooped into Alvord's Washington office to demand that Pace be placed in charge. Conducting himself like a politician at a national convention, Pace said that he was being drafted and denied that he had done anything to set the *Putsch* in motion. Still, he did nothing to stop it. Alvord defended his friend as best he could, but the task was hopeless and on April 29, at a short board meeting, General Dynamics' founder was voted out of office. More important to those who remained, the last chance that a new man would be brought in from outside to whip the company into line abruptly vanished. Pace was to take the reins and business would continue much as before, the divisions safe from the threat of intrusive supervision.

It is probably unfair to say that any one thing can kill a man who is already dying of cancer, but in this case, the timing speaks for itself. The next day Hopkins left his apartment for a hospital bed, where he had his last conversation with Alvord, his friend and chief defender. "My usefulness is gone," he said. Three days later, amid professions of grief from those who deposed him, Hopkins gave up the fight and lapsed into a coma; he died soon after.

While Hopkins grappled with death, Convair's misadventure with Howard Hughes was rapidly degenerating into a running joke. At one point, Hughes decided it would be nice if the new planes were finished with bronzed aluminum, since this would allow him to call them "TWA's Golden Arrows." The name had a little more zest than the "Skylark" designation Convair had originally chosen for the plane, but the idea proved impractical and both parties settled by default on the "880" tag, a name that reflected an ability to carry 88 first-class passengers and graced with an extra zero, at least according to one former airline executive, "because they didn't want it to sound like a piano."

Hughes, however, was now only part of the problem. At just that moment, when the need for research and development dollars was most pronounced, Washington began a temporary cutback in defense spending that crimped the division's cash flow. And further to complicate the problem, the Pentagon delivered a second blow by ordering the Atlas missile production line moved to another, more secure assembly site.

The worst news concerned the division's inability to find buyers, a deficiency magnified by the often bizarre decisions of the department heads. United Airlines, a large potential customer, with the ability to make or break the 880 program, had indicated it would be interested in the plane only if it could handle six-abreast seating. Convair's engineers made it just wide enough for five.

The finances, too, presented a series of peculiarities. The 880 was being sold to Hughes for $3.5 million apiece, yet no one in San Diego knew how much it would cost to complete until later that year, when a relatively junior executive analyzed the available data and discovered that there was no present or future break-even point. The line representing costs never came close to the one depicting sales. Losses would begin with the first aircraft in the series and mount steadily with every one thereafter. When the man brought this information to his superiors, he was relieved of his post and the program continued as if nothing had happened. How Naish and the rest could have chosen to ignore the mounting evidence against the scheme will always remain a mystery. But ignore it they did, even though their own, extremely optimistic sales figures indicated that the 880 would be only marginally profitable at best. Originally, Hopkins had been told that sixty-eight sales were necessary to put the plane in the black. Since the market was put at a generous 358 potential sales, the break-even point appeared a modest objective, particularly with Hughes' and Delta's orders already in hand. The next official estimate, compiled at about the same time Hopkins retreated to Palm Springs, envisioned a potential market of just 150 sales and a break-even point well on the way to seventy planes. Meanwhile, total sales were fluctuating around the forty-plane figure, small new orders balancing larger cancellations.

McNarney had recently retired, and by the time of Hopkins' death the future of the 880 fell into Naish's hands. The airframe veteran had been skeptical about the program from the beginning, but now, with no choice but to proceed or declare an immediate $50 million loss, he elected to plunge ahead. The division's future was at stake and neither he nor the other executives were going to let it go down without a fight. Failure now, so soon after Hopkins' death, would invite dismemberment; Fort Worth would be given its independence and San Diego subjected to the close scrutiny of the head office.

Hopkins would have been horrified by the "rubbery" estimates and sketchy progress reports emanating from The Rock, but Pace allowed the affair to proceed to new depths. With one scorned jetliner already tying up a large part of the plant, Naish received permission to build a second and substantially larger passenger plane. This was the 990, and the folly of its conception was even more pronounced than that of its smaller sister.

It began when Convair asked American Airlines president C. R. Smith if he would be interested in a slightly altered version of the 880, equipped with the new fan-jet engines Fort Worth was fitting to its B-58 Hustler bombers, the successor program to the B-36. This new jetliner would boast much greater power at takeoff and permit substantially larger payloads. Better yet, in an era fascinated by jets and speed and sonic booms, the re-engineered version would be faster than any comparable aircraft in the skies. Smith was receptive but demanded that Convair follow his instructions to the letter. The plane that emerged from the negotiations was forty thousand pounds heavier and fifteen feet longer than the 880, higher, wider, and generally so different that Smith could rightly claim to have designed it himself.

The financial terms of the contract continued in the same vein. With the 990's additional research, development, and tooling costs, the break-even point for both planes soared to a total of two hundred sales, surely an impossible figure in the light of the available evidence. Desperate to get the business, Convair accepted a fleet of old DC-7 piston aircraft instead of a cash deposit. There were twenty-five of them in all, and each was assigned a trade-in value of $1 million, at least 50 percent more than they were worth on the open market. Even before the first construction jigs were set up, the 990 was $12.5 million in the hole.

From this unfortunate beginning, the 990 contract progressed through an expensive list of fringe benefits drawn up at Smith's insistence. American would not have to pay for its spare parts until they were actually needed, nor would it be required to pay for storage costs. Finally, there was the most costly concession of them all: Smith extracted a promise that pledged Convair to return part of each plane's purchase price if the 990 failed to beat the best of its competitors' point-to-point times between New York and Los Angeles. No engineer would have made the promise, dependent as it was on a score of technical variables, but Convair's negotiators went ahead regardless.

The contract was presented to the board in New York by Earl Johnson, Pace's former under secretary from his Army days, who had been brought into the company not long before Hopkins' death. Ellsworth Alvord would later maintain that the other directors were given no chance to consider the proposal or even to voice their opinions. It was "a contract written to American specifications with an American delivery date but the plane was not even on paper," Alvord lamented. "The 990 was signed, sealed and delivered without board approval. It was just a *fait accompli.* An announcement was made to the board that there would be "a slight modification of the 880."

History soon began to repeat itself. Orders for the 990, originally put at upwards of 130, pegged out at the paltry figure of 37, while the 880, now released from Hughes' sales moratorium, had managed to amass just 50 confirmed sales. Even at this late stage Convair might still have backed out, swallowed a $100 million loss and then tightened its belt until the

Pentagon came up with something to get the division back on its feet. That possibility was never even considered, and soon it was too late to retreat at all. Once again, it was Howard Hughes who pushed the division beyond the point of no return.

Inconceivable as it may seem for a man whose reputation then as now rested upon an apparently bottomless reservoir of private wealth, Hughes was facing a strange and unfamiliar problem: He lacked the money to pay his upcoming bills. The mainspring of his fortune had always been the Hughes Tool Company, with its patented drilling bit perfected by his father back at the turn of the century. The bit was a breakthrough at the time and it had remained of such vital importance that the dips and rises in its monthly sales were treated as near-infallible indicators of the energy industry's future health. Just so long as the tool company kept bringing in money, Hughes was free to continue ruining each of his other possessions, one after the other. By the late 1950s, as Convair raced to complete the first 880, Toolco ran into hard times. The demand for new bits was going through one of its cyclical downturns and it left Hughes well short of the $120 million he needed to make the planes his own. He had just taken delivery of some of the 707s he'd ordered from Boeing and was now so short of liquid cash he was at the point of selling half a dozen to Pan Am, chief rival to TWA.

A sane man faced with a similar dilemma would have gone to a bank and asked for a loan. Hughes, however, could not bring himself to broach the subject. His mismanagement of TWA was legendary throughout Wall Street, and the one sure thing about a successful loan request was that it would be conditional on a new, and hopefully more rational, man being brought in to operate the airline as a business rather than a private toy. By any conventional standards of behavior there were no two ways about it. Either Hughes surrendered his voting stock and day-to-day control of his beloved airline, or he admitted to the world that he could not satisfy his creditors. True to form, he found a demented alternative.

Just after the noon lunch break on October 5, 1959, a convoy of large sedans, each packed with more than one thousand pounds of well-muscled human flesh, pulled up at the gate to the Convair plant. The convoy paused briefly at the gatehouse while one of the passengers explained that he and his associates were Toolco inspectors on their way to examine the first plane, now within a few weeks of its maiden flight. There was not a Convair employee who did not know Toolco was synonymous with the division's chief customer, and the barrier quickly went up. If the millionaire's odd little ways demanded that he send what appeared to be a gang of longshoremen to attend to business on his behalf, then so be it. Any client spending that sort of money was entitled to any and every diplomatic indulgence he desired.

The convoy moved swiftly away from the gates and accelerated to the main construction hangar, the largest of the assembly areas gathered around the taxi apron of the division's private airfield. The lead car came to rest no more than a few feet from the huge sliding main doors; the men piled out as their vehicles slowed down, and sprinted inside. If they had carried guns openly—and there is no doubt some carried concealed weapons—an observer could have been excused for thinking some sort of paramilitary operation was taking place. And in a sense, it was. The world's first and strangest hijacking was under way at the express orders, and under the personal direction, of Howard Hughes.

A few yards inside the assembly area, a large cross-section of Convair's engineers, interior fitters, and production workers bustled about the No. 5 plane like ants attending to the carcass of a giant silver bird. It was the first production model—Nos. 1, 2, and 3 were marked for flight testing, while the fourth was destined for Delta—and was complete down to the floor fittings and interior paneling. Had Hughes not interfered, TWA would have announced its first 880 service sometime early in the new year.

Fanning out from the entrance, the new arrivals formed a human wall around the plane. Convair workers outside the cordon were warned to stay away, while those inside were rapidly ejected. The only explanation was as brief as it was mystifying: "This plane belongs to Mr. Hughes and he wants all work to stop." And stop it certainly did. The plane was dragged outside into the open, where Hughes' men stood guard unsmilingly throughout the afternoon and on into the night, refusing to let anyone near the plane even to retrieve tools and hastily abandoned plans. Nor was there any point in trying to reason with the intruders. Several days later, Convair production supervisors tried to have the engines draped with covers to protect them from the corrosive sea air, but the guards still refused to move aside.

Hughes' strategy was as simple as it was unhinged. If Convair could be stopped from completing the planes, they would not be delivered. And if Hughes was not called upon to accept them, he did not have to pay. There was consternation at Convair and also at TWA's corporate headquarters. Both parties knew that Hughes was having liquidity problems, but they had been led to believe he would shortly sign the long-awaited loan agreement. The same day that the guards towed plane No. 5 out of the construction hangar, one of TWA's directors, Emmett O. Cocke, had told a fellow airline executive over lunch at 21 in New York that Hughes would soon have his financial problems straightened out. Now there was this latest outrage, which neither Convair nor TWA knew quite how to handle.

Predictably, Convair chose to accept the situation without putting up a fight. The workers driven away from the No. 5 plane were reassigned, and production on both the TWA and Delta planes continued more or less as usual. As TWA would argue when it later sued its disenfranchised chief stockholder for damages, in what was to become the longest and most expensive case in U.S. legal history, there was really nothing the manufacturer or the end user could have done. Although the 880s were due to fly with TWA, they were actually on order to Toolco, which reserved sole right to negotiate with the builder. Both TWA and Convair tried to exert whatever pressure they could, but all concerned realized that it was a hopeless pursuit. Impervious to logic, Hughes had retreated to a secluded Hollywood bungalow, where, to the exclusion of almost every other subject, he fretted about the germs and unsanitary conditions which he saw as the chief threat to his life, health, and fortune. On the rare occasions when he did address the jetliner problem, he was more likely to dictate long, rambling, and entirely pointless memos describing possible ways to finance the fleet without surrendering his voting stock.

For almost six weeks after the October seizure, an unnatural calm hung over San Diego. Some of the evening shifts were cut back, but there were no layoffs, and flight tests on the first four planes continued more or less on schedule. Then Hughes struck again, this time seizing not one plane but two, which were dragged out and roped off next to the first, now showing initial signs of exposure damage. By February, he had seized No. 9, taking his tally to four, and in March he set a new personal record by commandeering another five, all in less than three weeks.

It was the beginning of the end for Hughes and what was left of his empire. Hijacking the planes had done nothing to alter the facts or postpone the inevitable. TWA still needed a new fleet, there was no money to pay for it, and the banks, as always, represented the only solution to the airline's slow strangulation at the hands of its better-equipped rivals. Still, Hughes refused to respond for the best part of a year. The planes were moved to a locked hangar on the other side of the airfield, where the guards still remained much in evidence.

TWA and Convair tried to work out some sort of solution among themselves, but without Hughes' active cooperation there was really no point. At one point, Convair succeeded in finishing one aircraft, the twentieth on the production line, before Hughes had a chance to drag it out of the shop. TWA sent over a flight crew and actually managed to get it into the air, once. The moment it returned from the test flight, Hughes' people arrived and towed it away to join the others.

After six months of further temporizing, Hughes did arrange financing, and just as he expected, it was conditional on his agreeing to step down. Not long after that, the Hughes Tool Company changed hands, the coffers long since rifled by its owner's profligate airplane and movie expenses. Hughes would remain a rich man, wealthy enough to continue living in the bizarre seclusion that had become his norm; but the days when his fortune was large enough to defy even his own inspired depredations were gone forever.

It was also the end of an era for General Dynamics. The most visible sign of the company's changed circumstances was evident on Wall Street. In April 1957, shortly before Jay Hopkins' ouster and death, the stock price stood at an all-time high, just short of seventy dollars. By December 1960, when Hughes finally swallowed his pride and signed the financing agreements, it had fallen to a bare forty dollars and was poised to plunge down a Himalayan gradient that would see it languishing in the mid-twenties by the end of the following year. Only then, as all the "forgotten" write-offs and fresh losses were totted up, did the full extent of the disaster become apparent. The previous record loss for a U.S. company had been chalked up by Ford in the aftermath of the Edsel disaster. General Dynamics beat that figure at a trot, going on to write off some $425 million worth of losses.

Everything that could have gone wrong had done just that. The 990, for example, was built without a prototype, in an attempt to save both time and development costs. The approach had worked well with the 880, which performed creditably on its maiden flight and required very little remedial attention. The 990, however, proved to be a compilation of aeronautical oversights and contractual blunders. It was just a little slower than the speed demanded by American Airlines president C. R. Smith—a mere six miles per hour, according to one estimate—but that was enough to see Convair forced to return a portion of the purchase price, since the airline could no longer claim that the 990 was the fastest passenger plane in the world. Worse, the wings on the first production model displayed an alarming tendency to "flap" at high speeds, a phenomenon diagnosed in subsequent wind tunnel tests as the result of the larger and much altered flight configuration. The wing flutter was solved by the addition of a short vertical stabilizer midway along each wing and the installation of vibration dampers. They were relatively simple steps, but by that stage, they were also enormously expensive, since each of the alterations had to be retrofitted to every one of the partially completed 990s now sitting idled on the production line.

But perhaps the the most telling blunder concerned the 880 and Convair's response to Hughes' series of seizures. When the planes were returned to the factory and work began once again, production supervisors found their suspicions about weather damage confirmed. All of the engines on the No. 5 plane and several from some of the others had to be sent back to General Electric. The cost of repairs was charged to TWA, but the other, less obvious consequences of the episode were borne solely by Convair.

When they were dragged from the workshop, Convair lost track of the amount of work needed to complete each plane. The result was massive confusion. Some planes had all their wiring removed and reinstalled, on the premise that it would be better to begin again rather than risk leaving something out. Worse, with so many different planes at so many diverse stages of completion, the economy of scale to be found on most production lines vanished completely. Under the original plan, the last plane off the production line should have been built with as little as one third the man-hours devoted to the first—a "learning curve," in the parlance of the trade. Now each had to be finished as an individual project at a cost of many additional millions.

Frank Pace blamed Convair's penchant for secrecy and indicated his displeasure with Jack Naish and the other members of San Diego's executive corps. "Convair resisted all along any attempt by General Dynamics to do anything," he said later. "There is not the slightest bit of doubt that the effort of General Dynamics to dominate Convair was completely resisted. The number of times I personally sat down with Jack Naish did not exceed seven or eight in a period of three years." The criticism was warranted and accurate. Convair had always played its cards close to the chest, and there was a considerable body of evidence to indicate its accountants had buried unflattering financial estimates beneath strata of cheerfully bland projections. Why Pace failed to exert his authority, insisting that Convair toe the corporate line, is a matter for conjecture. He was, after all, Hopkins' successor and very much in control of the board, something evident in the way he allowed the 990 program to go ahead despite the misgivings of Alvord and others. If Convair remained an intractable, fiercely independent kingdom within a kingdom, the reason would seem to be that it was given the latitude to follow its worst inclinations.

And of course, Pace did have a lot of other things on his mind. He was "a very small dark horse" in the race for the Democratic presidential nomination, according to one commentator who added that "some Democrats have talked vaguely of trotting him out." Apart from Pace's political ambitions, there were other areas for concern within General Dynamics. Fort Worth was working on its latest bomber, the B-58, and things were not going as well as planned. The Air Force was unhappy about the electrical and flight control systems, as well as the basic design. With four fan-jets slung beneath the delta wings, it had a natural tendency to execute cartwheels whenever one of the outboard engines stopped working. In public, the Air Force was warmly enthusiastic about the aircraft; privately, it was becoming increasingly clear that no new orders would be placed when the last of the 116 under construction were delivered.

Finally, there was the public relations problem, not just General

Dynamics' but the entire defense industry's. Back in Hopkins' day, the arms game had been both profitable and patriotic; now an unsavory tint was creeping into public perceptions of the nation's corporate defenders. Evidently a magnet for criticism, Pace also played a minor part in remodeling Roosevelt's "arsenal of democracy" as Eisenhower's ominous "military-industrial complex."

In 1959, with trouble looming on all fronts, General Dynamics found its practice of hiring former military and political leaders coming under investigation. F. Edward Hébert, chairman of a House Armed Services investigating subcommittee, conducted a series of hearings to determine the extent to which procurement decisions were influenced by indirect and applied pressure from contractors and their lobbyists. General Dynamics was not the only company whose employment rolls were scanned for the names of retired military personnel, but it certainly produced the richest haul. In addition to Pace and his deputy Earl Johnson—once the two top men at the Army Department—Hébert learned, the company had a full 187 retired officers above the rank of major on its payroll, twenty generals, and seven full admirals.

Pace's further contribution to the controversy took the form of an ill-advised invitation asking several aerospace industry leaders to attend a testimonial dinner honoring Lt. Gen. Bernard S. Schriever, the man in charge of all Air Force research and development programs. A copy of the invitation found its way into the hands of Hébert's researchers and soon became the focus of another round of unsettling disclosures in Congress and the press. In an accompanying letter signed by executives of two other military contractors, Aerojet and Martin, Pace described the event as "a small off-the-record party" at which "Gen. Schriever would like to tell you about his plans and some of his problems at the Air Research and Development Command." The gathering was abandoned to a chorus of apologetic explanations from the thwarted hosts that it was to have been nothing more than an innocent conference convened in the best interests of a strong defense.

It was to be the last miscalculation of this, the first stage of Frank Pace's business career. While General Schriever and his embarrassed contractors denied that there was anything unusual or unethical about their get-together, less conspicuous corporate officers at General Dynamics' New York headquarters were working toward a deal that would bring the Hopkins era to an official end. At the start of 1959, Pace had begun discussions about a possible merger with a sand and gravel company owned by Chicago millionaire Henry Crown. The talks had progressed slowly at first, but now, as the stock price continued to slip in step with worsening news from San Diego, the pace was picking up.

Like Jay Hopkins, Colonel Crown was a self-made man. Unlike Hopkins and those who would succeed him, the self-styled sand and gravel man was not a product of Harvard, the bar, or, despite his title, the upper levels of the military. Crown's background was strictly South Side Chicago, and his rise from the slums spectacular even by that city's colorful standards. His father was a Jewish immigrant, a weaver of men's suspenders who turned his hand to the match trade and succeeded in burning down the family home. That story may be a fiction—there are so many concerning the Crown family, it is often hard to tell where fact stops and fancy begins—but even the official record indicates a man with an eye for a good investment.

At one stage, Crown owned the Empire State Building and New York's Waldorf-Astoria Hotel, as well as the fountainhead of the Crown family fortune, Material Services Corporation, favorite supplier of sand, gravel, and general construction matériel to the Corporation of the City of Chicago and its many administrative limbs. Success in Chicago, particularly when it stems from a close relationship with City Hall, has always invited speculation about the political connections of the recipient. So has it been with Henry Crown and Material Services. The Colonel angrily rejects any hint of favoritism in his dealings with the city. "Honest politicians far outnumber the dishonest ones," he once said. "A man does not have to pay bribes to operate a business successfully." Such sentiments make his own triumphs all the more remarkable.

In the early 1920s, Henry and his brother Sol raised twenty thousand dollars and went into the quarrying business. Sol died of T.B. a few years later, and Henry continued on his own, completing his first major coup when he bought a thousand acres of industrial land from City Hall at an average price of sixty-four dollars an acre. Chicago journalist Ovid Demaris reported in his book *Captive City: Chicago in Chains* that the deal turned gravel into gold: "These properties had mountains of earth and rock deposits on their surface—spoil banks rich in limestone used for crushed rock and cement—which were the residue from channel widening and deepening at the turn of the century. They saved MSC the expense of quarrying for years."

Later, Crown branched out into the brick business and acquired a coal mine, both companies finding special favor with the city's school board. The coal went into the school's boilers, while the bricks went into new classrooms and auditoriums. Demaris explains: "When James McCahey, chairman of the School Board, was accused of favoring MSC, he announced that in future the board would split up the business among several companies. A man of his word, McCahey then bought bricks for six new schools from three different firms: MSC; Howard-Matz Brick Company, owned by MSC; and Garden City Sand Company, a subsidiary of MSC. The bricks, worth $28 a thousand on the open market, were sold as low as $33 a thousand."

When the war was over, Colonel Crown—he received his commission not long after Pearl Harbor and served throughout the duration as the chief of procurement for the Great Lakes Division of the Army Engineers in Chicago—began looking for investments outside Chicago. By 1959, having picked up the Empire State Building and a large interest in Hilton Hotels, Crown was to make his next and by far his biggest investment to date: General Dynamics.

The jetliner program was discouraging but not yet disastrous when the negotiations began. The absence of orders made it clear the plane would never be a big money-spinner, but the rest of the company appeared sound, and it was consistently prominent among the top three military contractors. It was, in fact, the kind of business Crown knew best—an enterprise filling the needs of a large public institution, staffed in this case by admirals and generals rather than Cook County public officials. General Dynamics' boardroom was a long way from Chicago, but the Pentagon's style of business was not far removed from that of City Hall.

For Pace and the rest, a merger with Material Services promised a fresh source of much-needed liquid cash to compensate for the millions being consumed by San Diego. MSC was responsible for sales of at least $100 million every year, and its profits were consistently impressive. The deal was finalized in late 1959. Crown merged MSC with General Dynamics, acquiring a 20 percent interest in convertible stock and the right to nominate at least one fifth of the directors on the board. It was, the financial press agreed, a perfect marriage. But then the jetliner losses began to emerge and the picture changed dramatically.

In 1960, Frank Pace announced a write-off of $91 million, commenting optimistically in his May letter to shareholders that while the amount was substantial, "We have every reason to believe the program will be one of our most successful ventures." The letter was scarcely in the mail when further bad news began rolling out of San Diego. Far from marking the end of the write-offs, it was just the start. Naish "discovered" another $40 million worth of expenses, and then, in September, a further $96 million worth of red ink had to be added to the balance sheets.*

As the tally began to draw closer to the eventual $425 million loss, Henry Crown grew increasingly restive. MSC was his life's work and his

*Convair's self-destruction became a prime talking point in the aerospace industry. "Watch for dramatic upheaval at General Dynamics," *Aviation Week* reported with a degree of gusto more commonly associated with gossip columnists and faltering Hollywood marriages. "One sign of the coming retrenchments was the sudden cancellation last week of all advertising in the media for the rest of the year."

proudest possession; if General Dynamics went to the wall, it would end up in the hands of bankers and other creditors. Never reticent about financial decisions, Crown took the company's future firmly in his own hands. Naish was the first to go, taking most of Convair's top people with him. Pace was next, although his departure was a little more dignified. Explaining that his ambitions were political rather than commercial, he announced that he would leave in the near future. In view of the fact that Pace presided over the largest loss in U.S. corporate history, his eventual destination was more than a little ironic. After serving with the Kennedy administration and holding a number of other public posts, he became the chief executive of a nonprofit organization that encourages retired executives to share their expertise with businesses in developing countries.

Now firmly in control, Crown reformed the company's bookkeeping system, imposing common reporting standards on the different divisions. He also found one last chore for Pace and Earl Johnson, who were sent off to renegotiate American Airlines' 990 contract in a series of talks with the hard-bargaining C. R. Smith. The pair succeeded in persuading American to decrease its order—a vital cost-cutting move since the jetliner program was to be abandoned entirely—but only at the cost of further immediate losses.

Finally, having convinced the banks to provide a $200 million revolving credit system to allay the liquidity problem, Crown found his new chief executive, former Pan Am chief and onetime Air Force secretary Roger Lewis. The company was still far from over the last repercussions of its ill-conceived decision to do business with Howard Hughes, but now at least it was in no immediate danger of sinking. There was ready credit, a few profitable enterprises like Pomona's family of missiles, and of course, just as had always been the case in Chicago, there was a good chance the public purse might finance the company's salvation and eventual rehabilitation.

In fact, that was exactly what happened. Salvation took the form of an Air Force contract for what was supposed to be the airborne marvel of its age, a fighter-bomber called the F-111.

8. The Birth of an Aardvark

On December 12, 1962, Air Force II swept in over the sprawl of Dallas–Fort Worth, to settle gently onto the main runway of Carswell Air Force Base. The vice-presidential jet taxied past the official reception point and continued in a broad arc to the other side of the tarmac, where its journey from Washington came to an end before a cheering crowd of three hundred General Dynamics workers. The day was sunny, if a little cold by Texas standards, yet Lyndon Johnson seemed totally unconcerned as he stood, arms raised atop the boarding steps, luxuriating in the warmth of the crowd. It was the kind of occasion he loved, the local hero surrounded by adoring fellow Texans, and for the next five minutes or so he posed happily for photographs with various General Dynamics executives and an assortment of home-state congressmen.

There was nothing unusual about the arrival of dignitaries at General Dynamics' Fort Worth plant. Politicians of all mainstream persuasions, foreign military delegations, presidents, and even a few members of the world's more obscure royal families had in their turn pulled up on that same concrete apron. One of the first visitors had been Vice-President Harry Truman, who arrived not long after the official opening in 1942, when the government-owned plant was being run by one of General Dynamics' corporate antecedents. He left less than impressed. Late that day, after a nit-picking tour of the mile-long main construction hangar, the future President concluded his visit by berating the plant's executives for allowing their workers to loaf on the job and for the poor safety record of the B-24.

This latest vice-presidential visit was altogether different—ironically so, in fact. Johnson had come to bestow a $6.5 billion contract, the largest sum ever spent on a military program by any government anywhere in the world. The order was a watershed in the history of American weapons procurement and its importance was reflected, albeit from a more personal perspective, by the banners and placards waving above the jubilant crowd. "We Have Jobs Thanks to LBJ," they said, or more commonly, "LBJ Has Saved the Day." Unlike Truman, who had come to root out waste and demand an improvement in the quality of the product, Johnson was on hand to mark the approval of a contract for a warplane that would soon be denounced in Congress and the press as a scandalous boondoggle, and later, as the least capable military aircraft ever to be lost in battle. If the road to today's gold-plated hangar queens is paved with theoretical abstractions and an unworldly acceptance of stupendous sums, the most appropriate path marker should be wrought in the shape of the F-111 fighter-bomber.

By any reasonable expectation, the celebrations of that bright and cold December day should never have taken place. The new decade had found General Dynamics suffering miserably through the consequences of the 880/990 fiasco, and the prevailing mood was one of abject gloom. The share price had fallen to around twenty dollars, almost fifty dollars below the peaks it had reached during the last years of Jay Hopkins' reign, and the indications were that it would drop even further before the jetliner program could be considered to have run its full, disastrous course. Frank Pace announced that he would resign as soon as a suitable replacement could be found, explaining that public service was his forte rather than business. His replacement, whoever that might be, certainly had his work cut out. Somehow the company had to be removed from the field of civil aviation altogether. There were simply too many producers and too few buyers to warrant the high-risk investments essential to restore Convair's competitive edge in any future battles with Boeing, Lockheed, and Douglas. Nor was there much hope of obtaining additional military orders, since the backlog of existing work placed General Dynamics high among the Pentagon's top three suppliers, with little room for further improvement. The immediate future was so bleak, many professional observers of the aerospace industry had long since concluded that Frank Pace's last official duty would be to announce the closure of at least one major manufacturing operation.

There was a good deal more to the story than mere rumor. On August 3, 1960, Pace received a letter from Roswell Gilpatric, a senior partner with the law firm of Cravath, Swaine and Moore, and the man who had brokered the merger negotiations with Henry Crown one year before. Pace and Gilpatric's friendship went back a long way. Both had held senior positions in the Department of the Army under Harry Truman and each moved easily through the same business and social circles in New York and Washington. Their easy familiarity was reflected in Gilpatric's blunt appraisal of General Dynamics' future. Certain difficult and delicate decisions would have to be made, he wrote after examining the key financial statements and internal reports. If the company was to regain its momentum, it would have to rationalize operations, preferably by closing either San Diego or Fort Worth. Of the two, Gilpatric deemed Fort Worth the more expendable, for reasons that melded subtle politics with basic economy theory. "While political considerations may play a part—Nixon for California and Johnson for Texas—objectively viewed, the plant complex selected for continuance would appear to be San Diego," he opined, adding that "the general preeminence of the southern California area for growth in aerospace and electronic industries" would prove to be a major factor in the quest for future civilian and military orders.

Nor did Gilpatric find any reason to believe the upcoming presidential election would work to the company's advantage. It was true that the rhetoric of the campaign had been laced with references to an alleged "missile gap"—a spurious threat that was to be disavowed by its chief propagator not long after Inauguration Day—but even assuming Washington moved to redress the purported imbalance, Gilpatric predicted General Dynamics' share would be no more than the crumbs left over by its smaller rivals. He pointed out that Douglas had been recently named prime contractor for the second stage of the Saturn rocket, and went on to identify the main reason General Dynamics could expect to be ignored. "In view of General Dynamics' present No. 1 position in defense order backlog, Convair cannot realistically expect to be assigned any major new aircraft program in the immediate future. . . . The most that Convair should count on is some major sub-contract work such as the B-70 wing."

Although never officially announced or confirmed, the possibility that General Dynamics was about to place its Fort Worth operation in mothballs prompted an outbreak of Texas-wide anxiety that was telegraphed through an informal network of elected officials, suppliers, subcontractors, and workers' families. Only slightly less pronounced than Seattle's legendary dependence on Boeing, Convair had been one of the local economy's chief props since the boom years of World War II, when as many as thirty thousand people reported for work every day. The years that followed saw the work force dip and fly through a mountain range of fluctuations, rising almost to a new peak during the Korean War before declining sharply at the conclusion of the B-36 program, on its way to an all-time low in early 1960 of just twelve thousand full- and part-time employees.

In the eyes of its local partisans, Convair just could not get an even break. The fate of the B-58 Hustler, the first supersonic bomber, was typical of the division's run of bad luck. Even though the plane went on to set nineteen individual world records, including one transatlantic dash from New York to Paris in just three hours and nineteen minutes, the Air Force chose to curtail the program early in the production run. Fewer than one hundred B-58s entered service, and even that small return on Convair's considerable investment was blighted by a later series of congressional hearings in Washington that saw General Dynamics accused of adding unjustified markups to the cost of electrical components obtained through its subcontractors. The Hustler's early demise had been bad enough. Now, or so it seemed, Convair was to be subjected to another injustice. Fort Worth had been a minor, unwilling accomplice in the jetliner debacle, yet it was about to become the chief casualty of the affair.

Johnson's presence less than two years after Gilpatric's gloomy appraisal was nothing but astonishing. By rights, the complex should have been largely empty, populated only by a skeleton maintenance staff and a few specialist teams crating the last of the division's chattels for shipment to the West Coast. Miracles are never easy to explain, but Congressman Jim Wright, one of the home-state backslappers on hand to welcome the Vice-President, did his best. "You have got to have friends," he told the crowd. Johnson beamed at the veiled compliment and the crowd roared anew. Their jobs had been saved by this godsend; there was every reason to celebrate. Outside the state, however, in places where Fort Worth's future prosperity was not of pressing concern, the F-111 was being dubbed "The Plane That Changed the Rules." Wright's moment of candor added further to the speculation that they had been changed for the explicit benefit of General Dynamics.

Before plunging into the events that led up to General Dynamics' unprecedented and unexpected coup, it is necessary to introduce the central character, the Kennedy administration's secretary of defense, Robert Strange McNamara.

When McNamara testified for the prosecution during General William Westmoreland's 1984 libel action against CBS, the newspaper accounts of the day's proceedings did their best to sum up the witness' career in a few parenthetically placed paragraphs. "The quintessential public servant" was a popular epithet, as were variations on the observation that McNamara was "the man who ran the Vietnam war for Lyndon Johnson." The summations were brief but more than adequate for someone who really needed no introduction at all. McNamara has become one of the great tragic heroes in the developing mythology of the sixties, a good and gifted man defeated by a war in which he was but a technical adviser. The narrow focus on the Vietnam years, just one short episode in the career of a man notable for his associations with spectacular failures, is to be expected; the war casts that sort of shadow over the lives of those who were its chief spokesmen. But it is also unfair. Long before the demise of his theory that war, like business, could be made to conform to certain ineffable rules, that the road to victory was a finely plotted curve bisecting body counts with body bags, there had been an earlier failure and perhaps, from today's perspective, a more significant one. For the first time in the history of the republic, an administration had installed a defense secretary committed to reforming the way the Pentagon did business. The tragedy was that McNamara made the F-111 the cornerstone of his campaign and thus cast the entire endeavor into disrepute.

The term "organization man" was much in vogue throughout the 1950s, and McNamara might have been its personification. He gave up a small fortune to join the administration, relinquishing a job as chairman of the Ford Motor Company as well as a share in the executive option plan that would have earned him at least three million dollars over the next two years. The speed with which McNamara agreed to join the administration was no less surprising than his selection. A registered Republican, he met John F. Kennedy for the first time on the morning his appointment was announced, to near-universal acclaim. Even President Eisenhower had criticized the military's disdain for economy, and McNamara, who was once described as "the strong-minded and insensitive tyro from Ford," seemed to be an inspired choice for the job. He even looked the part. With his severe hairline, academic spectacles, and thin, grim mouth, he gave the impression of a man whose authority would not be open to challenge.

Kennedy's faith in McNamara was rooted in more than an appreciation for the new cabinet member's public image. As a former PT boat skipper and something of a war hero, John Kennedy took the kind of personal interest in the services and their equipment not seen since the days of Theodore Roosevelt. It was Kennedy who established the Green Berets and the Navy's equally elite Seal units, approving and in some cases designing the distinctive headgear and regimental badges that set them apart from other, more conventional servicemen. The fabled Kennedy fascination with heroes—astronauts, commandos, and other men of action—is well documented. The less photogenic side of his commitment to a more potent military was represented by McNamara's appointment.

The Pentagon of 1960 was an affront to everything McNamara represented. In the thirteen years since the armed forces were brought together under the one roof, there had been seven secretaries of defense, none lasting longer than three years. McNamara identified the problems in his first official interview, two days after the presidential inauguration. His aim, he said, was to bring "efficiency to a $40 billion enterprise beset by jealousies and political pressures while maintaining American military superiority." Other men had found the job impossible, but then, as he went on to explain, they had pursued the wrong methods. Rather than acting as an unquestioning conduit for the transfer of money from the Treasury to the Pentagon, McNamara planned to put his own stamp on all major procurement decisions. From now on, every purchase would have to conform to the creed of cost effectiveness, the philosophy of management he had espoused from his early career on the faculty of the Harvard Business School through his fourteen years with Ford.

The question of uniforms, one of McNamara's first and best-known assaults on the system, staked out the battle lines for future confrontations with the service chiefs. The new defense secretary reasoned that significant savings could be achieved if the various branches of the military would agree on common standards for the more ubiquitous items in a typical service wardrobe. It was a reasonable suggestion, but one that proved incapable of full execution. Each service refused to forsake its traditional belt buckle, shoes, or style of shirt for those of a rival, while all were united in their opposition to the idea that airmen, soldiers, marines, and sailors could perform their duties with equal dispatch in trousers cut and sewn to a common pattern. This aversion to "commonality"—one of the McNamara era's most oft-repeated catchwords—extended well beyond nickel and dime items.

Debates about the merits of triple-stitched seams were annoying in their own way but of little importance in the grand scheme of things; the next stage of McNamara's drive for commonality prompted a reaction of undisguised horror. In 1960, after seven years of development, the Navy was at the point of introducing the new F-4 Phantom jet fighter to its aircraft carrier battle groups. McNamara looked at the statistics and concluded it might be made to serve equally well with the U.S. Air Force.

Then as now, the strategy of fleet defense rested upon the ability of carrier-borne planes to locate and destroy incoming enemies before they could come close enough to launch their missiles. This task requires an on-board radar system, which, in its turn, demands a reception dish of impressive proportions. In most Navy planes the radar dish is fitted into the nose and the rest of the plane built around it. The end result is a flying compromise, a pig-nosed aircraft, like the A-6 Intruder, that sacrifices streamlining, and hence speed, for the sake of enhanced surveillance capabilities.

The Phantom was very different. Perhaps because it began life as an unsponsored design project free from the restrictions of the Navy's official performance specifications, the McDonnell engineers were able to strike a new balance between the desire for speed and the need for a farsighted radar system. Through the use of the latest, solid-state electronics, not to mention the design equivalent of a shoehorn, the F-4 emerged as a fast, sleek package with radar capabilities second to none. In a further break with naval tradition, they also equipped the plane with two immensely powerful General Electric engines that substituted brute force for the aerodynamic lift of long, troublesome wings. The engines' one major drawback—they produced volumes of dark exhaust gases that made the plane easy to spot with the naked eye—was mitigated by the new ease with which the plane could be maneuvered on and off the flight deck elevator.

The Phantom's potential as a dual-service aircraft was obvious. A single plane could carry more bombs than a World War II B-29 Superfortress and do so at speeds of up to 1,600 mph. No less a fighter than a bomber, it was agile enough to hold its own in dogfights, and if the situation turned nasty, there were those twin jet engines to rocket the plane and its two-man crew away from trouble. Of course, close-quarter air-to-air fighting was considered unlikely in any event, since the pilot, alerted by his radar, would long since have dispatched the intruders with long-range missiles. The theory, together with the Sparrow and Falcon missiles on which it depended, were both tested and found wanting in the skies over North Vietnam, but in those early years of the era of the transistor and solid-state circuitry, it still seemed an entirely reasonable proposition.

McNamara's argument was not vindicated until late in 1963, when the first U.S. Air Force version of the F-4 was placed on active service, rapidly winning the loyalty of American aviators and their counterparts in the ten foreign countries that adopted the Phantom as their frontline fighter-interceptor. Yet to the Air Force commanders whose objections had been ignored, the plane's performance was not the main issue. What mattered most was the appalling break with convention. For the first time since the service won its independence from the U.S. Army in 1947, a defense secretary had taken it upon himself to define a military need and then acquire a weapons system to meet it. At first it had been easy to dismiss McNamara's talk of "commonality" and "cost effectiveness" as so much business school prattle. The F-4 victory did away with the illusion that he could be confounded and subdued by the mere complexity of the system. The man chewed through paperwork like a machine; his will to prevail was intractable.

The most visible signs of McNamara's presence were the teams of systems analysts, cost accountants, and fiscal planners who roamed through the Pentagon's project offices, eyes ever alert for examples of the administrative inelegance their master so abhorred. At first the military men had been inclined to treat the smart young business school types with patronizing condescension. But now, as the strength of McNamara's will became ever more apparent, the "whiz kids" became "bow tie bastards," their requests for information regarded as a sure sign that trouble was on the way.*

*SAC chief Curtis LeMay is said to have summed up Air Force sentiment in a cutting remark about one of McNamara's acolytes, future defense secretary Harold Brown. "Why that little son of a bitch was in a junior high school while I was out bombing Japan," he said.

And indeed it was. Throughout 1961, McNamara and his team focused their attention on a theoretical project that was soon dubbed the TFX, Tactical Fighter (Experimental). The Phantom had provided the defense secretary with the opportunity to demonstrate the savings that could be achieved by splitting the development and production costs of a new aircraft between two or more services. Yet from McNamara's point of view, the result was less than satisfying. While the Phantom had the makings of a good land-based plane, it had been designed primarily for duty at sea and the change required the replacement of hundreds of major and minor electrical and structural components. The TFX, later to become General Dynamics' F-111, was to be a dual-service aircraft from its inception, a shining example of an enlightened procurement system in action.

By one of those happy coincidences, the Air Force and the Navy were both casting about for new frontline aircraft just as McNamara entered office. Their respective needs were very different—the only common thread was that both were to be the first planes of their kind—but that did not daunt McNamara in the least. As much a believer in advanced technology as a disciple of managerial efficiency, he accepted his analysts' assurances that a little additional research would overcome the services' objections and allow the plane to fulfill its many and disparate functions.

The Navy had its eye on a supersonic airplane that could also loiter at low speeds high above its mother ship, a farsighted sentry with the capacity to identify and destroy its enemies at long range. This had been one of the duties originally assigned to the prototype Phantom, but it soon became clear that later production models would be unequal to the task. Designed with an eye for speed and power, the Phantom used so much fuel climbing to its patrol station that there was precious little left to do anything but come down and refuel. This chink in the fleet's armor had been partially plugged by the Grumman S-2 Tracker, a slow, twin-engine propeller plane that could remain aloft for as long as eight hours at a stretch. When an enemy approached, the Tracker's crew were to summon the Phantoms into the air and direct them to an interception point, where the fighters could use their own on-board radars and Sparrow missiles, moving in to finish off any surviving intruders with short-range, heat-seeking Sidewinders.

In 1957, a restless faction in the Navy air corps broke with tradition and launched an attempt to turn the decidedly unglamorous Tracker into a fighter, with the hope of freeing the Phantom for purely offensive missions. If a Tracker or something like it was equipped with long-range

Eagle missiles, the progenitor of all standoff weapons still so much in vogue today, it would be able to identify airborne enemies and shoot them down at the same time. Dubbed the F6D-1 Missileer, the extensively reworked surveillance plane almost became a reality. Development funds were set aside in the 1958 budget, and for a little more than eighteen months Douglas worked feverishly to cram the computers, radar systems, and fire control electronics into a dumpy, inelegant, and entirely theoretical fuselage. Unfortunately, the Missileer proved even more expensive than it was complex, and in December 1960, outgoing defense secretary Thomas Gates acted on the objections of traditionalists who ridiculed the very idea of a "slow" fighter and canceled the airframe sections of the contract. The experimental target selection computers, the Eagle missile, and the multidirectional radar were allowed to continue, in the expectation they would later be grafted onto another advanced but more orthodox warplane.

The Air Force's quest for a new aircraft was also something of a stroll through the realm of technological speculation. Even more than the Navy, there was a strong perception at the upper levels of the Air Force that the existing order of battle—heavy bombers, smaller tactical aircraft capable of delivering so-called theater nuclear weapons, fighters, and interceptors—no longer reflected strategic reality. Improvements in Russian radar and missiles were about to render the time-honored doctrine of high-altitude bombing obsolete, something that would be established beyond doubt when Francis Gary Powers' U-2 spyplane was shot down over Sverdlovsk in May 1960. By that time, the Strategic Air Command was already attempting to write a new book of tactics for its B-52s, although at that stage, the effort promised to provide nothing more than a stopgap until something better came along. The big bombers were built for the thin air of the upper atmosphere, and while they could be brought down to the deck for cross-country dashes, they would always be slow, easy to spot, and bereft of the maneuverability needed to foil even the most ham-fisted pilot of a Russian interceptor.

A possible solution was laid out during the summer of 1960 in a document entitled "Air Force Specific Operating Requirement 183," a compendium of performance and engineering criteria that was circulated through the aerospace community, Congress, and the office of the secretary of defense. McNamara was captivated by the possibilities.

"A truly enormous advance in the art of military aircraft," was the way he later described the project, "the greatest single step forward in combat aircraft in several decades." It was easy to understand his enthusiasm. "SOR 183" amounted to a technical portrait of a plane that could do everything and anything. It was to be both an adequate fighter-interceptor and an example of the new, ground-hugging nuclear bombers that would replace the B-52. At high altitudes, the Air Force wanted a top speed of 1,700 mph, not much slower than a bullet fired from a small-bore rifle. Down low, at steeple height, performance was to be even more impressive. Since it was likely the plane might one day be required to strike targets up to 400 miles inside the Soviet Union, it had to be capable of "skating" over hills and down valleys at heights of no more than 250 feet and speeds in excess of 1,000 miles per hour. This was clearly beyond the competence of any human pilot, so "SOR 183" went on to outline the kind of electronic guidance equipment that would be required. Like the plane itself, the avionics' successful development was highly speculative. Five on-board radars, each programmed to double-check the others, would scan the terrain ahead, calculate the necessary adjustments, and issue the appropriate instructions to a robot pilot. The pilot, at least in theory, would be required to do little more than sit tight and passively watch his instruments.

The designers' revolutionary ambitions were most readily apparent in any comparison with the F-105 Thunderchief, the Tactical Air Command's frontline fighter-bomber at the time the project was announced. Where the F-105 required a 9,000-foot runway in order to put down with any assured degree of safety, the plane described in "SOR 183" was supposed to land and take off in one third that distance or less. The Thunderchief had a mission radius of 800 miles; the theoretical newcomer promised 1,600 miles at speeds substantially higher than anything previously thought possible.

All these miracles depended upon the wings, by far the strangest of the proposed plane's many peculiar twists. When it was taking off, landing, or maneuvering to link up with an airborne tanker, they were to jut straight out from the body at an angle of almost ninety degrees. At the other end of the scale, when approaching the speed of sound or cleaving through the thick, turbulent air at the bottom of the troposphere, the wings were to be tucked back along the fuselage, dovetailing neatly with the rear horizontal control surfaces in a close approximation of the classic delta wing configuration employed by the B-58 and many of the era's other transonic warplanes. Any pilot, indeed any layman with a rudimentary knowledge of aerodynamics, could identify the unique advantages.

At low speeds, a plane behaves like a man on a tightrope, struggling to stay in the air while simultaneously using its wings as a balancing pole to counteract the loss of lift and stability caused by the drop in airspeed. At high speeds, when the pressure of air rushing over the control surfaces is enough in itself to guarantee a measure of stability, the primary goal of all designers is to reduce drag. Thanks to its movable airfoils—variable geometry, in the parlance of the trade—the proposed aircraft would be able to enjoy the best of both worlds. A slow and stable weapons platform one minute, it could be transformed at the flick of a few switches into a streamlined killer.

The Air Force was typically keen to explore and exploit the new technologies, but it also had doubts, both practical and political. It was true that the computations and technical projections had been formulated, revised, and polished many times by NASA's aeronautical engineering research laboratory at Langley Air Force Base in Virginia, arguably the most brilliant collection of aviation theorists in the country. But even the assurances of John Stack, the NASA team's leader and the primary instigator of "SOR 183," could not dispel reservations born of experience.** Every development encountered snags; they were inevitable, the nature of the game. Since the future of the Tactical Air Command might well be determined by the success of the program, there seemed no point in rushing to production ahead of time. The risks were simply too great. Small, overlooked details that might later prove expensive, or even impossible, to correct had the potential to humble a project that promised to revolutionize aerial warfare like nothing since the introduction of the jet turbine. And there was also another reason to proceed with caution. For more than two years, the Air Force had been battling to win the funds needed for the B-70 bomber, and regarded this latest weapon as a potential threat. If the aircraft lived up to even half of John Stack's predictions, opponents of the larger bomber would be justified in claiming that the B-70 had become obsolete, and there were too many careers tied up in the project to risk that unfortunate turn of events. The best policy was the traditional one—nurse the prototype through its trials and use each well-publicized success to build congressional support for the appropriations needed to begin full-scale production when the plane was more or less ready to be put to work. In the meantime, the B-70 would be shepherded through Congress and well on its way to replacing the B-52.

In the office of the secretary of defense, the proposed plane was recognized as a weapon of many virtues. It meshed perfectly with McNamara's recent departure from the doctrine of massive retaliation that had dominated the Air Force's war plans since the start of the 1950s. Now, with a small, low-flying aircraft like the one outlined in "SOR 183," it seemed possible to deliver just one or two nuclear weapons to targets inside the Soviet Union rather than obliterating the whole country. The degree of retaliation should match the provocation, in McNamara's view, and the F-111 offered a degree of flexibility impossible with ballistic missiles or a massed bomber attack.

**Stack also designed the supersonic wind tunnel that led to the development of the F-106 and the Super Sabre. He won the Collier trophy for his work.

Even more important than the proposed plane's role in future nuclear wars was its potential as a weapon that could be put to good use in McNamara's escalating wars with the services. The Pentagon had always regarded commonality and McNamara's other objectives as distasteful civilian fantasies, to be treated with the contempt they deserved. The attitude had been in force since a few days after his appointment and had hardened with every fresh, infuriating intrusion. Within weeks of joining the administration, McNamara had asked for a comprehensive report on the feasibility of developing dual-service aircraft. The proposal had been given desultory attention before being dismissed with predictable disdain. The services' respective needs were so different, he was told, no common ground in the design of major weapons systems was possible. The attitude was to crystallize in a farcical episode that same year when the Air Force agreed to adopt the Navy's F-4 but insisted on labeling its version the F-110 Scepter even though the planes were almost identical. The original designation was restored only at McNamara's insistence when various congressional committeemen found themselves discussing the merits of one plane only to be informed that the subject had already been considered under another name.

With his own civilian analysts' enthusiastic appraisals of the proposed plane before him, McNamara decided the time had come to dispel any doubts about who was in charge. If the Navy and the Air Force could not settle their petty jealousies and agree on a common plane, he would have to do it for them. And if the NASA evaluations were correct, "SOR 183" was the perfect starting point.

It all seemed so simple on paper. The Navy wanted most of all a plane that behaved so well at low speeds it could be flown on and off the decks of even the small aircraft carriers left over from World War II and Korea. Variable geometry promised all that and much more besides. On land, where long runways provide a margin for pilot error, short landing and takeoff characteristics are always desirable but seldom essential. On the pitching deck of an aircraft carrier, perhaps at night or in the midst of a midocean storm, there has never been any room for negligence or bad luck. The theoretical Air Force aircraft seemed ready-made for duty at sea. Quite apart from the advantages of being able to make a carrier landing with the wings in the forward, low-speed position, the pilot would not be troubled by any loss of forward visibility, one of the worst traits of fixed-wing delta aircraft, whose basic configuration obliges them to come in with noses upthrust at sharp angles in order to avoid stalls.

There were, of course, a few problems, but McNamara's whiz kids believed all could be easily solved. The plane outlined in "SOR 183" was essentially a tactical bomber, certainly nothing like the utility fighter the Navy wanted. Once again, technology was tapped to make it fill the bill. If the plane was equipped with the Eagle missiles left over from the

Missileer experiment, it would not need the finely honed agility essential in traditional close-quarter air-to-air fighting. Moreover, if it was also given some form of rapid-fire cannon, there was no reason it could not be assigned to beating up ground targets and ships.

The Navy was far from convinced, and did all that was diplomatically possible to dampen the defense secretary's misguided enthusiasm. McNamara's analysts were unmoved, countering the Navy's arguments with extracts from the mission statistics compiled during the last years of World War II and Korea, when the primary mission of both land- and sea-based fighters was ground attack. Why, they reasoned, was it necessary to indulge the services' sentimental attachment to dogfights and dogfighting, when future squadrons might never be required to perform this mission, at least not on a regular basis? Annoyed that its objections had been rejected out of hand, the Navy soon withdrew into a sullen, offended silence. No less dismayed by McNamara's seizure of its pet project, the Air Force soon followed suit.

Much later, when the F-111 investigation had become one of the longest-running shows on Capitol Hill, Assistant Navy Secretary Robert A. Frosch made one last, doomed attempt to portray the Navy version as a reputable weapons system. "Our experience with the F-111 is giving us new insights into the writing of specifications for aircraft," he commented in late 1967. Since he was addressing a conference of aerospace and electronics executives, there must have been quite a few who found it hard to suppress a smirk at the assistant secretary's unfortunate choice of words. Indeed, there was a lesson to be learned from the F-111 project: This was no way to design an airplane.

Without the assistance or encouragement of the services, McNamara's team drafted their specifications like hungry men at a groaning buffet, picking through the technological advances laid out in "SOR 183" with a taste for extravagance rather than balance. Since the plane was to be an all-altitude fighter, a long-range bomber, and a ground-attack aircraft, there were ample opportunities to mix and match every latest breakthrough. Whenever a problem presented itself, the solution was always a little more high-tech hardware, much of it as yet theoretical. The F-111's cockpit, or more correctly, the Self Contained All Altitude Crew Survival and Escape Module, remains one of the best examples of the way the process ran away with itself.

Not long after it began, someone realized that no pilot could bail out into a 1,000 mph slipstream and hope to survive, particularly when the plane might be only 200 feet above the ground at the time. Far from regarding that as a problem, the Department of Research and Engineering recognized a welcome challenge to further advance the state of the art. If a conventional ejection seat was out of the question, which clearly it was, why not call for an entire cockpit assembly that could be blown out of the plane, and just for good measure, add a few extra touches that would make it the most sophisticated working environment any pilot could hope for? The theoreticians' ambitions were glowingly described in a publicity release put together by General Dynamics' PR office not long after the company was named prime contractor:

> Air conditioned and pressurized, the module provides a "shirt-sleeve" environment for the F-111's two crewmen, allowing them to operate without fatiguing pressure suits or other special flight clothing.
>
> Equally important, the module provides F-111 fliers with an unprecedented escape and survival capacity. It can eject in a zero-speed and zero-altitude condition, or at any point in the F-111's performance spectrum.
>
> In an emergency an explosive cord shears the module from the aircraft and a rocket motor powers it clear. The module, with crewmen inside, descends by parachute.
>
> For an underwater escape, flotation bags raise the module after it has been severed from the aircraft.
>
> The module, *which resembles a spacecraft rather than a conventional cockpit* [author's italics], includes the entire crew compartment and a portion of the wing to stabilize the module in flight.

And if that was not enough, this flying, floating, rocket-propelled marvel came equipped with a control column that did double service as the handle of a built-in bailing pump!

Like so many of the other much-vaunted improvements, the cockpit helped to compromise the plane's performance while doing little to safeguard its occupants. The rocket ejector, flotation systems, air conditioning, and detachable wing stumps all helped to increase the weight until eventually, when all the other frills and gewgaws were added in, the plane was too heavy to achieve the levels of performance that had made an ejection cockpit necessary in the first place. As for the pilots and their navigators, they received only a false sense of security. Of the F-111s lost over Vietnam during the plane's two short tours of duty, only two crewmen survived.

Before officially unveiling his vision, McNamara and the head of the Department of Research and Engineering, future Carter defense secretary Harold Brown, did the rounds of the Pentagon for a series of talks that both men later portrayed as dispassionate consultations on a variety of fine, technical points. General Frank F. Everest, commander of the Tactical Air Command, SAC's LeMay, and various admirals recalled the meetings in a different light in testimony before a congressional committee two years later, in 1963. McNamara had come to demand concessions, they explained. Intrigued by the choice of words, the panel sought clarification. Surely a compromise implied some mutually satisfactory agreement, while "concession" could be interpreted as a synonym for capitulation. LeMay and Everest both explained why the plan was fatally flawed. The Navy needed a short, fat fuselage in order to fit the proposed plane on its flight deck elevators. The Air Force, on the other hand, was prepared to accept an aircraft that was at least 70 feet long, with a fully laden takeoff weight of up to 77,000 pounds, more than 20,000 pounds beyond the Navy's maximum tolerance. McNamara's insistence that the Air Force accept the lower gross weight became the first concession and, in the witnesses' estimation, by far the most damaging. Thanks to its wider, foreshortened fuselage, wind resistance and fuel consumption both increased to such an extent that the requirement for a 400-mile low-altitude, high-speed dash had to be slashed by half. This effectively reduced the number of potential targets inside the Soviet Union by a factor of 60 percent and led the B-70's supporters in the Strategic Air Command to question the defense secretary's motives in continuing to push the project. After all, was not deep penetration of Russian territory one of the chief reasons the plane was being built?

The question of weight cast its shadow over other aspects of the plane's performance. Much of the permissible 55,000-pound gross weight was represented by the structural bracing needed to help the Navy version withstand the tremendous strains imposed by regular short-stop landings at sea. To the Air Force, the additional bracing was counterproductive, since it swapped the capacity to carry additional fuel and weapons for dead, useless weight.

There was scarcely a corner of the plane that was not subjected to concessions of one sort or another. The Navy, for example, had set its heart on a radar dish roughly four feet in diameter. This was no frivolous yen for luxury but an essential part of the original plan to build a far-sighted, high-flying fleet defender, and its importance was reflected in the Navy's willingness to accept an airframe stressed to only 6.6 g's ("g" is aerodynamic shorthand for the force of gravity, or in this instance, a measure of the plane's ability to hang onto its wings when pulling tight turns). Now, 6 g's is uncomfortable but tolerable, and most pilots can take the strain without experiencing a temporary loss of consciousness. By the standards of other fighter planes, it is still not terribly high. The reason for this voluntary concession was simply that the Navy planned to use the proposed plane as a launching platform for long- and medium-range missiles, a role in which the ability to outturn an enemy was a secondary consideration. However, by the time McNamara and Brown had imposed their combined will, the plane was equipped to do neither job with any particular degree of efficiency. Saddled with the three-foot-diameter radar dish favored by the Air Force, the Navy version was no better equipped to spot distant enemies than the simpler and much cheaper Phantom, which it was supposed to replace as the aircraft carrier battle groups' chief protector.

Even apparently trivial considerations like the configuration of the landing gear became the subject of bitter wrangling. The Air Force wanted individual landing wheels large enough to roll over the bumps and tussocks of unpaved runways without pitching the plane on its nose. The Navy, on the other hand, sought small wheels mounted in pairs, since this would have helped to lower the plane's center of gravity and thus make it easier to stop with standard flight deck restraint equipment. Even more important, twin-mounted wheels provided a measure of insurance against the possibility that a single blowout might send two human lives and several million dollars' worth of technology sliding across the flight deck and into the sea. Once again, these technological arguments were lost on McNamara's civilians, who justified their insistence on single wheels by pointing to the potential savings.

If the rulebook had been torn up during the TFX's conception as a theory, the contract competition that followed saw the tatters flung rudely in the face of convention. McNamara had found it easy to reject the pleas of service chiefs, whom he regarded as rival lobbyists bent on defending their own turf. However, his refusal to heed two allies, Navy Secretary Paul Fay and Air Force Secretary Eugene Zuckert, remains harder to understand. In August 1961, a mere six weeks before the project was opened to bids for contractors, the two secretaries informed their boss that the plane would fall far short of its goals. According to Fay, the more outspoken of the two, the Navy version of the TFX "could not be considered technically feasible since it was likely to meet only 37 percent of its stated mission requirements." McNamara simply ignored the advice and continued with his preparations to select a builder.

The second phase of the scandal opened officially on October 1, 1961, when McNamara unveiled the list of compromises he and Brown had wrung from the services and announced that the contract was open to bids from potential builders. Like the plane itself, the contract competition reflected the McNamara approach to business. The ten competitors, some choosing to increase their chances by teaming up with potential rivals, were given just two and a half months to prepare submissions and deliver them to a high-security compound on the grounds of Wright-Patterson Air Force Base. Each aspiring builder was to provide a team of engineers and a contract specialist to explain their submissions at a series of hearings before the official selection panel. And then, when the manufacturers had all been shown to the gate and sent home for Christmas, a three-hundred-member appraisal team drawn from the Navy, Air Force, and NASA would sift through the mountain of charts, plans, cost projections, wind tunnel models, and evaluations of prospective subcontractors. All going well, the quest should be wrapped up by February, with the first prototype's maiden flight scheduled for early 1965. That was the official projection of events, although those who took part in the whirlwind procedure remember it rather differently. "It was McNamara's version of the Miss America Pageant," one former Air Force analyst recalled many years later.

The competition began in early December, when the manufacturers' liaison teams descended on the Wright-Patterson compound for nine days of oral submissions. With a contract worth perhaps $6.5 billion at stake, the atmosphere inside the isolated barracks was electric, so much so that the companies had to draw straws to establish the order in which they would go before the selection panel. McNamara wanted a total of 1,700 aircraft, 500 of them in the slightly modified form that had been imposed on the Navy. By itself, the U.S. component of the contract promised to keep the winning bidder hard at work for a decade, and that was but one part of the potential rewards. Britain and one or two other favored allies had already indicated their strong interest in the TFX contract, and there were hopes that overseas orders would see the production run increased to perhaps 3,000 aircraft. Finally, there was the promise implicit in all major military contracts, even in this latest approach to the game, which McNamara had dubbed his "Total Procurement Package": When the production of planes for the U.S. and overseas markets finally came to an end, the manufacturer could still expect to reap the benefit of staged improvement programs, alterations, and retrofits of the kind that had marked the history of the B-52.

Ten contenders soon emerged, each pursuing its own strategy. Northrop, for example, took the measure of the opposition and soon retired from the fray altogether. Boeing, on the other hand, sank more than $1 million into a submission that included not only wind tunnel models, computer performance analyses, and detailed cost projections, but also a draft copy of a final contract that had taken one hundred accountants more than a month to prepare. Of the other competitors, three decided to take on junior partners, for reasons that were both practical and political. By signing up Grumman early in the piece, General Dynamics aligned itself with the Navy's favorite aircraft supplier of more than thirty years, while simultaneously adding the strength of the New York congressional delegation to its own Texas boosters.

Six weeks later, in mid-February 1962, the seven-man appraisal and selection board concluded its deliberations and announced the verdict: None of the proposals promised a plane that could fulfill the assigned mission, although the Boeing sketches came closest to the mark. Of the rest, only the General Dynamics–Grumman entry was deemed worthy of further consideration. Even so, as the panel pointed out in some detail, it was a poor second.

Boeing's early lead was entirely predictable. The company's engineers had been mulling over "SOR 183" for more than a year, combining trade gossip and items of intelligence gleaned from NASA's Langley laboratory with their own estimations of the kind of plane likely to win Air Force support. McNamara's decision to bring the Navy in on the deal prompted a series of frantic last-minute alterations to the working sketches, but it was not enough to wipe the shine from the company's aggressive optimism. If any airframe manufacturer ever mounted a more determined, or less successful, campaign to win a plum contract, Boeing's preliminary groundwork for the TFX was it.

Over the previous eighteen months, Boeing's teams had roamed America and the world, studying the Tactical Air Command and the way it operated. Pilots, ground crews, and commanding officers were interviewed, and their opinions rendered, down to a long list of desirable traits that any future plane should display. "We thought we had done everything right," one former Boeing official said later. "At one point we even put together a book that described TAC and the way it operated. Well, the Air Force liked it so much they made it one of their training manuals, and that was about all we got—a vote of thanks followed by a kick in the face."

Even more than its rivals, Boeing had a lot riding on its bid. Halfway across the country from its Seattle headquarters, another Boeing plant, in Wichita, Kansas, was facing many of the problems confronting Convair in Fort Worth. Existing orders were almost at an end, the labor force was falling, and unless something new turned up to keep the production lines operating, there was little chance the plant would remain open for more than a year or so, at best. And there was also a question of company pride. Although better known for jetliners and heavy bombers, Boeing had achieved its first successes as a builder of pursuit planes during the 1920s. World War II and the later contract for the B-52 banished any inclination or opportunity the company might have had to return to its roots. But now, with one of its principal plants in serious trouble, the TFX competition presented an opportune moment for the long-awaited return to the fighter market. The rewards extended well beyond short-term security for the Wichita work force. With its swing wings and other advanced features, the TFX was looked upon by Boeing as an invitation to establish an unassailable lead in the application of the new technologies.

The breadth and depth of Boeing's preparations were evident in the pile of documents, charts, and wind tunnel models submitted for appraisal. Where many of its competitors were clearly ill at ease with the applied principles of variable geometry, Boeing demonstrated a thorough familiarity and a fine eye for detail. In terms of performance, it was also the design that came closest to McNamara's ideal. Boeing promised a range of 4,630 miles, sufficient to span the Pacific in a single hop, and more than 700 miles better than its nearest rival, the joint effort conceived by General Dynamics and Grumman. The General Dynamics design fared no better at low altitudes, where the fuselage and wing designs proposed by Boeing were judged to be almost 20 percent more efficient in terms of wind drag and fuel consumption. In its high-altitude role as a Navy surveillance-fighter, Boeing's entry also left General Dynamics in the shade. The Seattle plane could not only lift a heavier weight to a greater height, it could also "loiter" at low speeds for up to four hours and ten minutes, while General Dynamics offered only three hours and thirty-five minutes.

In opting for the Boeing proposal, the Source Selection Board arrived at a unanimous, though hardly unqualified, decision. The degree of commonality—in this case, the number of component parts and subsystems shared by the Air Force and the Navy models—amounted to only 50 percent, well below McNamara's specified minimum. And there was also an embarrassing problem with the twin General Electric turbofan engines that Boeing had nominated as its ideal power plant. The designers had originally called for Pratt & Whitney engines but changed their minds when General Electric announced that it had just perfected a promising prototype that would soon go into full production. The Air Force, swayed by the masterfully orchestrated campaign of publicity and promotion that accompanied the announcement, let it be known that it was keen to install the new engines in the TFX. Unfortunately, by the time the evaluation team assembled, late in 1961, the Air Force had taken a closer look at the General Electric unit and concluded that it required many major refinements and at least two years of further development. Boeing could hardly be blamed for its decision—several other bidders had also chosen to build their planes around the untried turbines—but the result was still enough to cast a shadow over a proposal that was, in most other respects, clearly superior.

The low levels of commonality were, however, all the company's own doing. Together with its friends and informants in TAC, Boeing had concluded that McNamara's scheme was impractical and perhaps impossible, at least with the high percentage of common parts demanded by the TFX prospectus. So, reassured by precedent, Boeing decided to limit the emphasis on commonality and build a plane that would appeal to pilots rather than misguided public officials. Other defense secretaries had had similar, if less grandiose, brainstorms, and at that early stage it was still possible to imagine that McNamara's crusade would also lose its way, bogging down under the weight of institutional opposition. The folly of this approach soon became apparent.

There were, however, few things about the TFX competition that could be considered normal and the sanctity of the panel's opinion was not one of them. Rather than ordering its Wichita plant to begin tooling up, Boeing found itself skewered by three bodies of opinion, each for its own reasons pressing for the award to be suspended pending a second review.

The chief advocates of postponing the award were McNamara's whiz kids, who confounded Boeing's early hopes by reviewing the selection panel's findings and concluding that the winning entry "was not one plane but two." The people in charge of the Air Force propulsion department were also much cheered by the delay, since a Boeing victory would have placed them in a most embarrassing position. It was only four months since they had placed their stamp of approval upon the General Electric engines, yet now, if the Boeing proposal was adopted in its original form, the final contract would have to include a proviso that extensive modifications be undertaken to redesign the plane around the larger, heavier, and less powerful Pratt & Whitney jet turbines. It would be much better, the propulsion people argued, if Boeing was given some money to finance a sweeping redesign and told to report back in a couple of months' time.

And finally, there was the Navy. Throughout the seven-week appraisal period, neither Rear Admiral Frederick L. Ashworth, the service's sole representative on the panel, nor Vice-Admiral Robert Pirie, the deputy chief of Naval Air Operations, had displayed any great enthusiasm for the TFX. The Navy still wanted a lighter, slower, low-performance plane—"a big black Packard," in the words of one NASA designer—and Boeing's proposal was definitely not what it had in mind. Even more unsettling was the realization that Boeing had never built a carrier-based aircraft or, for that matter, ever handled a major Navy contract. If the admirals were going to accept a plane they did not want, then it was going to come from a tried and tested builder like Grumman, which could be counted on to make the most of the fighter's already limited possibilities. Still clinging to the hope that McNamara might yet respond to reason and cancel the entire project, the Navy added its voice to those calling for a second runoff between Boeing and General Dynamics–Grumman.

Much later, after General Dynamics had been declared the winner and cries of "foul" were echoing around Washington, *Fortune* magazine described the events that followed the first decision to overturn the selection board as a test of endurance. "Boeing and General Dynamics faced the 12 week run-off like a pair of exhausted runners who had put all they had into a sprint to the finish line only to find that it had been moved a mile down the track." In fact, Boeing's ordeal was to continue for ten months rather than twelve weeks, and the finish line was to be moved three more times before the race finally ran its course.

While the news was a morale-sapping blow to Boeing, the reprieve was the source of much rejoicing in Fort Worth. General Dynamics had entered the competition months behind Boeing and had never closed the gap. Boeing's problems with its General Electric engines were considered an act of divine providence. Convair's designers had wanted the Pratt & Whitney engine from the beginning and were now free to spend the best part of three months fine-tuning their proposal. Boeing, on the other hand, was faced with redesigning its entire plane while simultaneously striving to increase the incidence of parts common to both Navy and Air Force versions.

Despite the difficulties and the additional expense—estimates of the contestants' expenses were widely put at one million dollars per month in preparation for the second submissions—both proposals were lodged well before May 1, 1962, the official deadline. To the amazement of industry watchers, the chagrin of the Air Force, and the immense satisfaction of the Navy, history repeated itself. Once again, Boeing was judged the winner, only to have its prize snatched away at the last moment. This time it was almost entirely the work of the admirals.

When first dragged screaming into the TFX quest, the Navy adamantly refused to consider any plane with a gross weight of more than 40,000 pounds. McNamara and Brown had forced the ceiling up to 55,000 pounds by promising that the final gross weight would not exceed that figure under any circumstances. But now it was up to 62,000 pounds, an increase of more than 50 percent over the original, "ideal" figure. To Admiral Pirie and the hard core of senior Navy aviators, Boeing's apparent disdain for the service's needs came as confirmation that it would be saddled with an aircraft that was first of all "an Air Force plane." They dug in their heels, branded this latest winning proposal "entirely unacceptable," and demanded yet another review.

While the Air Force had its misgivings about the TFX, it was generally inclined to view the project much more favorably than the Navy. This was hardly surprising, since the concept was a direct outgrowth of its own "SOR 183." One year earlier, the Air Force had been forced to accept the Navy's Phantom, and the affront to its dignity still rankled. Now, thanks to the TFX, it had the chance to repay the compliment with interest. The entire project was to be overseen by Air Force Systems Command, the administrative instrument designated by McNamara's office to handle the selection of a contractor, approve changes in design, issue payments, and administer the test programs. The Navy's input would be filtered through this rival bureaucracy, and the admirals did not like it one bit, particularly since the TFX project represented the Navy's only chance to acquire a new frontline aircraft until the late 1970s.

McNamara professed amazement at the vehemence of the Navy's objections. "After eight months of work we'd come up to May and the Navy was still saying neither design met its requirements," he told an interviewer one year later. "That was a terrible setback. My first reaction was one of disappointment, and then disbelief, because in mid-1961 we had finally got the services to agree on a set of requirements that were basically similar." All the same, McNamara ordered the competition into its third round of redesigns and appraisals, using the Navy's opposition as a springboard to make several demands of his own.

As with each of the previous submissions, McNamara considered both Boeing's and General Dynamics' estimates of the final cost to be "completely unreasonable." Each company was hawking its version of the plane with what McNamara regarded as "hungry bids," prices so low they made a mockery of his Total Procurement philosophy. "It appeared that they were following a practice that is evident elsewhere in our society of trying to entangle a customer by a low initial bid, keeping the thought in the back of the mind that it can be raised later." His own analyses of previous major weapons contracts indicated that the final price was, more often than not, at least 300 percent in excess of the initial estimates and sometimes as much as 1,000 percent higher. Obviously, both companies were playing their same old games and McNamara, pledged to ushering in a new era in defense procurement, was adamant that they would not be allowed to get away with it.

Boeing concealed its annoyance with this latest postponement beneath a veneer of stoic perseverance. Privately, however, its executives were growing increasingly paranoid. Ever since the TFX competition began, there had been persistent rumors that the contract had General Dynamics' name on it. The Texans in Congress were said to be pulling all the strings they could find, including the one that led to the office of Vice-President Lyndon Johnson. Easily dismissed at first, the pieces were now starting to fall into place. The Navy's repeated assaults on the Boeing proposal could be interpreted as a ruse aimed at winning Grumman the time it needed to develop a superior design. Even some of the country's most senior public officials were not above suspicion. Navy Secretary Fred Korth hailed from Forth Worth, while the assistant secretary of the Air Force was none other than Roswell Gilpatric, the same New York lawyer who had so recently served as General Dynamics' special counsel and evaluator of the company's dim prospects. Increasingly suspicious that the deck was loaded against it, Boeing established an intelligence file of all the little hints gleaned from political contacts and friends in the military indicating that its best efforts would not be enough to carry the day.

The third evaluation, like the previous two, concluded with a Boeing victory and yet another decision to postpone the final award. Boeing was distraught, and more suspicious than ever. As usual, Admiral Pirie and his fellow airmen led the charge, complaining bitterly that these latest proposals were no more acceptable than their predecessor. At this point, McNamara decided he had put up with as much as he was going to take. By his estimation, the latest proposals, while still far from perfect, now fell close to the Navy's goals; if there was a problem, it lay with Admiral Pirie and his cabal of obstructionists. Moving with speed and the full support of the White House, McNamara shunted the admiral into early retirement and dispersed his most vocal supporters to jobs where their talents for procrastination would find less provocative outlets. Meanwhile, Boeing and General Dynamics summoned up their last reserves of stamina and plunged into the fourth and final leg of the marathon. This time, however, the race was to be conducted in accordance with a slightly different set of rules.

By this stage, both competitors realized they were doing something wrong. Their common problem was that no one would tell them exactly what. This was not an example of bureaucratic indifference but a deliberate decision intended to keep the rivals on their toes. If Boeing, for example, had learned that the table of drag coefficients submitted with its first proposal came close to the desired levels of performance, the company could hardly have been expected to continue striving for an even more attractive set of figures. Thus, by keeping each of the competing design teams equally ill-informed, McNamara believed they would be obliged to continue working on the entire project rather than dedicating their best efforts to polishing isolated specifics.

The notion made sense in theory but fell apart in practice. Like backward children incapable of making their own informed deductions from the available evidence, the competitors needed to have the rules for this strange new game spelled out in painstaking detail. McNamara, who was now coming under increasing Air Force pressure to hasten the preliminaries and get the proposed plane into production, decided to enlighten both General Dynamics and Boeing by revealing the three key attributes that had always carried the most weight in the eyes of the selection panel. The first two, commonality and a realistic estimate of the total cost, were so straightforward they should have required no explanation at all. The third revelation, the rather vague issue of overall structural design, was, from Boeing's point of view, fraught with danger.

Ever since the first round, General Dynamics had lagged consistently behind. But now, armed with the selection panel's own vision of what the plane should look like, the company's design team had a yardstick to measure its own compliance. Of course, the same could be said of Boeing, although, as soon became clear, the advantages to the front-runner were much less pronounced. Boeing's goal was still the fulfillment of the panel's ideal. General Dynamics, on the other hand, merely had to close the gap between its own best effort and the one from Seattle. This task was made a good deal easier by the panel's newly ordered candor. Now that its members were free to guide the designers toward the ideal configuration and performance, their advice on the placement of air intakes, landing gear systems, and the hundreds of other points of aerodynamic arcana had the inevitable effect of "standardizing" both designs, of maintaining the superficial differences while drawing both sets of performance and pricing projections closer to a common mean.

The inherent benefit to General Dynamics in the new system of rules was further enhanced by its own recent breakthrough in the applied science of wind tunnel testing. Until the third round of the competition, General Dynamics had made its test models from stainless steel, handcrafting each lump of raw metal in a painstaking process that often took two months or even longer. And since each of the hundreds of preliminary variations on the final submissions had to be observed in the wind tunnel, the job entailed literally thousands of draftsmen, craftsmen, and computer technicians organized into independent teams, each requiring a minimum of ten weeks to turn their preliminary sketches into three-dimensional scale models. In mid-June 1962, however, Fort Worth's Research and Engineering Department had a brainstorm that was to revolutionize the entire field of aerodynamic test modeling. Instead of stainless steel, the designers began to cast their models from fiberglass, a breakthrough that shrunk the lead time from three months to a mere ten days. One of General Dynamics' vice-presidents, Robert Widmer, explained the advantages to a *Fortune* reporter not long after the contract was awarded. "It permitted us to go back and look at the whole commonality problem again, instead of just making small changes," he said, adding that he had informed the Air Force that his people planned to "take another look at everything." Widmer lived up to his word, confounding the panel's expectations that General Dynamics had taken temporary leave of its senses. "We built and tested more models in that July–September period than in the previous nine months all put together."

The TFX competition had long since acquired the spectator appeal of an Olympic marathon, and now, as the test of endurance led the runners into the arena for their last and final lap, the crowd came to life in a surge of partisan cheering. Lined up behind the flagging front-runner, one contemporary observer noted, "was undoubtedly the most glittering array of top brass since the Japanese surrender ceremonies aboard the battleship *Missouri.*" Robert Widmer's move to fiberglass models appeared to have given the plucky challenger a second wind within sight of the tape. But would it be enough to close the lead?

The answer was a resounding "no." On November 8, 1962, the Wright-Patterson selection panel announced its unanimous decision: Boeing's was still the best proposal. And this time not even the Navy was prepared to demur. Cowed into meek submission by McNamara's head-lopping, the service had tendered formal notice of its capitulation in early October, when Admiral George Anderson, the chief of naval operations, informed McNamara that both proposals were now considered equal to the assigned mission. Cynics were inclined to smirk at the timing of the announcement and its diplomatic wording. By agreeing to accept either plane well before the final decision had been made, the Navy was pledging its fealty to the secretary of defense while leaving room for some discreet lobbying on behalf of Grumman's variation on General Dynamics' basic theme. Not that its quiet preference for Grumman did any good. As the contestants came to the wire, Boeing still retained its narrow lead on points.

"There is a clear and substantial advantage in the Boeing proposal over the General Dynamics proposal which is magnified by the environment found under the austere conditions usually inherent in limited-war actions," the Air Force Council announced in a ponderous endorsement of the Source Selection Board's decision. General Curtis LeMay, the charismatic wartime bomber commander, now head of the Strategic Air Command, was a Boeing adherent, as was General Walter C. Sweeney, the new chief of the Tactical Air Command. Those in charge of Logistics Command and Systems Command all praised the decision; so, too, did a surprisingly large collection of admirals, although not with the same degree of public conviction that dominated the Air Force wing of the Pentagon.

According to the Source Selection Board report, Boeing promised distinct advantages in all the most sought-after areas of performance. It would be able to fly farther and higher and carry a greater payload of bombs and general weaponry. It was to be more maneuverable at low altitudes and consume less fuel when dawdling at forty thousand feet as a fleet defender. Designed for shorter, rougher landing strips than any previous tactical aircraft, it weighed less than General Dynamics' entry and was marginally cheaper. And if all those attributes were not enough, the plane itself was to be the most wickedly beautiful package of lethal technology ever put together. In their attempts to minimize both wind resistance and the plane's radar signature, Boeing's designers had dispensed with convention and mounted their twin jet engine air intake atop the fuselage like supersonic sugar scoops. General Dynamics, on the other hand, placed their intake beneath the bulky "glove" that housed the swing-wing mechanism. The result was a dumpy, boxy airframe with a long beaklike nose that would later lead pilots to dub the machine the "Flying Aardvark," an appropriately ironic nom de guerre for a warplane with such diminutive talents for aggressive action.

All that Boeing required to make its victory complete was a final word of approval from McNamara's office, and after four successive triumphs, that appeared to be a mere formality. "We were looking at eggs and seeing chickens," one former Boeing worker ruefully remarked some years later. At Wichita, where the local press and city officials were waiting for the phone call from Washington that would mark the official start of the long-delayed celebration, few even considered the possibility that the prize would once again be snatched away. But incredible though it seemed at the time, that was exactly what happened. When the telephone rang, it was not in Wichita but in the home of Fort Worth congressman Jim Wright. The caller was a junior member of the Air Force Legislative Liaison Office, and his news was that Secretary McNamara, Navy Secretary Korth, and Air Force Secretary Zuckert had overturned the selection panel's final verdict and decided to give the contract to General Dynamics. Wright, his face behind a lather of shaving cream, dialed Convair, where he identified himself to the switchboard operator. "Did we get it?" she shrieked, and cut off the connection to spread the good news as soon as Wright uttered his one-word reply.

The second stage of the TFX scandal was about to begin.

9. Skewered on the Hill

Back when the competition was dragging toward its third postponement, in the late summer of 1962, Senator Henry Jackson of Washington placed a call to Deputy Defense Secretary Roswell Gilpatric. Certain disquieting rumors were doing the rounds, Jackson said, persistent whispers that Boeing did not have a snowball's chance in hell of securing the TFX contract. Speaking for himself, Jackson said that he found such stories hard to believe. A Kennedy supporter and an admirer of McNamara, he was sure the administration could be relied upon to take the honorable path. All the same, and just for the record, he felt obliged to raise the fears of his constituents at an official level.

Gilpatric dismissed the rumors by reminding Jackson that loose talk was Washington's stock-in-trade. The TFX program was going to be one of the mainsprings of McNamara's drive for an efficient, effective military; it was inconceivable that such an important undertaking would be compromised so early and for such base, political ends. Reassured but still professionally wary, Jackson relayed the news to Seattle, with the advice that the best policy was to continue striving for a technological edge over General Dynamics. McNamara, or so Jackson maintained, could be expected to keep the competition on an honest footing.

Thus it was that the announcement on November 24 that General Dynamics would build the TFX struck both Wichita and Seattle with the force of an unexpected disaster, a bolt from the blue that defied comprehension or explanation. On four separate occasions over the course of a year, a panel of experts had pronounced Boeing's proposals to be by far the best. Yet now, and for reasons that were not at all clear, McNamara had taken it upon himself to overturn the jury's opinion and award the contract to the runner-up. Like Boeing's own workers and executives, Senator Jackson moved quickly from incredulity to hard-edged anger. No matter how he looked at the decision, it just did not make sense. Boeing had not only submitted plans for a warplane that would be technically superior; it had also promised to do the job for roughly half a million dollars less than General Dynamics. Economy should have been a virtue, yet according to the joint statement prepared by McNamara and assistant secretaries Korth and Zuckert, the lower price was a liability, an example of "unrealistic cost estimates."

Commonality, that other great McNamara battle cry, constituted another and, as far as Jackson was concerned, an even more spurious justification for the decision. Boeing's plane had 14,423 individual components, of which some 60 percent were common to both the Navy and Air Force versions, while General Dynamics boasted 84 percent commonality. The figure was impressive at first glance, but it did not survive close scrutiny. In order to achieve high levels of shared components, Fort Worth's engineers had been forced to add some 1,400 pounds of additional weight to an aircraft that was already showing a decided tendency toward obesity. Moreover, as hearings before the Senate's Permanent Subcommittee on Investigations would soon establish, General Dynamics' higher figure offered very little in the way of financial advantages. They were "parts we do not normally stock and buy as spare parts," Air Force Major General Robert G. Ruegg disclosed early on in the proceedings. "In other words," the committee's chairman, Senator John L. McClellan, asked, "these were parts that are seldom replaced because if they get broken the plane is destroyed, is that right?" McNamara's justification that commonality would lead to substantial savings on the purchase of spare parts evaporated with Ruegg's affirmative reply.

McNamara was not the only one prepared to break with convention. Seething at the perceived injustice and smarting from the insult of McNamara and Gilpatric's ingenuous assurances, Boeing was about to deviate from another of the arms industry's time-honored traditions. Up until the TFX controversy, contract competitions usually ended with a nod to the winner and a shrug of sportsmanlike resignation on the part of the loser. To carp and complain or call for congressional investigations was not done, since the resulting outbreak of public acrimony stood to benefit none but scandal-hungry newspapers and grandstanding politicians. It was far better, or so the industry believed, to keep quiet and wait for another contract competition and the opportunity to regain lost ground. This time, however, Boeing's sense of outrage was enough to dispel all customary reservations. It was "discouraging and disillusioning," Boeing's president, William M. Allen, remarked, to have the selection panel's decision overturned by the "arbitrary judgment" of a group of ill-informed civilian meddlers.

Despite the mounting furor, General Dynamics took possession of its Christmas present on December 21, when the company's new chairman, former Pan Am chief Roger Lewis, and its Fort Worth boss, Frank Davis, put their signatures to the first of the "final" contracts. The initial agreement demanded only twenty-two prototypes, five of them for the Navy, with the first to be ready for flight testing and evaluation by November 1964. If these first proved satisfactory, General Dynamics was given to understand, it would be asked to provide as many as 1,700 additional aircraft, worth roughly $3 million apiece. Those who recalled that the plane's original premise had been to save money received a short lesson in the ways of the Defense Department when they pointed out that the TFX, soon to be rechristened the F-111, was by far the most expensive warplane of its kind. You have to spend money to save money, they were told.

If that concept was somewhat obscure, the contract's rejuvenating effect on General Dynamics' prospects was readily apparent. At the conclusion of 1960, the company had been obliged to write off a $27 million loss, which would be increased by a further $143 million the following year. Coming on top of the losses already posted, these latest setbacks brought the company to the brink of receivership. Under its agreements with the various banks that had agreed to put up emergency capital, General Dynamics was obliged to maintain certain minimum levels of liquid working capital. By the start of 1961, the situation had deteriorated to the point where its creditors were entitled to invoke the insurance clauses in the lending contracts and demand that the company effect emergency cost-cutting measures. And that, as Roswell Gilpatric had pointed out in his letter to Frank Pace, meant the closure of either San Diego or Fort Worth. The only thing that was likely to save it, as *Fortune* magazine pointed out in a lengthy two-part series published over January and February 1962, was a favorable decision in the TFX competition.

To the growing legion of critics, the deus ex machina nature of the award and the apparent haste with which McNamara moved to bestow it after rejecting the Source Selection Board's recommendation added spice to an increasingly hot scandal. On the same day that the papers were signed, for example, Senator McClellan lodged an official request with Gilpatric in which he called for a moratorium while his committee investigated not one but two apparent conflicts of interest among senior Defense Department appointees. From a diplomatic point of view, Gilpatric's reply was most ill-advised. Rather than address the issues McClellan had raised, he fell back on the traditional justification for every major weapons procurement program since the end of the Second World War —the F-111 was needed to counter an increasingly sophisticated enemy that would soon be in a position to threaten the nation's security and its territorial integrity. The deputy secretary might have done better to explain his own business and personal connections with General Dynamics, because even as he dictated his reply, a chain of circumstantial evidence was unfolding that further strengthened the convictions of those who believed Boeing had been given less than a fair deal.

When Gilpatric left to serve the interests of the national defense in Washington his place as General Dynamics' favorite source of legal advice was taken by M. Tex Moore, a close friend and partner in his former law firm, who was also a director. The coincidence seemed too neat for mere happenstance, and there were many in Congress who wanted to ask what part Gilpatric might have played in the decisions.

The second apparent conflict of interest concerned Navy Secretary Fred Korth, a man whose ties to General Dynamics were even more intriguing than Gilpatric's. Before he replaced Texas' future governor John Connally as the head of the Navy Department in January 1962, Korth had been the energetic president of Fort Worth's Continental National Bank, a job that brought him into frequent contact with senior Convair officials, several of whom he numbered among his close friends. In fact, Korth had been in constant consultation with the division as recently as October 1961, when he personally approved a loan of $400,000, a small but nevertheless important chunk of the $20 million the company had borrowed in its efforts to recover from the jetliner debacle. The banker steadfastly maintained that he had severed all connections with his former employer and now held nothing but a parcel of stock worth $160,000, an amount he characterized as too small to call his integrity into question. Cynics on the Hill were inclined to disagree, since there was a growing body of evidence that indicated Korth still took a proprietary interest in the bank's affairs, even to the extent of using the perquisites of office to attract fresh clients. In any case, even if those suspicions were not enough in themselves to call his integrity into question, there was the matter of the $400,000 loan. If General Dynamics' Convair division went to the wall, the bank would not only suffer with the rest of Fort Worth's business community, it would also have trouble collecting one of the largest loans on its books.

With innuendos, suspicions, and talk of general malpractice serving as the currency of the day, it was not long before the battle was joined by a squad of Republican senators who questioned both the propriety and the practicality of the entire exercise. Senator John Williams, for example, a Delaware Republican, and several other congressmen asked the Justice Department to determine if any of the participants in the contract negotiations were open to charges of unethical conduct. Under normal circumstances, complaints such as these would have been dismissed as predictable manifestations of political partisanship. On this occasion, however, the demands for a wide-ranging inquiry gained much credence from the active participation of Democratic senators Jackson and McClellan, the two most senior members of the Investigations Subcommittee.

Nor did the Vice-President's taste for public adulation help matters. On his December 12 visit to the Convair plant, Johnson provided the plane's foes with additional ammunition by hinting that he had indeed exerted discreet influence in the final decision. "The twentieth century is one of the most exciting, challenging times in the history of the world," Johnson said. "Texas is going to have a little chunk of the century." It was hardly surprising that critics of the contract and the administration took to calling the plane the LBJ rather than the TFX.

Questions soon began popping up at presidential press conferences. Although Kennedy was spared the grilling that awaited McNamara, Gilpatric, and Korth, his endorsement of the decision and his brother Robert's refusal to order more than a perfunctory investigation in his capacity as attorney general did much to enhance the atmosphere of conspiratorial conjecture. Could the Oval Office have been involved? The evidence was purely speculative, but it was intriguing all the same. Texas and New York, Convair and Grumman's home states, had made vital contributions to Kennedy's narrow victory in the Electoral College while Kansas and Washington had both backed Richard Nixon. And if the contract was both a payoff for past favors and a down payment on a second presidential victory in 1964, might it not also be a sweetener intended to charm a valuable ally? Colonel Henry Crown, General Dynamics' new chief stockholder, was a political patron in the grand tradition of Cook County, and his support, both financial and personal, was generally reckoned to outweigh any possible advantages to be had in wooing the prominent citizens of Kansas and the Pacific Northwest.

The administration could hardly claim ignorance of the apparent conflicts of interest. Quite apart from the increasing number of calls for an investigation coming from the House and the Senate, Robert Kennedy had received several confidential briefings from Democratic figures and a particularly candid appraisal of the situation from one of the most perspicacious members of the Washington press corps. Clark R. Mollenhoff, a Pulitzer Prize winner and the then Washington correspondent for the *Des Moines Register,* remembers calling on the attorney general with a five-page synopsis of the scandal and a summation of the latest charges. "I told him that the paper contained everything I had been able to find out and that once he had held it in his hand he would not have any excuse for not taking action," he said, adding that he knew Robert Kennedy well enough to step outside his professional role as an objective observer. "It was all McNamara's business, he told me. He said he had no intention of getting involved, under any circumstances."

By mid-January, the burgeoning growth of rumors and the desire for revenge had reached a critical mass that began its chain reaction when McClellan announced that his panel would begin taking evidence the following month. As an indication of his serious intentions, he let it be known that McNamara, Gilpatric, and Korth would be among the first and most closely questioned witnesses.

At the Pentagon and in certain mutinous corners of the defense secretary's own office, the hearings presented an opportunity to even the score. McClellan's staffers and the committee's chief investigator, Tom "Tiger" Nunnally, were soon being supplied with the kind of background information and technical advice that promised one of the most entertaining spectacles seen in Washington since Senator Nye's probe of General Dynamics' corporate forebear more than a quarter of a century earlier.

The investigators' only disappointment was what might best be described as an outbreak of cold feet in Seattle. Those who were close to the committee recall that Boeing's executives began to have second thoughts about their belligerent stance, in the face of veiled threats from various Defense Department contacts. "It was made clear that any trouble they caused over the TFX would be kept in mind the next time Seattle was due for a contract," a former Pentagon systems analyst recalled many years later. "Boeing's people asked Jackson if he could get McClellan to back down, but he refused. By that stage, of course, Boeing's input was largely irrelevant anyway, because the investigation had taken on a life of its own."

The committee began its deliberations by acceding to an Air Force demand for secrecy that would later seem farcical in the light of the F-111's many failings, agreeing to go about its business behind closed doors and limit the release of information to censored transcripts that became public property one, two, or even three days after the event. Excluded from the action at center stage and not terribly impressed by the delayed testimony of the first few witnesses, the press was forced to scout the flanks of the fray, gleaning partisan snippets from contacts with their own individual barrows to push. Senior Air Force personnel who were publicly committed to the program spoke predictably of the plane's advantages and future savings, at the same time that colleagues were blaming McNamara for the gradual degradation of the performance projections that had taken place during the course of the evaluation process. As the junior partner in this shotgun marriage, the Navy was no less eager to have its private misgivings broadcast to the widest possible audience. The F-111 was going to be too heavy; it should be revamped before the development of the first prototypes locked the service into the purchase of a plane that promised to be more trouble that it was worth. Since less than five months had passed since Admiral Anderson's announcement that both aircraft were considered acceptable, these latest outbreaks of anonymous sniping amounted to a curious about-face. Strange though it seemed, there was method in the madness. The Navy had always been disposed toward any plane that carried the Grumman nameplate, although in this case, the company's involvement as General Dynamics' subordinate did nothing to dispel the pall of skepticism that hung over the entire project. As it happened, the only sailor on the Source Selection Board cast his vote in favor of Boeing, a surprising decision but one that meshed neatly with this latest strategy to obtain a different aircraft entirely of its own choosing. With its first official choice eliminated by executive fiat, the Navy felt free to denounce the General Dynamics–Grumman aircraft in terms unfettered by any concern for the appearance of hypocrisy.

However, it was neither the Navy nor the Air Force that generated the first rash of front-page headlines. That job fell to McNamara's own supporters.

In mid-March 1963, the *Washington Star*'s Pentagon correspondent, Richard Frylund, obtained a confidential Air Force memo that criticized the committee's investigators for using "Gestapo tactics" to extract information from witnesses. The dramatic tone was further enhanced by an assertion that might have sent the committee into guffaws had its members not felt their integrity under attack. Harassed Air Force officers were cracking under the pressure, the memo's anonymous author maintained, making reference to heart attacks, ulcers, and in one extreme case, a total nervous collapse. Chairman McClellan was moved to incredulity. "Good grief," he said, "if we have Air Force officers collapsing all over the place before they appear before this committee, what are they going to do when the shooting starts?"

Like any seasoned Washington hand, McClellan and fellow panelist Senator Karl Mundt looked beyond the leak to the motives of its anonymous propagator, concluding that McNamara's people had circulated the document in an attempt to strip the committee of public support. If that was indeed the case—and it was something McNamara insistently denied—it was an example of yet another brainstorm gone wildly astray, because the defense secretary was about to take the stand for the first time just as the early-afternoon edition of the paper hit the streets. The resulting terse exchange as the chairman pressed his witness for an explanation produced the kind of semantic debate that was to be repeated many times before the hearings were done.

CHAIRMAN: You don't approve of the statement being released?

MCNAMARA: I do not. I don't wish by that, however, to indicate that I doubt the truth of the statement.

CHAIRMAN: You are saying it is true, the statement?

MCNAMARA: No, sir; I just don't wish to indicate that I doubt the truth.

Given the widespread expectation that McClellan and Jackson would draw blood in this, their first official encounter with McNamara, the defense secretary emerged from the hearing room in surprisingly good cheer. He had explained his decision to overturn the Source Selection Board's recommendation by citing Boeing's allegedly "unrealistic" cost estimates and comparing them with General Dynamics' "highest dependability." To scotch any suspicions that his decision had been motivated in part or wholly by a personal financial interest in General Dynamics, Grumman, or one of the numerous subcontractors, he had tabled an itemized statement that put his family's holdings of stocks, bonds, and real estate investments at roughly $1.3 million. Senator Mundt had thanked McNamara for the courtesy, but added that there had never been any suggestion of his acting for motives of personal profit. "You have lived in a goldfish bowl so long," Mundt said, "that there was no need to check up on your background."

By contrast, Gilpatric's background was ripe for exploration. During his first sojourn at the witness table in mid-March, Gilpatric intimated that he might just as easily have been accused of bias toward Boeing if the Source Selection Board's original verdict had not been overturned. Sure, he said, Frank Pace was a good friend and a senior client, but so, too, was Boeing's Bill Allen, a man with whom he had "a very close personal relationship." Allen had visited his home, knew his family, and on various occasions throughout the 1950s, had asked him to help in the preparation of several legal actions involving both civilian customers and the Department of Defense. By October, when Gilpatric returned to the stand, this picture of an evenhanded relationship with both the TFX competitors had come under close scrutiny by the committee's investigators, and the panelists were keen to press the witness on what they perceived as major inconsistencies.

General Dynamics' own records indicated that Gilpatric had been a regular speaker at meetings of the company's board of directors, as well as an occasional recruiting agent for retiring Air Force personnel and assorted civilian executives. When McClellan asked about the remuneration he had received for his work on behalf of the company, Gilpatric replied that it amounted to no more than $3,000. Surely, he was asked, a senior member of one of the nation's largest law firms earned more than $1.60 an hour, the figure obtained by dividing Gilpatric's admitted earnings by the number of hours he had spent working on the account. Gilpatric's reply was a rambling statement of the vagaries of law firm accounting practices, which shed no further light on his earnings and obliged the committeemen to float their own estimates of General Dynamics' total contribution to the Gilpatric family fortune, none below

$45,000 a year. By way of comparison, the lawyer's work for Boeing had produced little more than walking-around money.

Gilpatric's relationship with General Dynamics was unusual in that it went well beyond the limited function of a mere lawyer. As late as March 8, 1960, for example, he had interviewed a certain retiring brigadier general from Wright-Patterson Air Force Base before recommending him to General Dynamics' board as a valuable future recruit. On another of several similar occasions, Frank Pace asked Gilpatric to ascertain the reputation and competence of a former power company executive then being considered for a position with the company's civilian nuclear power division, General Atomic. And finally, there was Gilpatric's extensive involvement in the merger negotiations with Henry Crown's Material Services Corporation of Chicago, a relationship that went well beyond the casual affinity implied by the testimony of six months earlier. In the period between May 28, 1959, when the merger was first mooted, and May 26, 1960, several months after it became official, he had been invited to attend every one of General Dynamics' board meetings, sometimes venturing as far afield as San Diego, Montreal, and Chicago. So active had he been on the company's behalf, it was hard to imagine where he found the time to serve as a senior adviser on national security with candidate Kennedy's campaign team.

The lawyer's standard defense against accusations that he had a vested interest in General Dynamics' financial rehabilitation never varied: He had severed all connections with his former law firm on entering the administration. Since he had given no thought to returning to the company, he could not be considered a candidate to profit from the TFX decision. Once again, the committee's investigators unearthed a file of documents that cast grave doubts on his assertions. In a round of correspondence between the law firm and its insurance broker, Gilpatric's departure was consistently labeled as a "leave of absence" rather than a resignation. The implication was clear: Gilpatric had come to Washington not as a patriotic servant of the public but as a ticket puncher looking to add another credential to his already impressive résumé.

It was all a simple mistake, Gilpatric explained when he was summoned to reappear before the committee on November 19, 1963. Although no written record of a formal resignation had been found in the law firm's files, the committee was forced to accept the witness' assurances that some petty clerical error was to blame for the misunderstanding. All the same, as devoted observers of the TFX affair were to note some two months later, even the most duty-bound public officials can sometimes change their minds. Not long after resigning from the Defense Department, on January 9, 1964, Gilpatric returned to his old office in General Dynamics' Park Avenue headquarters to assume a "new" job as one of the Cravath firm's most active senior partners.

Fred Korth, the third signatory with McNamara and Air Force Secretary Eugene Zuckert to the five-page memorandum that overturned the selection panel's decision, was not destined to escape quite so lightly. Like Gilpatric, the Fort Worth banker always claimed to have cut all his ties with his former employer. Senator Mundt had his doubts and decided to ask McClellan's investigators to inspect the bank's correspondence files. The findings obliged Attorney General Kennedy to call for Korth's resignation.

Far from turning his back on the world of banking, Korth was found to have exploited the office of Navy secretary in his attempts to whip up new accounts. "I note on today's new account list that you have secured a $25,000 account from Neiman-Marcus for us and only a few days ago another $25,000 savings account the name of which I don't recall at the moment," Leon Jordan, the bank's comptroller, wrote to Korth on September 14, 1962, adding that "it was probably more business than the people primarily responsible for new business have gotten in the past two or three months." Korth's methods in wooing potential investors also drew the committee's fire. In a succession of letters bearing the official seal of the Office of the Secretary of the Navy, Korth not only praised the Continental National Bank as a reputable and solid financial steward of its depositors' funds but also invited a number of important clients to join him for pleasure cruises aboard the presidential yacht *Sequoia.* The amounts involved were generally small, seldom more than $50,000 and more often than not a good deal less. If such paltry sums elicited the Navy secretary's personal attention, the committee was inclined to wonder how the bank looked upon its business with Convair. The airframe company held one of the largest loans then on issue, as well as various corporate accounts amounting to some half a million dollars in cash and credits. Just as Gilpatric would do a few months later, Korth packed his bags and returned to his old job, all the while denying any and all suggestion of wrongdoing.

With Korth's recent departure and the fresh revelations about Gilpatric's relationship with General Dynamics, the hearings appeared to be on the verge of unleashing a major national scandal to rival the celebrated Teapot Dome affair of the 1920s. But then, just as it all seemed set to come tumbling out into the open, the story was swept from the front pages by a national tragedy.

On November 22, 1963, after stopping off at Fort Worth to press the flesh and make a speech in which he praised the F-111 as "the best fighter system in the world," John F. Kennedy continued to Dallas and his appointment with an assassin's bullet. The investigation did not stop, but it did lose its momentum amid the outpouring of national grief. There had been almost nine months of entertainment, but now, in the space of a single day, the pursuit of further revelations became an exercise of academic interest. Korth was gone, Gilpatric would resign within two months, while the man most often identified as the chief villain in the piece was the new occupant of the Oval Office. The inquiry was not over; indeed, it would continue in fits and starts until 1972, when McClellan released his final report. But to all intents and purposes, the quest for the F-111's hidden patrons was at an end. From that point on, it was to be the plane's technical failures that kept it as a running sidebar story to the greater and growing drama of the Johnson administration's escalating involvement in the Vietnam war.

Back when General Dynamics' victory was first announced, McNamara justified the decision by alluding to confidential staff estimates that put the projected savings at one billion dollars. McClellan made great sport of the documents submitted to justify what amounted to a highly dubious contention. Replete with overly optimistic predictions and glib assumptions about the ease with which the F-111 could be developed and flown, the slipshod logic of the postdated "report" was rounded out by several glaring arithmetic errors. Even Albert W. Blackburn, a systems analyst whom McNamara had described as an enthusiastic proponent of General Dynamics' version of the TFX, could not bring himself to endorse the reputed billion-dollar savings. Early on, long before the ten original competitors submitted their proposals to the first round of appraisals, Blackburn admitted, he had indeed done some preliminary calculations that indicated considerable financial advantages if the Navy and the Air Force could be persuaded to use the same plane. But he also pointed out that his analysis was purely theoretical, "a ballpark figure" applicable to a general concept rather than to any individual competitor's final design. As Blackburn explained it, no competent systems analyst would risk labeling the savings as a hard estimate, since prior experience with other advanced aircraft indicated that unscheduled development expenses would figure inevitably in the final cost. His estimate, for example, was based partly on the premise that the winning plane would be able to take off from an aircraft carrier's deck at the speed specified in the original list of performance demands. If takeoff speed proved higher, he said, part of the basis for his conclusion disappeared, since the plane would require new, high-powered catapult gear to get it into the air, as well as additional structural bracing to help it survive the tremendous strain.

Blackburn dropped his most embarrassing bombshell when he revealed the panic that had overtaken the Office of the Secretary of Defense when it became clear that McClellan would want documentary evidence to support McNamara's decision in favor of General Dynamics. Two teams, code-named Red and Blue, had been instructed to act as devil's advocates, each producing paper records to "prove" that Boeing's scheme had been rejected for valid reasons. For his own part, Blackburn confessed to a decided preference for Boeing, which he praised for the sophistication of its design, the modest advances in construction technology needed to make it a reality, and the company's innovative use of production line techniques. Described in an early session of the hearings as "a mathematical William Shakespeare," he explained that while McNamara's concept of common parts made sense as an abstract theory, it had little basis in fact. Boeing, for example, planned to manufacture slightly dissimilar parts for each version with the same tools, an innovation he identified as the basis for a much cheaper aircraft. General Dynamics, on the other hand, opted for the unimaginative expedient of simply cramming as many common parts as possible into the same basic shape, an approach that compromised performance while doing nothing to lower the final cost, since in many instances it would have been both cheaper and more effective to follow Boeing's lead.

"When the announcement was made that General Dynamics would be given the development program, it was clear to all those involved that this decision could be justified only on the basis of a broad, high-level policy of the administration, and could not in any way be associated with the merits of the two proposals," Blackburn recorded in a memo written in early 1963. Continuing with his criticisms of the award, he contrasted Boeing's "superior" management with that of its rival. ". . . the handling of such major Air Force weapons systems as the B-47, B-52 and the Minuteman as well as the KC-135/707 development must be considered superior to the management given to the F-102, F-106 and B-58 series and the notoriously poor management exercised in the General Dynamics jet transport program."

Blackburn gave his evidence in mid-1963, concluding his testimony by revealing that he had tendered his resignation to protest a decision "that was not in the best interests of either the Secretary or this country." By mid-1965, when the Air Force prototype took to the air for the first time, Blackburn's protest had been vindicated by events in the skies over Fort Worth. In order to reap a promised cash bonus of $875,000 for sticking to McNamara's twenty-five-month production period, General Dynamics pulled out all the stops to transform the wind tunnel model into a seventy-foot-long, fifty-thousand-pound flying test bed.* It had not been an easy job, especially when an independent appraisal by NASA's Langley laboratory in March 1963 identified higher than anticipated levels of transonic drag that required further, extensive modifications to the "final" design outlined in the fourth round of the competition. All the

*The snub-nose Grumman version was shorter.

same, General Dynamics managed to come up with the goods twenty-four days before deadline. Unfortunately, as the old adage maintains, haste makes waste, and the F-111's first fully operational test flight proved to be no exception.

After taking the plane off the tarmac and folding the wings back to the maximum high-speed sweep position of 72.5 degrees, Fort Worth's chief test pilot, Dick Johnson, seemed to have everything under control. But then, just as he was about to breach the sound barrier and collect his second bonus for fulfilling a design goal ahead of time, catastrophe struck. The aircraft shuddered and both engines went dead, victims of a phenomenon known in the trade as compressor stall. Johnson kept his head and brought the plane down to the thicker air of the lower atmosphere, where he managed to restart the twin turbines and bring the aircraft in for a safe if somewhat subdued landing. While the source of the problem soon became apparent in a further series of wind tunnel tests, the solution proved both expensive and elusive.

All jet engines depend upon an assured supply of incoming air. This may appear to be a statement of the obvious, but to a pilot or an aeronautical engineer it is an elusive objective determined by the interrelationship of literally hundreds of variables. With the F-111 prototype, the problem was quickly identified as a consequence of the air intake's position beneath the "glove" that housed the vital swing-wing mechanism. At high speeds and at certain angles of attack, air turbulence created by the forward sections of the fuselage and overhanging wing roots disturbed the airflow and led to weird entry angles, high-speed eddies, and gross fluctuations in the pressure of incoming air. The result of a stall on the hurtling internal components of a straining jet engine can be imagined. At best, it cuts off power with a sudden, unpredictable shudder and a loud *whump* like the sound of a million-dollar truck backfiring. At worst, it might tear loose a compressor fan, wrecking the engine or even the entire plane. In just about every area, the prototype proved to be a textbook example of the way not to design and install a jet engine.

In order to get the most out of the Pratt & Whitney turbines, General Dynamics' engineers had cut back the length of the inlet sleeves that guided the air into the engine and smoothed out random turbulence. This might not have been so bad if the engine had been a little more forgiving, but like so many other parts of the plane, it had been stretched to the limits of its theoretical performance. Originally conceived as a sedate, economical power plant for a scrapped midrange jetliner, the engine had been "specced out" to meet the military's need for speed and enormous power. In addition to an afterburning fan and a variable nozzle to focus the force of the exhaust, the compressor geometry—the angle at which the rotating blades bite into the airflow—had such a low tolerance for disruptions that the mishap which stopped General Dynamics from collecting its bonus occurred when the plane was flying a level, stable course in still air. Obviously, this was no way for a purported fighter to perform. Somewhat shamefacedly, McNamara's office announced that flight testing of the F-111 would be suspended pending investigation. Over the years to come, the term "pending investigation" was to become an increasingly common refrain among the corps of Air Force spokesmen assigned to act as the plane's official apologists.

While General Dynamics and Pratt & Whitney struggled to remedy the problem of unpredictable compressor stalls, the best-known of the F-111's shortcomings began to emerge. It was not, as cartoonists and editorialists maintained, that the wings fell off;* it was just that the swing mechanism did not always work as well as it should. The culprit was a totally unexpected outbreak of metal fatigue. In the first few series of test flights, the early F-111s all moved their wings without problem, something that led McNamara's people to conclude that at least one part of the aircraft would not need expensive revision. By late in 1968, however, fatigue cracks began appearing in the pivot box and surrounding structures, and once again the planes were ordered back to the research hangar. The compressor stalls had been bad enough, but the wings proved to be a disaster. For the next three years, the vital, stress-plagued wing boxes were redesigned, reworked, and replaced at enormous cost. And just to make a difficult technical exercise even more so, McNamara's earlier insistence that production continue unabated even while the first prototype was being extensively revamped meant that all of the essential alterations and improvements had to be performed on not one aircraft but many.

There was only one bright light in the F-111's early development, and that also bore the heavy stamp of McNamara's circumspect involvement. When the TFX project was first hawked around Washington, McNamara and his point men made much of the potential for overseas sales. And for a while it looked as if they might be right. In October 1963, the Australian government of Prime Minister Robert G. Menzies signed up for twenty-four slightly modified versions, and early the following year the British government ordered roughly seventy aircraft in yet another variation on the basic design. Both agreements, together with a rumored third overseas sale, to West Germany, were trumpeted as vindications of the McNamara approach: If foreigners were prepared to buy before the first plane had even left the factory, then the program must be doing something right. In both cases, however, the plane's purported virtues had

**Later, as metal fatigue joined the list of problems, wings did start to fall off. This was not until 1968 however.

little to do with either buyer's early commitment. In Britain, the decision was an outgrowth of the perennial debate about the future of the nation's ailing, and often addled, aerospace industry. The government of Harold Macmillan had left office in 1963 after making a qualified commitment to build a British fighter-bomber, the TSR-2. Harold Wilson's incoming Labor government took one look at the anticipated total cost and rapidly scrapped the entire project. As Defense Minister Denis Healey explained, the market for such a plane simply did not exist. Britain would not be able to sell in Europe, where Dassault of France, Sweden's Saab, and the ever-present phalanx of American salesmen dominated the field by virtue of geographic proximity or sheer weight of numbers. America offered no better hope, since the U.S. Air Force, like every other wing of the Pentagon, had an institutional aversion to any weapons system not designed and built within the Lower Forty-eight. As for the third world, that was clearly out of the question. As it was then envisioned, the TSR was to be both too expensive and too complex for nations whose tastes and budgets ran more toward Northrop's cheap, utility warplane, the F-5 Tiger.

In a speech that marked the official end of Britain's place among the world's leading aerospace manufacturers, Healey announced that the F-111 would become the new, deep-strike arm of the Royal Air Force. The announcement plunged the native industry into the deepest gloom, which was not dispelled when Healey attempted to explain his reasoning. While the production of high-technology jet warplanes was beyond Britain's means, there were many other, more simple weapons systems, like mortars, small arms, and tanks, that could, if promoted properly in the international marketplace, keep British workmen at their lathes for years to come. In return for Britain's pledge to purchase fifty F-111s, Healey believed he had taken the first step toward establishing a two-way flow of arms traffic across the Atlantic. This proved to be a false hope. The U.S. made no significant purchases of British weaponry, while the F-111's escalating cost and interminable setbacks soon eroded what little support it commanded in the Wilson cabinet. Early in 1967, less than six months before the U.S. Navy finally managed to extricate itself from the F-111 project, the British announced that they were canceling their order, to concentrate on yet another, even more wonderful warplane, to be built by a pan-European consortium.

If Britain's flirtation with the F-111 was, as some maintained, a clever ruse intended to justify the demise of the TSR, Australia's experience with the plane still brings to mind the plight of those simpleminded folk who sometimes buy the Brooklyn Bridge. In 1963, Conservative Prime Minister Menzies cast an anxious glance at the leftist regime of Indonesian president Sukarno and began checking the national arsenal. The U.S. embassy in Canberra let it be known that the Royal Australian Air Force could, if it so wished, purchase several squadrons of recently retired B-47 jet bombers at minimal cost. There was nothing wrong with the aircraft as such, just that they now seemed rather dowdy and outdated in comparison with the latest generation of U.S. Air Force planes, like the B-58 Hustler and, of course, the F-111. But for Australia they would have made perfectly adequate symbols of the bristling mistrust of Sukarno and his policies.

Menzies' problem was that while the B-47s might be equal to the military task, they stood to do his party little good on the stage of domestic politics. An election was due in a few months and Menzies wanted some vote-catching emblem of his government's will to stand up to the perceived threat of Indonesian aggression. Aged, obsolescent by the standards of the superpowers' arsenals, and decidedly unglamorous, the B-47 simply did not fill the bill. RAAF appraisal teams had already considered and rejected the British TSR on the grounds that its development was likely to be halted without warning. They had also toyed with buying Mirages from Dassault and had given serious thought to the F-4 Phantom and the A-5 Vigilante, when McNamara entered the picture. The U.S. embassy in Canberra was told to push the F-111 as the only plane capable of giving the Australians what they sought.

U.S. attachés who argued that the B-47s were equal to the Australians' needs soon found themselves back in the United States, their places taken by the likes of one Alexander Butterfield, an Air Force attaché who would earn his own niche in history during the Watergate hearings, when he revealed the existence of the White House tapes. Those who worked in and around the Pentagon at the time maintain that Butterfield's chief mission was to make sure the Australians added their small order to General Dynamics' list of orders. Local politicians were informed that only the F-111 would be able to hit any island in the Indonesian archipelago and still fulfill a purely defensive role as a fighter-interceptor. The Menzies government bought the sales pitch without reservation or hesitation, and in October 1963, Defense Minister Hugh Townley committed his government to the largest and most curious arms purchase agreement in the country's history. With only a rough idea of the final price and no clear indication if the as yet theoretical warplane would work half as well as planned, Townley agreed to purchase twenty-four aircraft at a tentative price of $125 million. The decision proved immensely popular, at least in the short term. Two months later, when the Menzies government was returned to office with a substantially increased majority, many local analysts saw the victory as an expression of the electorate's approval of the government's willingness to stand up to Sukarno and the Asian hordes.

In fact, as soon became clear, the decision in favor of the F-111 proved to be one of the most costly mistakes an Australian government has ever made. Late in 1972, when the first F-111 finally arrived, more than five years and 700 percent over budget, it had long been the parliamentary opposition's weapon of choice in a seemingly endless round of attacks on a succession of defense ministers, as well as the butt of innumerable jokes. During the late 1960s, the Australian press took to calling it the Flying Opera House, in grim reference to the now famous cultural edifice on the shores of Sydney Harbor, which had to be reworked many times and at enormous cost before the original architect's impressive but impractical design could be made to stand up. The situation became so bitter that the U.S. Air Force was forced to lease a squadron of Phantoms to the Australians as a temporary stopgap.

Although they eventually dropped from the headlines, the RAAF's problems with the F-111 were never entirely settled. In early 1985, twenty-two years after the Menzies government's hasty blunder, Australian strategists were still at a loss to know how and where to fit the aircraft into the framework of the national defense. It remains too short-winded to reach potential enemies—Indonesia's threat disappeared in 1966, when Sukarno's regime was toppled by a right-wing coup—while its attributes as a fighter-interceptor are negligible. So pronounced are its deficiencies that the RAAF recently played host to a visiting U.S. Air Force strategic planner who spent three months in Canberra attempting to create a local mission to justify the aircraft's existence.

Domestic politics in London and Canberra are all part of the F-111's strange saga, but of course, they are mere sideshows to the galloping controversy in the U.S. In 1968, the Navy finally removed itself from the project for good, pointing to a staggering 253 individual deficiencies that made the aircraft unsuitable for Navy surface. This time, the admirals had the statistics they required to give their argument the force of indisputable fact. When first conceived, the naval version of the F-111 was supposed to be able to take off with an eight mph tailwind—a prodigious achievement, since it is the flow of air over the forward edges of the wing that lifts an aircraft into the sky. By July 1967, the last, much altered version needed a nineteen mph headwind, something that would have obliged an aircraft carrier to steam at full speed every time an F-111 was ready to fly a mission. Airborne performance suffered a similar marked decrease. Now weighing almost 70,000 pounds, the plane could not remain airborne on one engine alone, and even with both engines working it could not drag itself higher than 44,000 feet, far too low to serve as the loitering sentry originally envisioned. Ironically, the cumulative effect of all the essential performance-sapping changes destroyed McNamara's chief rationale for the original TFX by reducing commonality from a promised 80 percent to a mere 29 percent, and boosting the price to $9.5 million.

In keeping with the tenor of the whole sorry saga, the Navy's departure from the program came as a total surprise. Only a few months before the admirals persuaded the House and Senate's Armed Services Committees to overrule the Defense Department's civilian secretaries and withhold funds for the project, Robert A. Frosch, assistant Navy secretary for research and development, made an enthusiastic speech in which he claimed that all the major problems had finally been eliminated. "When all of the elements are considered," he said, "we find that it meets our fleet air defense requirements better than any competing system available for study."

The admirals must have been beside themselves with mirth. Not only were they preparing the final, damning report that would scuttle the F-111 for good; they were also hard at work on another theoretical aircraft, this one entirely their own. While struggling to iron the bugs out of the F-111's trouble-plagued wings, Grumman began developing a slightly different, much stronger "swing" box that eventually became the most important feature of the Navy's F-14 Tomcat. As one former NASA engineer commented, "The Navy always wanted its own plane and it made damn sure that was what it got." Even people like Bill Gunston, an aerospace writer who has championed the F-111 in articles and a well-researched book, was forced to concede that the Navy version was an unqualified disaster. "The F-111B succeeded merely in consuming $377.7 million, degrading the Air Force F-111 versions and delaying the F-14 six years," he lamented.

Today, if you ask an Air Force flack or a General Dynamics official about the F-111, the response is likely to be a recitation of glowing statistics. The plane has a safety record second to none. It is versatile, beloved by its crews, and, quite possibly, the single most impressive aircraft of the latter twentieth century. The speech will probably be followed by some glowing reference to the Terrain Following Radar, the plane's infallible accuracy in delivering its bomb load, and, inevitably, a ringing refutation of the charge that the F-111 is a lemon. There are reams of statistics to support all the claims, including some highly imaginative figures that depict the F-111 as the most reliable bomber ever fielded against an enemy. Impressive as they are, the "facts" exist as perhaps the most cutting indictment of the aircraft and the procurement system that produced it. In the words of an unimpressed Pentagon analyst who worked closely with both the Air Force and General Dynamics during the latter stages of the F-111 program: "All they prove is that the Pentagon has a fertile imagination."

While its pitchmen point to the F-111's low abort rate—the incidence of mechanical and electrical malfunctions that force a crew to return to base before dropping their bombs—the plane's actual history paints a very different picture. Its performance in combat, the bottom line for any weapons system, was little better than deplorable. In 1968, for example, six aircraft from Nellis Air Force Base were sent to Takhli Field in Thailand to take part in the bombing campaign against North Vietnam and Laos. Two disappeared without a trace within five days, and a little more than a week later, a third aircraft went down. As the 1969 edition of *Jane's Aircraft* noted, "the type was grounded shortly afterward."

In 1972, after four years of immensely expensive static testing to ascertain the incidence of metal fatigue in the wing pivot boxes and tail assembly, a second group of F-111s went back to Thailand, where history repeated itself. Flying in an F-111 was like sex, one enthusiastic pilot remarked at the time: "You don't know how good it is until you've tried it." If the aircraft did warrant comparison with the procreative act, it would seem to have had more in common with certain varieties of spiders, because once again, this latest batch of much-repaired F-111s proved to be extremely lethal to their own crewmen. Four aircraft and eight crewmen disappeared without trace in just two months, and early the next year the surviving planes were called home for the last time.

Why did the planes prove so unreliable in combat? The answer is anyone's guess. The ground crews and pilots who flew the planes out of Takhli speculated that the short, sharp, and dense rain squalls common over North Vietnam may have confounded the Terrain Following Radar. Back in the United States, analysts far removed from the war zone were inclined to suggest structural flaws as the main culprit. Of course, one guess was as good as the next, because the aircraft flew too low to radio their positions, and no wrecks were ever found. The most plausible, though most often ignored, explanation was supplied by the North Vietnamese, who maintained the missing F-111s had simply been shot down. Even though the Air Force rejected these claims out of hand, they still demand serious consideration. A former Air Force pilot who now serves as a Pentagon consultant explains why:

> First, there were all those radars sending out signals all over the place. What that meant was the plane was really telling the North Vietnamese, "Hey, here I am. Come and get me."
>
> Second, if it did come under attack, the plane could not maneuver out of its own way. The moment it had to bank, or turn or climb to take evasive action, it lost airspeed because there just wasn't the power to offset the tremendous weight loading on those little wings. It is the same reason why the F-111 is still forbidden to do any dive bombing today—because it doesn't have the wing area to pull out at the bottom.

So, if you accept that the North Vietnamese knew the F-111s were coming long before they reached the targets, it is safe to assume that they had time to prepare a reception. And any plane with the F-111's handling characteristics that flies into the teeth of an alerted antiaircraft defense system just has to be a sitting duck. If the SAMs or MiGs didn't get it, the pilot's attempts at evasive action would have been enough to punch a hole in the jungle.

Of course, if the final versions of the F-111 had lived up to McNamara's demand for an aircraft that could streak away from danger at the speed of sound, the fatality rate might not have been so high. However, that quality, like so many others, had long since fallen victim to the series of enforced alterations and dubious improvements. Where "SOR 183" had called for a high-speed, low-level dash capability of 400 miles and McNamara's compromised Air Force–Navy proposal demanded 220 miles, the planes that flew over Vietnam could manage only 30 miles of low-level supersonic flight before running out of fuel.

Ironically, it was the plane's deficiencies that made the greatest contribution to the security of the United States, something originally predicted in jest by the chairman of the Senate Appropriations Committee, Senator Richard Russell, during the 1969 Defense Department budget hearings.

> **SENATOR RUSSELL:** Would it be a very serious matter if one of these planes were recovered by an enemy in reasonably good condition?
> **GEN. MCCONNELL (Air Force CoS):** Yes, we have quite a few things in it we would not want the enemy to get.
> **SENATOR RUSSELL:** Of course the Russians got a B-29 when they were our allies. They fabricated a great many of them as nearly compatible to the B-29 as they could. I was hoping that if they got an F-111 they would fabricate some of them as near ours as they could and see if they had as much trouble as we did. It would put their Air Force out of business.

Senator Russell's best hope appears to have come true. The U.S. built only 520 F-111s in various configurations, while the Soviets, seduced apparently by the volume of official disinformation, made swing wings a prime feature of the next generation of warplanes. Early in 1973, the year Fort Worth finally wound up production, the Russians unveiled the first in a series of variable-geometry fighters, fighter-bombers, and ground-attack airplanes. Despite glowing Pentagon descriptions of its prowess in combat, the first Russian swing-winger, the MiG-23, has demonstrated a singular inability to get the best of the more conventional U.S.-supplied warplanes fielded by the Israeli Air Force, while the SU-17 provided little sport for the U.S. Navy's own swing-wing fighter, the F-14 Tomcat, in their short, one-sided confrontation in the skies off the coast of Libya in

1981. But the Soviet's supreme vote of faith in variable geometry remains the SU-24. An F-111 look-alike, the Fencer goes its inspiration one better by boasting a power-to-weight ratio and a wing-loading factor that are respectively the lowest and highest to be found in any frontline aircraft anywhere in the world.

III MODERN TIMES

10. The Combative Colonel

General Dynamics left the sixties much as it had left the fifties—uncertain, financially troubled, and with a much-enhanced reputation for always making the worst of a bad situation. Thanks to the F-111 and the timely appearance of Henry Crown, the company had survived its encounter with Howard Hughes, even managing to prosper for the few short years between the announcement of the TFX contract decision and the discovery of the first serious flaws in the design. But the price had been extraordinarily high. The plane itself might not have been "the unsafe and defective deathtrap" depicted by Senator William Proxmire in his repeated assaults on the program, but even by the most charitable standards, it could hardly be rated a success. Try as they might, neither General Dynamics' spokesmen nor official Air Force endorsements of the aircraft could dispel the widely held impression that the company owed its escape to an agreement formulated by old friends intent on saving a clumsy giant from its self-inflicted wounds.

Success might have mitigated the lingering impression of corruption, but there was precious little about the F-111 that warranted praise. Even General Dynamics' chairman, Roger Lewis, was forced to admit that the plane was a multibillion-dollar example of Murphy's Law. "The only thing worse than having the F-111 contract," he once said, "would have been not having it." Back when it all began, McNamara predicted a production run of more than 2,000 aircraft at a total cost of roughly $6 billion. In 1972, the year work came to a halt on all but a few experimental variations and retrofitting programs, the total price had surpassed $17 billion, while the number of planes had shrunk to a mere 520. And even these few examples of the breed presented the Air Force with a logistic dilemma. In their efforts to produce at least one version that made some sort of sense in the inventory of combat-ready equipment, General Dynamics' engineers and the Department of Defense had come up with so many different types—electronic warfare versions, the aborted prototypes for Britain and the U.S. Navy, as well as a couple of "stretched" models envisioned as potential replacements for the B-52—that it was almost impossible to equip any one operational wing with a full complement of the same kind.

And of course, at every turn of the story there were the ever-present charges of favoritism and featherbedding. Consider, for example, the terms under which the Navy departed from the project in early 1968. Since each of the delivered prototypes was more than eight tons overweight and thus totally unsuitable for duty at sea, the Navy could simply have washed its hands of the deal and left the contractors to sort out the mess as best they could. However, that did not happen. In a series of meetings between Lewis, various Grumman officials, and Under Secretary of Defense Paul Nitze, the parties arrived at an agreement that gave every appearance of rewarding failure.

Many years later, when Nitze was about to present the Reagan administration's case in arms limitation talks with the Soviet Union, one critic of the appointment labeled him "the high priest of the hard line." There was, however, no trace of the confrontationist in his negotiating position with the contractors' representatives. As the overseer of the Navy version, Grumman was awarded all its fair costs plus a percentage of lost profits, and best of all, an assurance that the next defense budget would include a generous allocation for further work on the F-14 Tomcat, the aircraft born of the F-111's many insoluble problems. General Dynamics also did better than it had any reason to expect, since Nitze not only agreed to restructure the surviving Air Force component of the contract but also awarded Fort Worth an additional one million dollars to remove the last traces of naval influence.

And so the scandals continued. As late as 1972, the year prime production came to an end, General Dynamics' good name was still an open question. This time, and not for the last time in the company's history, the charges concerned misconduct on the part of a principal subcontractor.

The Selb Manufacturing Company of Walnut Ridge, Arkansas, had been given the job of producing top and bottom bracing plates for the swing mechanism, as well as several hundred sixteen-foot spars that formed the rear fuselage and helped to keep the tail in place. It was demanding work, vital to the structural integrity of the finished product, yet evidence produced before a federal court indicated that General Dynamics had allowed Selb executives to get away with fraud and deception on a monumental scale.

Charged and later convicted of conspiring to conceal defective parts, Selb's owner, Saint Louis businessman Harry Bass, confessed that it had been an easy matter to foil Fort Worth's lackadaisical quality-control inspectors. The serial numbers on rejected spars and plates were simply ground down, the flaws welded over, and the refurbished parts resubmitted as new. On other occasions, he admitted, the appropriate mixture of acids in a testing machine used to detect illegal welds was replaced with a kitchen-sink cocktail of water, vinegar, and honey. And whenever slight of hand was not enough to guarantee approval, Bass resorted to old-fashioned bribery. One General Dynamics official was found to have accepted a new car; a second walked away with quantities of cash and other items of expensive merchandise. Both quality-control inspectors were acquitted at later hearings, but the impression lingered that Fort Worth went about its business with scant regard for the safety of the pilots and crewmen who entrusted their lives to its products, a suspicion heightened by a number of Air Force crash investigation panels, which identified structural flaws in the tail assembly and wing boxes as the chief cause of the plane's tendency to roll onto its back and nosedive into the ground.

The F-111's greatest and most valuable quality may have been its ability to make converts out of confirmed skeptics. Richard Nixon's was perhaps the most remarkable about-face. Three weeks before the election in early November 1968, the future President swung through Texas, taking time, like Kennedy and Johnson before him, to touch down at Fort Worth and lavish praise on both General Dynamics' work force and the F-111. "The F-111 in a Nixon administration," he announced, "will be made into one of the foundations of our air supremacy." This was a curious comment indeed, because his earlier statements on the subject rang with the fervor of a righteous crusader against waste and mismanagement: "Against military advice, the F-111 was selected as a superior, yet economical, weapons system," the candidate had opined in a position released only six months earlier. "In view of the recent decision that the Navy version is unacceptable and substitute aircraft be initiated, the final cost of the program will increase enormously, coupled with years of delay."

One did not need to be a keen student of political motivations to catch the strong whiff of duplicity in the November reversal. No less aware than his predecessors of Texas' prime importance in the electoral college, Nixon found it convenient to ignore his earlier comment that the F-111 "did not meet Air Force operating requirements," and adopt it as a pillar of the national defense. But whom was he defending, the United States or General Dynamics? The vast weight of circumstantial evidence indicated the latter.

Less than three weeks before his plane touched down in Fort Worth, Standard and Poor's investment guide appraised General Dynamics' stocks as "a speculation in the success of the F-111 project," while

Moody's News was even more blunt, pointing out that the company had just swallowed a loss equal to $1.51 per share of common stock in the third quarter of 1968. In the first nine months of the financial year, *Moody's* continued, the company had amassed profits of only $9 million, a drop of 75 percent compared to the corresponding period of the previous years. One hard-bitten observer in the press commented that Nixon's change of heart marked the continued survival of General Dynamics as one of the few issues on which Democrats and Republicans were in full agreement.

And the company certainly needed all the help it could get. With the exception of the Pomona division's work on the Stinger antiaircraft missile and the family of Standard surface-to-air and surface-to-surface missiles, the other operations were almost all beset by crises of one sort or another. San Diego, for example, would have been amassing substantial profits from its prime contract work on the Atlas and Centaur rockets had it not been for a diet of piecework that included fuselage sections for McDonnell Douglas' first wide-bodied jetliner, the DC-10, and tail sections for Lockheed's trouble-plagued C-5 military transport. Neither project was going particularly well. The DC-10 subcontracts were running way over budget and lagging far behind schedule, while the number of C-5 tails on order had been recently cut from 250 to a mere 80 in the wake of revelations that the plane, already $2 billion in the red, could not do many of the things for which it had been designed.

On the other side of the country, Electric Boat's problems were even more severe. Caught in a lull between successive generations of nuclear attack and ballistic missile submarines that would not end until 1974, when the Navy accepted the first of the 688-class Los Angeles submarines and began preparing for the development of the much larger Ohio-class Tridents, Electric Boat was contributing nothing to the corporate coffers but cash flow problems. Thanks to a shortage of skilled labor and some grossly inaccurate cost projections, which had led to substantial losses on a number of boats completed during the late sixties, the yard's most notable feature was the air of gloom that dominated the corridors and offices of the executive administration building.

Farther to the north, the picture grew progressively worse. The Quincy shipyards had just finished the last of a fourteen-ship contract for the Navy that had never come anywhere near making a profit. Back in 1963, when Roger Lewis ignored Henry Crown's objections and insisted on buying the yard from Bethlehem Steel for what then seemed like the bargain basement price of $5 million, he defended the decision by arguing that Quincy would be able to draw on the vast reservoir of technical expertise available at Electric Boat. If the company could build ships that sailed under the surface, surely it could do an even better job with vessels that floated on top of it. The argument betrayed a costly ignorance of the shipbuilding industry, not to mention a misplaced confidence in Electric Boat's ability to set a good example. Rather than emulating Groton's good points, Quincy soon fell victim to all the same vices. By the time the last of the fourteen ships was handed over, Quincy's problems with unions, inflation, and the Navy had produced losses over three years amounting to more than $240 million.

Across the border in Montreal, Canadair was virtually withering on the vine. Although it continued to produce small quantities of aircraft for Canadian and Commonwealth customers, Canadair had become but a shadow of the thriving concern Jay Hopkins once described as General Dynamics' greatest jewel. Like San Diego, the Canadian operation had sought subcontracts for Lockheed's C-5, only to find that the job involved more trouble and expense than the project was worth. Unlike San Diego, the $10 million worth of nonrecoverable overruns on the subcontract work were entirely its own doing. So low had Canadair sunk, there were widespread rumors that it would soon be offered for sale in an attempt to alleviate some of the parent company's mounting cash flow problems. In fact, it remained within the General Dynamics' family until 1976, when the Canadian government completed the cycle begun thirty years earlier and repurchased the division for $38 million.

The defense conglomerate's lackluster performance was best reflected on Wall Street, where General Dynamics' stock was producing earnings of just 24 cents a share, an abysmal, all-time low. And there was worse to come. In the early months of 1969, the newly elected President reversed his position on the F-111 for a second time by approving Defense Secretary Melvin Laird's plan to cut short the FB-111 program, a nascent development and production project which envisioned yet another modified version of the original aircraft as a cheap alternative to both the B-52 and another, theoretical bomber that would later take flesh under the Reagan administration as the B-1. It was true that Laird had included a consolation prize in his defense budget, but as industry analysts and Fort Worth editorialists pointed out, the defense secretary's apparent generosity was not enough to assure Fort Worth's future for more than a year or so at best. General Dynamics was to continue producing the F-111s at an estimated cost of $599.8 million plus a further $209.2 million "to cover excess costs generated in 1968." Long-term critics of the program like Senators Proxmire and Curtis were quick to identify the reference to "excess costs" as yet another example of General Dynamics' ability to get administrations of either persuasion to pick up the bill for its own mistakes, although in this case, their comments were not entirely justified. It was true that the additional $209 million was slightly more than the sum Nixon had allocated for urban renewal in the wake of the riots that followed the assassination of Martin Luther King, but at the same time, it inspired no rejoicing in General Dynamics' New York Boardroom, where it was correctly identified as a lump-sum settlement preparatory to the project's assured and final cancellation in the military budget next year or the year after.

The start of the new decade produced its share of embarrassments and victories for General Dynamics' chief stockholders, Chicago sand and gravel men Colonel Henry Crown and his son, Lester. In 1972, the younger Crown joined with a number of other local construction and cartage companies to deposit roughly thirty thousand dollars in a suburban Lake Bluff bank account that would later be portrayed before a grand jury as a floating slush fund. The Crown family's Material Services Corporation, by far the largest contributor, would have done handsomely if the state legislators who received this covert largesse had succeeded in passing a bill to increase the load limit on heavy trucks carrying, among other things, sand and gravel.

Paul Sullivan, a reporter for Chicago columnist Mike Royko, looked into the incident more than a decade later, when it was once again the subject of much speculation, this time emanating mostly from Washington. "The highway engineers always said it would tear the roads apart and there were a lot of people going around calling it the Highway Buster Bill," he recalled. All the same, the sought-after amendment to the load-limit regulations was approved by the state legislature and sent on to Governor Richard Ogilvy for executive approval. That was when things started to go wrong. Ogilvy sided with the engineers and vetoed the legislation, amid growing allegations of corruption that were soon being investigated by various federal agencies. Five state senators were indicted on bribery charges in rapid succession, and from there the trail led directly to Lester Crown, who was to admit contributing some $23,000 to the slush fund.

The grand jury investigation of Lester's activities dragged on until mid-1974, when he was granted immunity from prosecution in return for his promise to testify against the indicted politicians. The trials did have their lighter moments, most notably when one of the defendants complained that it was most unfair to persecute those who accepted the bribes —a common enough custom in Cook County—when the man who authorized them was allowed to escape untouched. Untouched but hardly untarnished. "The two-bit cigar chompers were found guilty of taking bribes," Royko wrote later in his best acid style, "while Lester and his button-down flunkies went back to the business of getting richer."

His father's activities through the latter half of the sixties were also the subject of much attention. In 1966, the Colonel astounded Chicago and the financial world by announcing that he would shortly exchange his interests in both General Dynamics and Material Services for $120 million in cash. He gave no specific reason for his decision other than the comment that he planned to devote himself "to other businesses and philanthropies." Like Crown aphorisms on the maligned honesty of politicians and the value of hard work, this enigmatic remark needed to be read in context. Far from leaving voluntarily, Crown and his allies on the board of General Dynamics had in fact been ousted in a bitter palace revolt led by chairman Roger Lewis, the man he had appointed to the job five years before.

The full details of the maneuvering that went on between 1966, when Crown was shown the door, and his triumphant return three years later may never be fully known. Roger Lewis now refuses to speak of his years with General Dynamics, while Henry Crown has never been known for volunteering information about his business dealings. All the same, from the available records and reports, it is possible to reconstruct a story line that would do justice to one of the more improbable television soap operas.

The 1960 merger with Material Services Corporation was a marriage of convenience, which saved General Dynamics from a ruinous reckoning with its creditors only at the price of mounting in-law problems. From the days of Isaac Rice, the company had drawn its leading lights from among the well-heeled members of the New York bar, former secretaries and under secretaries, and of course, the corps of ubiquitous retired admirals, generals, and war heroes. It was not a patrician heritage, but it did come close. Henry Crown's problem, one of them, was that his did not.

There was about him the hint of Chicago and the peculiar brand of politics for which the city is famous. It was said that he had engaged in business dealings with several former members of Al Capone's inner circle, and there were stories of MSC's legendary ability to secure public works contracts, often at prices well above normal market levels. There were even questions about the way he had founded Material Services. A long-stalled legal case brought by a Mrs. Dora Greiver Stern claimed that she owned at least 21 percent of MSC. In 1919, Mrs. Stern alleged in papers filed with the Illinois Appellate Court, Henry Crown and his brother, Sol,* called on Mrs. Stern and convinced her to invest her mother's life savings of $4,250 in the firm, promising the women 170 of the original 800 issued shares.

"I remember Sol came over to our house very enthusiastic about going into the material business," Mrs. Stern informed reporter Drew Pearson in the early 1960s. "He said Henry had 'connections' with the

*Sol died of T.B. several years later.

city and would get all the city contracts. When we signed the papers, Sol put his arm around me and said, 'Dora, if this goes through you'll never have to worry.' I signed for my mother because she could not write." When MSC merged with General Dynamics, Mrs. Stern had been pursuing the matter for almost eleven years, finding it almost impossible to have the matter heard by a Chicago judge.

Crown rapidly demolished the faint hope that he would stay close to Chicago and limit his involvement in General Dynamics' affairs to overseeing the activities of Material Services. After luring Roger Lewis away from Pan Am and installing himself as chairman of the executive committee, Crown set about making his presence felt. He brought in several Chicago cronies, including one, Patrick H. Hoy, whose background was even more at odds with General Dynamics' traditions than that of his mentor. Before becoming president of Material Services and a General Dynamics senior vice-president, Hoy had run Chicago's Sherman Hotel, a job that brought him into regular contact with purported friends of organized crime like Charles (Babe) Baron, whom journalist Ovid Demaris once described as a former employee of Meyer Lansky, and "a Syndicate representative in Las Vegas." Several years later, when Hoy's investments in an insurance group turned sour and left him facing debts of $8.5 million with assets of only $500,394, Henry Crown was named in court papers as a joint guarantor of a $2.6 million loan then in default.

Although it was Crown who brought Lewis into General Dynamics, the relationship soon turned sour. "He was always on Roger's back about something or other," a former General Dynamics official commented in 1970. Another veteran of the early sixties explained why: "Henry figured that he owned General Dynamics and that gave him the right to lay down the law."

Lewis' reluctance to do Crown's bidding brought about the first of their major boardroom confrontations late in 1963, when the pair fell out over the proposed purchase of the Quincy shipyards. Crown had been adamant in his opposition from the beginning. If a veteran shipbuilder like Bethlehem Steel could not make the yard work, what hope had an innocent newcomer like General Dynamics? The U.S. maritime construction industry was in decline, he argued. Why else would Bethlehem be so eager to sell for a mere $5 million a division that had been part of its empire since the turn of the century, particularly when the deal also included 180 acres of waterfront land? The Chicago millionaire stuck to his guns right up until the last ballot, when, for the sake of appearances, he accepted defeat and agreed to vote with the other thirteen directors.

The strained relationship continued to cause problems for two more years, until Lewis and his boardroom allies made a dramatic move to rid themselves of Crown's troublesome influence once and for all. Their chosen weapon was found in the very same merger agreement that had brought Crown into the company back in 1959.

The boardroom maneuvering requires a little in the way of background explanation. The terms of the 1959 alliance granted Crown and various members of his family slightly more than 1.7 million shares, representing roughly 18 percent of General Dynamics' issued capital. Under normal circumstances that would have been more than enough to guarantee control of the company. In this case, however, the holdings were subject to a number of important qualifications. The first was that Crown's stake consisted almost entirely of convertible preference stock. This meant that Crown received an assured annual dividend equal to 5 percent of his shares' market value in return for direct control of only three seats on the board. Had the holding been acquired on the open market, a stake of that size would probably have guaranteed effective control.

Since the agreement also guaranteed the Crown interests' right to convert the preferred shares to common stock at a specified date in the future, this might not have been a problem if General Dynamics' stock had continued to rise through the years following the merger. If, for example, the market price for the preferred shares had been higher than the agreed conversion price, Crown could simply have exchanged one block of stock for the other while simultaneously pocketing the difference as an easy and substantial profit. Late in 1964, however, the agreed conversion price was still far in excess of the actual prices being paid on Wall Street. Lewis was quick to recognize the disparity as his window of opportunity.

While Crown was entitled to certain advantages, General Dynamics itself reserved the right to "call" the outstanding preference shares at two specified dates, the first in May 1965, and the second twelve months later. This enabled Lewis and his chief ally, financial vice-president John A. Sargent, to confront Crown with a choice between paying somewhere in the vicinity of $24 million to convert his preferred stock to common or allowing himself to be bought out at the current market value.

There was a certain logic behind the move, although as Crown attempted to point out in several counterproposals, there appeared to be various better, more financially sound ways of going about it. The chief motive, as it was later explained to shareholders, was that the preferred dividends of $2.90 per share could not be written off as a deduction before taxes, whereas interest on the $104 million that would have to be borrowed to finance the redemption of the preference stock could be subtracted from gross profits. By Sargent's own calculations, the financial advantages would amount to additional earnings of roughly 22 cents per share.

Crown's proffered alternatives included a promise to reduce his dividends by half for a period of three years, as well as an understanding that he would pay the commitment fees due to the lending institutions that had agreed to make the money available for the buy-out. As he explained it, this would have made it unnecessary to add a further burden of debt to the company's books, while simultaneously reducing the obligatory outflow of cash dividends specified in the preferred stocks' purchase agreements.

His opponents were unmoved, and in May 1965, Lewis and Sargent made their move; curiously enough, it was not the *coup de main* most observers were expecting. Instead of buying Crown's holdings in their entirety, the board voted to redeem only one half of the preferred stock, reserving the right to complete the eviction the following year. And that was exactly what happened. On March 14, 1966, Crown and his two nominees were forced to sit in silence while the board members entitled to vote on the question pushed them firmly out into the cold. Those who were there when the vote was taken maintain that the Colonel shrank visibly in his chair before regaining his composure and stepping silently away from the table. Then, still without saying a word, he walked out the door, flanked by his two equally mute allies, Pat Hoy and Chicago attorney Albert Jenner.

Despite what the press may have thought, the elderly gentleman had no intention of taking his $120 million and retiring peacefully to Chicago. Back home but far from idle, Crown was soon taking the first, small steps toward reclaiming his lost prize. Less than one month after the meeting, he bought a small parcel of General Dynamics stock and informed his brokers to be on the lookout for any other handy-sized blocks that might become available. Meanwhile, the board did just about everything it could to make his job that much easier.

Having risen to a high of $73 in 1967, the stock began a steady decline that saw it dip to a low of $23 in the last quarter of 1969. And at every downward notch on the chart, Crown was waiting to snap up further parcels of stock. The F-111 played some part in the fall, although not as much as might have been expected, since each successive setback saw the contract renegotiated, with constant provision for a profit of some sort or another. The real reason, just as the Colonel had predicted, was the growing turmoil at Quincy, where the fourteen-ship contract with the Navy had degenerated into an organizational disaster to rival the worst follies of the 1950s.

The decision to take on the work reflected the new management's gung ho enthusiasm rather than any realistic appreciation of the yard's capabilities. Since there was no room to build all the ships at the same time, the program's success was dependent on a degree of timing, efficiency, and good management seldom seen outside of wartime. When the first ships fell behind schedule, so too did all those waiting to enter the slipways behind them. And to make matters worse, Lewis and his fellow directors had agreed to deliver the vessels at a fixed price, which obliged the yard to absorb the full cost of its mistakes. Try as they might, the yard's executives simply could not extricate themselves from an increasingly harrowing situation. At one point, in an effort to catch up on lost time, design work was let to subcontractors, who proved no more successful than Quincy's own best efforts. Reflecting Quincy's $240 million drain on General Dynamics' working capital, the share plunged through 1968, losing some 60 percent on the big board price between January and December.

Crown himself was not averse to doing whatever he could to hasten General Dynamics' decline. Stripped of Material Services, the family company into which he had poured the best years of his life, the Colonel went home to Chicago and formed an alliance with one of MSC's chief competitors, Vulcan Materials, in which he established a five-million-dollar stake. Nor did he fail to prosecute his campaign with psychological weapons. Even though he no longer owned the building, Crown had maintained his chief office in the Chicago headquarters of MSC, where he regularly encountered former employees in the elevators and corridors. "Tell Roger," he would say, "that I have just bought another 250,000 shares."

Crown's needling was made all the more effective by the uncertainties surrounding his ultimate intentions. Was he, as some assumed, buying the stock simply because its depressed market value made for a good investment or because he was out to recapture the family jewels and avenge the insults of 1966? That question was not resolved until late in 1969, when he teamed up with an old friend, Nathan Cummings of Consolidated Foods. The pair were enjoying a meal in a Paris hotel when Crown raised the possibility of a joint attack. Cummings would later explain that while he knew little of the defense industry and even less about General Dynamics, "my confidence is in Henry Crown." By the time the pair were back in the United States, Cummings had agreed to put up as much as $15 million of his own money to help his old pal make the final push. And in December of that year, the pieces all fell into place.

On New Year's Eve 1969, Crown picked up the telephone and called Roger Lewis in New York to relay some news that must have thrown a heavy pall of gloom over the evening's festivities. Crown's chief broker and purchasing agent, Lazard Frères of New York, had just bought a further 70,000 shares of General Dynamics stock. It was far from the largest purchase Lazard's chief, André Meyer, had made on his behalf, but from a practical and emotional point of view, it was by far the most important. On that same day ten years before, Crown had ratified the original merger agreement with Frank Pace. Now, thanks to his latest acquisition, he owned a full 10 percent and was eager to do some serious negotiating. If the unexpected call caught Lewis and his fellow directors off guard, there was worse to come. Early in the new year, André Meyer obtained a further 526,400 shares from the Dreyfus Fund, to bring the Crown interests' total holdings up to roughly 18 percent.

Lewis and his people had been outflanked, outthought, and, quite frankly, outclassed. Short of prompting an expensive and divisive proxy fight for outright control, they swallowed their pride and agreed to sit down at the negotiating table to work out a deal. The compromise, if that word can be applied to the one-sided settlement that ensued, delivered just about everything for which Crown could have hoped. Stanley de J. Osborne, a Lazard partner who would soon join General Dynamics' board under Crown's auspices, thrashed out the settlement with the same Roswell Gilpatric who had negotiated the original merger a decade before. Once again, Crown became chairman of the executive committee, now called the executive steering committee in deference to Lewis' insistence that he would not be saddled with another clique of second-guessing overseers intent on meddling with his decisions. The change of name was, as most observers realized, largely cosmetic. The chief question was not what Lewis would be allowed to do, but the length of time he was likely to remain in office. Crown had already sent John Sargent, the other leader of the 1966 coup, on his way and few doubted that Lewis would be far behind.

The crunch came in October, after months of ever more rancorous internal friction which began when Crown demanded that Lewis surrender his position as president. Lewis agreed, but only on the condition that he retain the duties and title of chief executive officer. The mooted recruit, Semon (Bunkie) Knudsen, refused to consider the job under those terms. A former vice-president at General Motors and president of Ford until Henry II forced him out late in 1969, Knudsen was not prepared to step into the no-man's-land between Crown and Lewis. The last few months at Ford had been bad enough, but the situation at General Dynamics would have been even worse. Crown held five seats on the board and control of the executive committee, and was busy taking potshots at Lewis whenever opportunities presented themselves. "If the company goes well, Roger will get the credit," he said in June, adding ominously, ". . . if it doesn't do well, he'll get the credit for that too." *Business Week* described the atmosphere in the boardroom as "incredibly strained" and went on to list Crown's latest criticisms. He should be more open to advice, Crown asserted. He should redirect the company's resources away from military work and find fresh, productive enterprises in the civilian marketplace.

This last charge was both hackneyed and, judging by the events of the next decade, extremely ingenuous. General Dynamics officials have been bemoaning the company's heavy dependence on the military ever since World War II, but none has done less than Henry Crown to loosen the close ties with the Pentagon. No less than five major civilian divisions —Datagraphix, Canadair, Stromberg-Carlson, American Telecommunications, and General Dynamics Communications—have been sold since he resumed control, while the company has acquired two tank factories and steered the Quincy shipyards onto what ultimately proved a fatal dependence on the Navy. Roger Lewis' successor, David Lewis,* explained the sale of the telecommunications divisions in his 1982 letter to stockholders: "Frequent major technological changes in computer-based telephone systems required such a high level of research and development expenses that these operations could be profitable only in the best of times." Defense work, one gathers, would seem to be warmly profitable regardless of the economic clime.

Given the prevailing mood, it was almost inevitable that Roger Lewis would fail to see out the full year Crown had promised him. In October 1970, to no one's particular surprise, Lewis announced that he was leaving. Of much greater interest was the identity of his replacement. Once again, Crown's actions were at variance with his words. During the negotiations with Bunkie Knudsen, the Colonel had insisted that he wanted a man with a background in civilian manufacturing rather than another veteran of the military contracting game. David S. Lewis, a former president of McDonnell Douglas, hardly matched the description. It was true that McDonnell Douglas built civilian airliners, but the backbone of its success over the previous twenty years had always been jet fighters like the F-4 Phantom and, later, the F-15 Eagle.

White-haired, soberly tailored, and possessing a disarmingly gracious Carolinian accent, Lewis conveyed the style and easy charm of an avuncular dean from some minor southern college. General Dynamics had acquired a face not even a mother could love. Was Lewis capable of restructuring its much battered features into something resembling a pleasantly profitable countenance? Some sort of tonic was needed if the company hoped to regain its former luster. Unknown to Lewis and Crown, salvation was almost at hand. Its name was the F-16 jet fighter.

*No relation.

11. Too Good to Survive

When the F-111 program began to go astray in the early sixties, Fort Worth found itself holding a tiger by the tail. So many things needed so much remedial attention, the company had no choice but to build on an already swollen bureaucracy, taking on literally thousands of aeronautical and electronic engineers, draftsmen, computer analysts, and cost estimators, who often spent years shaping and reshaping just one or two small parts of the whole design. The burgeoning white-collar work force was not a financial burden since the contract guaranteed the reimbursement of all fair expenses plus a percentage profit. It was a major liability in other, less obvious ways. A former Grumman engineer who paid regular visits to Fort Worth during the mid-sixties still describes his first impression of the design and engineering department in terms tinged with awe: "It was the most awful, appalling, depressing, and infuriating building in the entire airplane business," he said.

The main assembly hangar was itself more than one mile long, and the Bullpen, as the design floor was known to its inmates, inevitably struck visitors as having been put together with the same truly Texan sense of proportion. "An endless vista of desk after desk after desk stretching off into infinity," was the way the man from Grumman described it. Split into their own individual working groups, each vying for money, talent, and greater preeminence in the design process, the Bullpen's inhabitants toiled in virtual isolation from all but the few of their colleagues who were working on the same problem. "We had layers of managers at Grumman, but I've still never seen anything like the situation at Fort Worth," he said. "The boxes in the organizational chart were translated into actual concrete walls. You absolutely could not talk engineer to engineer across a box. You had to go to some unbelievably high-ranking boss, who would talk to his equivalent, and then the word would come down."

The tenor of his comments was confirmed by a former Air Force project officer who now serves as an independent consultant within the aerospace industry. "When you are trying to get the landing gear to fit behind a door and the guy who is making the door can't talk to the guy who is working on the wheel, that is when you start to mess up the design," he said. If Fort Worth had been building cars or washing machines or even civilian jetliners, the situation would never have been allowed to develop; appliance retailers and commercial airlines do not order their wares through open-ended, "cost plus" contracts. But with a committed customer like McNamara's Defense Department not only paying for the constant redesign work but defending the additional expenditure at every turn, the pressure to let an already overblown managerial system run totally out of control became irresistible.

The consequences of this organizational chaos became apparent late in 1968, when General Dynamics joined just about every other major U.S. aerospace firm in bidding for the right to build the F-15 Eagle, today the premier Air Force fighter and by far the most expensive. By rights, Fort Worth should have been prominent among the front-runners. Crammed with the most sophisticated electronics and featuring the latest aerodynamic touches, the F-15 required a contractor with the human and physical resources to overcome a host of formidable technical barriers. Thanks to its hands-on experience with the equally complex F-111 and the Bullpen's vast reservoir of underemployed talent, General Dynamics was superbly placed to steal an early march on the competition. Much to the dismay of Roger Lewis and his board, it did not happen that way. Instead of dominating the field, the submitted proposal was, in the words of an Air Force analyst attached to the evaluation panel, "a bit of a joke." In any aircraft, particularly one displaying the F-15's leaning toward the baroque, performance is no more than the sum of the multitudinous systems and subsystems. General Dynamics' problem, one of them, was that while some elements of its design were laudable, the sum of the proposed plane's parts simply did not add up to a satisfactory whole. Once again, the Bullpen bore full responsibility.

The defeat could not have come at a worse time. Henry Crown was out there somewhere, stalking Lewis and the other usurpers with the indefatigable patience of a man waiting for his quarry to make its last, fatal mistake. And the mistakes—General Dynamics', McNamara's, and many others'—were coming home to roost. Safely in office, President Nixon had just reversed his position on the F-111 for a second time and allowed Defense Secretary Melvin Laird to eliminate the FB-111 altogether. Quincy was floundering and Electric Boat was its usual troubled self. If the Eagle had been landed, the board's defensive position would have been somewhat better; it would certainly have buoyed the company's flagging performance on Wall Street and calmed Fort Worth's

24,500 anxious workers. Instead, the year ended with Crown's triumphant telephone call and the disappointment of defeat on all fronts. The question now was what the Chicago millionaire proposed to do with such an uninspiring prize.

Like any good soldier, the Colonel first brought up reinforcements to secure his position. Along with Lazard Frères' Stanley de J. Osborne and Albert Jenner, he seated Nate Cummings on the board and installed him at his right hand in meetings of the executive committee. Robert V. Hansberger, the president of Boise Cascade, also joined the board, as did Crown's devoted financial adviser from Chicago, Milton Falkoff. In October, the consolidation became more or less complete when David Lewis arrived from McDonnell Douglas, an event that prompted Crown to remark, "Now we are finally getting somewhere."

The Colonel may have believed General Dynamics was coming back to life, but few others were inclined to share his opinion. The financial press agreed that the new board represented "a powerhouse team," but they faced the same problems that had humbled their predecessors. The company was caught between the proverbial rock and the hard place of its own mismanagement. Programs like the F-111 and the Atlas-Centaur launch vehicles, which were making money, faced the prospect of cancellation in the near future, while more permanent endeavors, like Groton's nuclear submarine operations, were blighted by an endemic inability to pay their own way. Not since Canadair stopped building its profitable and politically untroubled version of the Sabre jet had the company enjoyed the advantages of a firm financial base from which to proceed into the future. Somehow, from this assortment of weak reeds and lame ducks, Lewis was expected to bring forth a new and hopefully more prosperous era.

His first major action took most observers by surprise. On February 10, 1971, just six weeks after he officially replaced Roger Lewis as the company's president and chief executive officer, David Lewis announced that General Dynamics would soon be moving its headquarters from New York to Saint Louis. The stated explanations were adequate, sensible arguments of the advantages to be enjoyed in the city that bills itself as the Gateway to the West. "We have chosen a location where our headquarters personnel will have quick and easy access to our major operating locations," Lewis said, allowing a company flack to add that it was difficult to get "good people" to move to New York because of "housing and commuter difficulties."

If Crown's return marked the end of a chapter, the journey to Saint Louis was the frontispiece for a whole new book. Like Electric Boat before it, General Dynamics had always been prominent among New York's leading citizens. John Holland's *Fenian Ram* had been commissioned amid the boozy atmosphere of fantastic conspiracy that characterized the Irish bars in the area around City Hall and the city's citizens had turned out in their tens of thousands to witness his next boat's excursions beneath New York Harbor. Several generations later, their descendants hung from the windows over Broadway to cheer Admiral Hyman Rickover and the crew of the *Nautilus* after the boat's astonishing journey beneath the North Pole. Now the ties had been cut, and while the explanations sufficed, they did not suppress the financial world's tendency to speculate about the deeper motives behind such a radical break with tradition. Lewis' preference for Saint Louis was a matter of personal taste. He had lived in the city for more than a decade, first as president of the McDonnell Aircraft Corporation and later, as the $194,000-a-year favorite son of James (Mr. Mac) McDonnell, on the board of the new company brought into being by the 1967 union with Douglas of California. So fond was General Dynamics' new chief of his hometown, it was even rumored he had insisted on the change of address before agreeing to come aboard. If that was the case, Henry Crown may well have had his own reasons for agreeing. After his experiences during the mid-sixties, the Colonel had every right to conclude that New York was a breeding ground for boardroom mischief. Saint Louis, by comparison, was virgin territory, far removed from the old-boy networks of the East Coast.

The larger problems remained unaffected by the change of locale and the new chief executive officer's first small reforms. At McDonnell Douglas, Lewis had enjoyed a reputation as a shirtsleeve administrator, descending when the need arose to the shop floor for eyeball-to-eyeball conferences with middle-level managers and foremen. He was soon putting the same methods to work in San Diego, where he untangled the snarls that had caused so many problems with the DC-10 subcontracts, a job for which he was uniquely qualified since the jetliner program had been one of his chief responsibilities during his last years at McDonnell Douglas. With California more or less under control, he turned his attention to Massachusetts, where the Quincy shipyard was sinking ever deeper into an ocean of red ink. To someone with a reputation as an energetic, hands-on manager, Quincy existed as an affront to common sense and good business practice. Like Fort Worth's Bullpen, it was ensnared in a brier patch of petty internal politics and institutionalized inefficiency. The solution demanded a strong man, and in 1972 Lewis went north to Canada to woo the one person in the North American shipbuilding industry deemed capable of sorting out such a monumental mess. The potential recruit, P. Takis Veliotis, agreed to join the team, accepting Lewis' assurances that any man who could clean out Quincy's Augean stables was assured of a brilliant future within the company.

The moves confirmed Colonel Crown's faith in the new man he had chosen to lead General Dynamics, but still, as the stock market was quick to conclude, the reforms amounted to little more than an outbreak of housecleaning that was long overdue. Needed most of all were a few new contracts, and large ones at that. As it happened, that need was well on the way to being met. The pivotal event, or so the story goes, occurred not in New York, Chicago, or Saint Louis, but at a well-lubricated Washington cocktail party.

In January 1970, Air Force Colonel Everest Riccioni was assigned to the Pentagon for one of the deskbound tours of duty that are a feature of modern life in the U.S. military. His new job investigating new construction and design technologies was not part of the procurement mainstream but an insignificant backwater with a small staff, a low budget, and little "pull." Riccioni was not, however, a typical ticket-puncher. The story is told that during a later tour of duty in charge of Hawaii's air defenses, the colonel told his pilots that when all else failed, they were to ram incoming Russian bombers. The tactic was mentioned nowhere in the various air combat handbooks, but it was typical of the Riccioni approach, a mixture of pugnacious determination and a willingness to examine problems from the other side of orthodoxy.

As a fighter pilot with experience in test programs, Riccioni had taken on his new job with a distinctly different view of the qualities that constitute a winning combat aircraft. It should be small, highly reliable, and above all, cheap enough to permit the Air Force to buy and operate in large numbers. Part of a murmuring undercurrent of dissent against the prevailing high-tech philosophy that dominated the Air Force's budgetary and strategic planning, Riccioni's logic was simplicity itself: Expensive warplanes can be purchased only in small quantities—in the case of the Navy's F-14 Tomcats, no more than thirty a year—and even then they are apt to spend inordinate amounts of time on the ground, since their very complexity makes them liable to frequent breakdowns. "The number of weapons systems that can be employed in battle," he maintained, "varies inversely to the square of the level of sophistication." In other words, planes that break down don't count.

Riccioni's thinking had been much influenced by another, perhaps most justly celebrated, of the original cheap hawks, Pierre Sprey. Looking at the question of fighter plane design from a slightly different perspective, Sprey had done some independent calculations in his capacity as a systems analyst with the Office of the Secretary of Defense. Like Riccioni, he valued economy and the advantage of overwhelming numbers, but he took the theory one step further. Statistical analysis of air combat figures from World War II and Korea indicated that the much-vaunted avionics systems in planes like the F-111 and the F-15 were more trouble than they were worth, since they beamed out a signal that advertised their presence to enemies far and wide. It would be better, he reasoned, to limit onboard radar equipment to relatively simple reception systems, passive warning devices that emitted no signals of their own but identified and "vectored" those of the opposition. If such a plane was also small, highly maneuverable, and built solely for air-to-air combat, it could then sneak up on its indiscreet prey, swooping down to take the victim by surprise. In Vietnam, no less than in the skies over Flanders and Piccardy, the pilots who survived longest were those who fired the first shot.

"It is curious that a very high quality fighter can be built cheaply and easily if it is done correctly. Yet despite its great utility, no really superior fighter aircraft has been produced in the last 15 years by any nation," Sprey asserted in a paper published by *Interavia* magazine. "It is a remarkable result that the basic aspects or characteristics that make a superb fighter aircraft are not only *not* contradictory, but they all emanate from a germane central trait or philosophy of design. It is primarily because this trait is denigrated by most military services and designers that truly outstanding designs have been lacking in recent years."

It is not difficult to understand why. The United States Air Force had hung its future, and its budgetary aspirations, on the F-15 fighter and the B-1 bomber. Each supported a monumental pyramid of project officers, political liaison squads, and design departments, and neither conformed to the laws of combat economics being advanced by Riccioni and Sprey. The B-1, for example, looks, and apparently behaves, like a direct descendant of the F-111, which it resembles in a sort of gawky, overblown way. The wings fold back, the radar follows the contours of the land, and it comes complete with the usual promise of astonishing speeds at any altitude. It also makes the same large craters in the desert when something goes wrong. In 1984, the prototype B-1 went down because the crew neglected to transfer fuel from one tank to the next, upsetting the plane's trim. It might have happened to any plane, but with the B-1 there was little time and less opportunity to compensate for negligence and bad luck. With one of the world's highest wing loadings—the ratio between the square surface of the wing and the load it has to support—the B-1 is both more likely to get into trouble and less capable of getting out of it. The last parallel with the F-111 is the most obvious and the most distressing. Just as Kennedy, Johnson, and Nixon all discovered the F-111's strategic value in their bids for the White House, so did the campaign trail lead President Reagan to lavish praise on an equally dubious weapon that has the added disadvantage of being twenty years out of date. The bomber was "one of the western world's principal defenses against tyranny," this latest President maintained at a roll-out ceremony in the Rockwell plant. Not coincidentally, the long-delayed contract helped the

California company to weather a financial crisis similar to the one that afflicted Fort Worth in 1961.

While Riccioni and Sprey might have made some pithy remarks about the B-1, the chief object of their enmity was the F-15. Now, the Eagle is a remarkable warplane, as indeed it should be at something like $40 million a copy. Visually, it is striking. The wings meld seamlessly into the body, while the fuselage is surmounted by two rearing tail fins that sandwich the exhaust nozzles for the twin afterburning fan-jet engines. And external appearances are but the start of the plane's many marvels.

Where most warplanes have one "active" radar to seek out enemies, the F-15 has three, each broadcasting its own powerful signal at a different frequency. In addition to twenty-eight minor computers distributed throughout the fuselage and wings, there is a master unit with the equivalent of a 24,000-word memory that supposedly sharpens radar receptions, tracks three groups of intruders simultaneously, and guides the various short-, long-, and medium-range missiles to their targets. There is even a separate, defensive radar that can identify incoming surface-to-air missiles, analyze their radio frequencies to determine if they are SAM series one, two, three, or four, and locate the firing site on a relief map projected onto one of the cockpit display screens through the auspices of one more computerized radar system. That the F-15 manages to be a creditable air-to-air performer despite a gross takeoff weight of close to 70,000 pounds is more a comment on the equally debased standards of other U.S. and Russian planes rather than a tribute to the Eagle's designers and engineers.

Sprey rejected the quest for electronic infallibility represented by the F-15, writing that "no technological breakthrough or exotic material development is needed." His list of qualities that constitute a superb air-superiority fighter began with the weaponry. Long-range missiles, so-called standoff systems, had no place on a combat aircraft since they relied upon sophisticated, inbuilt "active" guidance systems that are themselves open to electronic countermeasures. "The lightest, cheapest, most effective, easiest to use, least refutable ordnance is the aerial cannon," he declared, adding that the most lethal missile also happened to be the cheapest, the heat-seeking AIM-9 Sidewinder. And if the weapons were to be used to the best advantage, they had to be mated with an airframe, engine, and fire control system designed for the exigencies of air-to-air combat rather than to satisfy the bureaucratic imperatives of a procurement system dealing in theoretical abstractions. Most combat aircraft were not only large, he asserted, but the wrong color. Painted in earthy browns, greens, and blacks to disguise them on the ground, they became visual beacons in the azure environment at ten thousand feet. Sprey's solution: Paint them pale blue.

He also laid the ax of statistical analysis to other cherished nostrums. "Another example of the tendency to avoid simplicity is the popular belief that twin-engined aircraft are more survivable than those with one engine, this despite the fact that in-depth analysis shows no such correlation in peace or war," he wrote. Further, prior experience indicated that it was not merely counterproductive to design a plane for more than one specific mission; it was disastrous, since the resulting hybrid could do nothing well. The first F-104 Starfighters, for example, entered service as "pure" interceptors designed to climb rapidly to a great height, shoot down Russian bombers, and return directly to base. The much-publicized problems that led to later Starfighters being labeled "flying coffins" came about only when the aircraft was extensively revamped to make it more attractive to the European nations then casting about for a counter to both the Soviet Air Force and the Red Army. Hauled down from the upper atmosphere into the turbulent murk of the European troposphere, laden with tons of additional equipment, and redesignated a dual fighter–ground attack plane, the resulting monster required extraordinarily gifted pilots to compensate for the aircraft's innumerable acquired deficiencies. "It is a curious result that a fighter aircraft which is an excellent performer in the air battle inevitably duplicates this fine performance as an air-to-ground aircraft," Sprey wrote. "The reverse has never been true."

Sprey's treatise also benefited from the practical experience of a third Pentagon dissident, veteran fighter pilot John Boyd. Though there were many among the F-15's legion of protectors and supporters who disagreed, Sprey argued that the F-86 Sabre jets which Boyd had flown over Korea were the last and by far the best examples of a practical effective air-to-air fighter. In addition to all of Sprey's other criteria, the Sabre was a true pilot's aircraft. Visibility from the large canopied cockpit approached 360 degrees, the power-to-weight ratio was high, maneuverability was excellent thanks to the large, low-loaded wings, while the on-board fuel supply was sufficient to permit reasonably lengthy patrols and still leave enough for the demands of dogfighting. The next U.S. fighter project, the three members of the so-called Fighter Mafia agreed, should exalt the principles that had made the ancient Sabres so effective. Riccioni, who at that stage had the distinction of having written the only new formal tactical doctrine to be accepted by the Air Force in fifteen years, began to nudge the concept of a cheap, lightweight fighter out of the realm of inspired conjecture and into the mainstream of the procurement process.

But first he needed something that would prod his superiors into making an initial commitment to what by now had been dubbed the Lightweight Fighter Program. He found what he was looking for among the small talk at a Washington social gathering attended by the usual assortment of military types, industry officials, and Defense Department public servants. Moved to candor by the convivial nature of the occasion, a senior naval aviator gloatingly informed Riccioni that the Air Force would soon have another fighter with a Navy heritage to park beside the F-4 Phantoms it had been obliged to accept more than a decade earlier. Riccioni confirmed the substance of the boast and began to sound a warning klaxon in the Air Force wing of the Pentagon. Reproduced in part by James Fallows in his book *National Defense,* his paper on the merits of a lightweight, pocket-sized air superiority fighter concluded with what proved to be the most telling argument of them all:

> Unless the U.S. Air Force thoroughly studies high performance fighters and is prepared to consider them as a necessary complement to other air superiority aircraft, the U.S. may be:
>
> A/ Outgamed by the Navy (again)
>
> and/or
>
> B/ Outfought by the Russians.

It was the mention of the Navy that did the trick. Despite the misgivings of the F-15 establishment, the paltry sum of $100,000 was made available for competitive design studies by two manufacturers, General Dynamics and Northrop. This was a radically different approach from the one that produced the F-111, a plane that cost the taxpayers more than $25 million in contractor reimbursements by the time the initial design and selection process had come to an end. This time, the only hard and fast rule was summed up by a simple but elusive equation: Performance in combat is the properly tuned relationship between a high power-to-weight ratio, a low wing loading, the correct wing sweep, and the smallest exterior surface. "They'd go through their design studies suggesting configurations and then we'd go through them again, saying things like, 'This is all right but you still need to do more work on this other bit over here,' " Pierre Sprey recalled during an interview in 1985, adding that it was "the only way to make sure we got what we wanted."

Although official Air Force spokesmen were at pains to point out that the project, by now dubbed the FXX, was merely an academic exercise intended to advance the frontiers of aeronautical theory, both of the contracted manufacturers had motives for pouring their best efforts into the work. General Dynamics' was a simple case of desperation. The production run on the F-111 was down to the last forty or so planes, with nothing else in sight. While the FXX was not, and had never been, proposed as a pilot design program for a major construction contract to be awarded at some later date, it represented at least a small repository for hope. Northrop's ambitions were more inclined to the grandiose. Convinced that Sprey, Boyd, and Riccioni really wanted an upscale, updated rehash of the design philosophy enshrined in its own aging F-5, the low-cost utility warplane often known as the Third World Fighter, Northrop recognized an opening that might well lead to thousands upon thousands of orders, not just within the United States but from all over the globe.

There were only two obstacles: the Air Force hierarchy and its passionate commitment to the F-15. One need not have been a class leader at the Air Force Academy to see the FXX as a broad indictment of all the Eagle's allegedly superior virtues. Here the situation was saved by one of the most beneficent strokes of good fortune ever to descend upon General Dynamics. The Nixon administration's new assistant defense secretary, David Packard, came to office with a staunch commitment to competitive prototyping by rival manufacturers that soon saw him put up a pot of $200 million, which he promised in part or whole to any service that could come up with fully operational weapons systems capable of being pitted against each other before a panel of judges.

The lure of all that cash was too much for the Air Force to resist. A panel representing the service's many branches was quickly convened to examine ongoing development projects and identify those close enough to fruition to qualify. "It wasn't as if they wanted one particular plane; in fact, they most definitely did not want the plane that later became the F-16," one current senior Air Force officer maintains today. "It was just that they could not stomach the thought of all that money going to the Army or the Navy." The panel identified two possibilities, the first for a medium-heavy transport that was soon to be labeled the C-17, and the second, General Dynamics' and Northrop's feasibility studies for the lightweight fighter. Packard's enthusiasm for pitting manufacturers against each other remains one of the most curious features of his tenure. He was certainly no hard-driving taskmaster in the McNamara mold. As Arnold Kanter explains in *Defense Politics,* the deputy secretary even sanctified one of the Pentagon's most divisive yardsticks by insisting that each service's share of the annual budget be included as a standard fixture in the briefing papers sent on to the White House and Congress. "We were just lucky that we were in a position to go," Pierre Sprey recalls. "If the Army had had its act together, the money might just as easily have been spent on a new tank."

Packard authorized a formal Request for Proposal (RFP) that soon drew responses from four of the industry's biggest players—Northrop, General Dynamics, Boeing, and Grumman—and one outclassed but plucky middleweight, Fairchild. The companies were given the technical essence of the original work done in the preliminary feasibility and design studies and told to put their own stamp on the basic physics. Once again, there were no long lists of performance criteria, and not the slightest suggestion that the contestants should prepare their designs in anticipation that the project would later be expanded to include missions other than air superiority. It was to be the best fighter in the world and that was enough.

The entries' evaluation proved a second deviation from the precedent set by the F-111. This time, the verdict was unanimous. General Dynamics' plans were by far the best. The plane promised a small, acrobatic airframe wrapped around a single, 24,000-pound-thrust engine large enough to send it rocketing vertically into the clouds. There were some nice design touches, as well. The body, for example, utilizing the same aerodynamic advances pioneered by the F-15, effectively turned the upper sections of the rear fuselage into an extension of the wings, a trick that greatly increased the plane's total "lift." If General Dynamics was the clear winner, Northrop's position as runner-up and second qualifier for the $25 million available to build two fully armed flying prototypes was not the subject of the same universal accord. While Northrop did indeed have its supporters, there were almost as many who believed Boeing should have been named as the alternative. That choice was never made. In one of the few appearances of an overtly political motive in the early history of the lightweight fighter, Northrop was given the decision, for reasons closely associated with its choice of General Electric engines, a company whose aviation division was then in need of a shot in the arm.

Throughout the course of these preliminaries, the official Air Force attitude was one of thinly disguised repugnance. It was quite all right to take Packard's money but another matter entirely to let the experiment produce anything of consequence. If these lightweight fighters had any future at all, the F-15's partisans informed anyone silly enough to ask, it was to be found among the other abandoned prototypes on display at the Wright-Patterson Air Force Museum.

Despite semiofficial assurances that neither contestant's entry would be allowed to reach maturity, David Lewis still pressed Fort Worth to take a long-overdue look at its operations. In fact, there was really no other option. The YF-16, the designation recently bestowed on General Dynamics' still-to-be-built contender, might be cursed by official disdain, but at least it was something. As General Dynamics had learned in the past, hope reigns supreme whenever a major contract is in the offing. On this occasion, however, there was no chance that the godlike hand of some latter-day McNamara would descend propitiously from the blue to pluck Fort Worth from the jaws of its latest predicament. The remedy for its ills had to be found within, and there seemed only one way to achieve the essential, hasty reforms: Imitate the methods of a notoriously successful rival.

During the late fifties and early sixties, Lockheed produced two of the most unusual and successful aircraft ever to fly, the U-2 spyplane and its even more radical successor, the SR-71 Blackbird. In each case, a small, tight-knit group of designers and engineers were plucked from their desks and sequestered in a secrecy-shrouded complex that came to be known as the Skunk Works, after the moonshiners' hideaway in Al Capp's comic strip *Li'l Abner.* The approach was not Lockheed's alone; France's Mirages were conceived and developed in isolation from the rest of the Dassault works, just as had once been the case in America with the World War II era P-51 Mustang. It was just that the Skunk Works and its legendary taskmaster, Kelly Johnson, were by far the best at the game. To the mixed astonishment and amusement of the rest of the aerospace industry, Fort Worth took note of the example and produced its own version of the Skunk Works.

"They actually built an entire new building and put a big fence around it to keep out the contamination," one former Air Force liaison officer at Fort Worth reported. "Then they grabbed two hundred or so people from the Bullpen, put them behind the wire, and told them that they could talk all they liked with each other, but they couldn't, under any circumstances, discuss anything with their old friends. It was easier to get into the place off the street than it was from the rest of the plant."

The plan worked spectacularly well. Unlike the F-111, whose myriad systems and subsystems had to be designed, checked, amended, and rechecked through successive layers of Defense Department bureaucrats, the F-16 proceeded from a paper theory to a flying test vehicle in a series of smooth, unencumbered steps. In February 1974, the first of the two prototypes was wheeled out onto a runway at Edwards Air Force Base for what was supposed to be a simple preliminary check of the landing systems and pilot controls. Everything went as planned for the first half minute or so until, about halfway down the runway, the plane was gripped by a series of violent, rocking convulsions that pitched and rolled it like a toy in the hands of an invisible giant. To those hunkered down behind the ground monitors, it looked like a disaster in the making. Test pilot Phil Oestricher had no time to wait for approval. Sensing that it would be more dangerous to back off than to proceed, he applied a little more power and pulled back on the short, stocky control stick nestled by his right knee. The oscillations stopped immediately as the plane rose serenely into the air. At first there was silence from the ground teams, but that quickly gave way to a rowdy chorus of whoops and cheers as Oestricher completed a low, smooth circuit of the field and brought his mount back for an uneventful landing. As Oestricher explained it, this new plane far exceeded even the most optimistic predictions. The pitching and yawing that had given the ground crew such a start began at a speed of only 125 knots, at least five knots below the takeoff threshold forecast by the computer projections. An enduring element of the F-16 had been born. The airplane not only handled and flew with consummate grace; it possessed a natural urge to leave the ground that could not be denied.

The maiden flight came as a massive, if anticipated, confidence booster. It had taken less than two years to put all the pieces together, and the results exceeded all expectations. Yet even though the test had been impressive, the technical achievement was but a small part of the victory. After the F-111's ten-year trail of shattered prospects, the realization that Fort Worth could actually do something right came as an exhilarating and encouraging revelation.

The boost to morale could not have been more timely, because Northrop was pressing confidently ahead with its own entry, the YF-17 Cobra. General Dynamics' California rival believed it possessed several distinct advantages, few of them aerodynamic. In one form or another, the idea for a new plane had been kicking around Northrop's project offices since 1965, when preliminary plans were drawn up for something called the P-503. The Cobra, much altered by time and Riccioni's emphasis on simplicity, was scarcely recognizable as a lateral descendant of that stillborn fighter, but its heritage, veiled though it was, remained of vital importance. With five full years of lead time, Northrop's engineers found themselves unable to accept the Fighter Mafia's emphasis on austerity. They had spent too long manipulating the design to pare back the layers of sophisticated extras, and as of late 1970, there seemed little reason to do so. After shepherding the lightweight fighter through the initial design phase and seeing his brainchild become the subject of a newly announced competition, Everest Riccioni had been told to clear out his Pentagon desk and ship himself off to Korea. It was all the encouragement Northrop needed. Convinced that what the Air Force really wanted was a *complex* simple plane, its engineers began to turn their entry into a pint-sized version of the F-15. The two warplanes even bore a superficial similarity, both featuring twin engines and futuristic dual tail fins. Unfortunately, that was where the comparison stopped, for while the Cobra replicated many of the F-15's limitations, it emulated but a few of the larger plane's better qualities.

Northrop would not be called upon to pay for its sins until later that year, when the F-16 trounced the Cobra in a series of realistic one-on-one dogfights at Edwards Air Force Base. In the meantime, Northrop's attitude was one of aggressive confidence that verged on the cocksure. Salesmen were already scouting Europe, singing the plane's praises and spreading the word among the company's well-placed and discreetly paid network of friends that it would be only a matter of time before the U.S. Air Force placed its first orders. There was much more to the promotion campaign than the hope of a few additional sales. Back at the start of the 1960s, Northrop had watched from the sidelines as Lockheed cornered one European country after another, eventually signing sales and coproduction agreements to supply the three thousand F-104 Starfighters that would constitute the backbone of NATO's aerial defenses for the rest of the decade. The Starfighters, those that had not crashed, were aging rapidly by 1970 and would soon have to be replaced. Northrop believed its Cobra was the plane for the job.

In the wake of the later congressional investigation into international corruption, it is easy to look upon Northrop's high opinion of its chances as a profession of faith in the utility of numbered Swiss bank accounts. It is certainly true that the company parceled out many millions of dollars in kickbacks to French assemblymen, German politicians, and a member of the Dutch royal family, but there were other, no less important sources to its optimism. The first was an entirely understandable supposition that Fort Worth would field yet another turkey. The second and most important reason revolved around a question of Air Force politics. As might have been expected, the Air Force's senior brass wanted to see the Europeans flying F-15s, since this would have increased the overall production run, assured McDonnell Douglas' continued stability, and, ideally if not in practice, lowered the total price for each plane.*

As far as Northrop was concerned, this happy scenario had little chance of coming true, for the simple reason that the F-15 was far too expensive. It was too much to believe that, at roughly $30 million each —the price Japan would agree to pay not long after the Europeans had made their own choice—the plane would find a place in the hearts of any but a few military men blinkered to the fiscal preoccupations of their political masters. U.S. Air Force generals could grumble all they liked about the degradation of NATO's air defenses, but it would not do any good. Once the F-15 had been discounted and the F-16 defeated, Northrop knew it could count on the U.S. government to add its weight to the company's own pitches, both covert and declared. It had to be that way. After their experiences with the Starfighter, many European politicians and union leaders believed the next generation of fighters should be built closer to home, a school of thought being assiduously promoted by France's Dassault and the aerospace division of Sweden's Saab conglomerate. If the United States hoped to maintain its technological

*In fact, the reverse tends to be true. Several GAO studies have pointed out that coproduction with another country increases the price.

hegemony over the NATO nations' air forces, it had no choice but to make the winner of the lightweight-fighter competition its official standard-bearer in this latest battle for Europe. All that was needed was a crushing victory over the F-16, and that, Northrop blithely assumed, was as good as in the bag.

Northrop's grand design crumbled in the fall of 1974, when the contenders met in their series of climactic jousts at Edwards Air Force Base. They strafed cloth targets strung between low poles in desert gullies, dropped mock bombs, and tested each other's mettle in dogfights that never lasted more than a few seconds. "There was no real comparison," Pierre Sprey noted much later. "The Northrop plane was shot down every time, and it usually happened very, very quickly." Designed solely for air combat, the F-16 also proved to be a superior ground attacker, just as the Fighter Mafia had predicted. But it was in the air that the difference was most pronounced. The F-16 climbed faster, pulled tighter turns, and switched from one maneuver to the next without loss of speed or control. "It was unquestionably the best fighter in the world; no doubt about it," Sprey enthused. Others shared the opinion. Among the crowd of U.S. and foreign military officials invited to witness the spectacle, there was an official delegation representing the European members of the NATO alliance. They were there at the invitation of the U.S. government, its first small act of lobbying on behalf of General Dynamics, and they left very much impressed. Out of a total of eight specific performance criteria, the Europeans judged the F-16 to be entirely satisfactory in five. What made the appraisal even sweeter was the fact that the evaluations were made not merely in comparison with Northrop's Cobra but also against Dassault's Mirage F-1E and Saab's Viggen.

Luck, the commodity that had forsaken Fort Worth so often in the past, now settled firmly in its corner. If the Air Force had been free to work its will, the F-16 would have remained no more than a "technology demonstrator" and the Edwards shootout an amusing sideshow that posed no threat to the future or reputation of the F-15. But once again, fortune intervened. Just as David Packard's belief in the value of competetive prototyping had moved the project from theory to flying fact, so now did another of the Pentagon's civilian administrators propel the plane into full production. The Ford administration's defense secretary, James Schlesinger, was an ardent admirer of the austere fighter concept and he spelled out his faith in an official memorandum to Air Force chiefs in February 1974, just one month after the premature maiden flight. "Lower-cost alternatives and/or complements to the first-line air superiority aircraft and weapons systems are needed," he wrote, adding that the Air Force and Navy should "investigate the development of less sophisticated aircraft/weapon systems to complement the F-14A and the F-15." It was a courteous, relatively diplomatic note, which neither challenged the F-15's place in the order of battle nor gave the Air Force's generals any reason to believe their favorite aircraft was in jeopardy. All the same, beneath the references to a lightweight fighter as a possible "complement" to the F-15, Schlesinger's intention was crystal clear. As McNamara had pushed the F-111 as a demonstration of his pet theories, the new man in the job seized upon the F-16 as a means to making his own mark on the Department of Defense. If the Air Force brass hoped to shelve the winner of Packard's shootout, it was clear their stated reasons would have to be very well thought out.

The depth of Schlesinger's commitment revealed itself in the winter of 1974, when the Air Force signaled that it would resist both the F-16 and the A-10, an inelegant, simple plane designed specifically to maul tanks, and another of the defense secretary's favorites. Just as Packard's $200 million worked wonders in 1970, so did Schlesinger's apparently paradoxical inducement. If the Air Force would agree to accept both planes without raising a fuss in Congress, he promised to make an unsavory package more palatable by increasing its size. The generals might have not have liked the F-16, but they were not about to reject the offer of four additional wings of combat aircraft. "Ten bent dimes still make a straight dollar," was the way one minor participant summed up the resulting bargain. The total "buy," originally set at 650 aircraft, soon grew to 1,184 F-16As, the single-seat version, and a further 204 F-16Bs, a two-seater envisioned as a trainer, and later as an advanced firing platform for medium- and long-range standoff missiles. The Air Force's game plan was becoming clear: Refine and "improve" this affront to the system, embellishing it to whatever degree was necessary to restore the F-15's superior edge. In January 1975, the same month that the official result of the Edwards shootout was announced in Washington, the change became official when both General Dynamics and the Air Force stopped referring to the F-16 as an "air superiority fighter" and began calling it a "multi-mission aircraft." The new emphasis was quietly explained as an attempt to make the F-16 more acceptable to the Europeans. Anyone familiar with the history of the Starfighter, however, could only have his doubts.

Now that it had the U.S. contract under lock and key, General Dynamics was happy to go along with whatever alterations the Air Force demanded. Every little change meant fresh business, and that, after all, was what the game had always been about. The small, disciplined task force that had done such sterling work on the original entry began to swell, increasing tenfold the initial hard core of two hundred engineers and designers. As always, the customer's logic and intent were never questioned. If someone wanted to add a bunch of radar and weapons control systems, the work was done even though the plane had been conceived specifically as a passive receiver of enemy radar transmissions rather than as an active sender of its own. Tactical Air Command decided it would be nice if the F-16 could drop nuclear bombs, and that, too, was done without protest. Like a small car jammed with a family of large, fat people, the airframe was soon loaded to capacity and sagging badly at the springs. The solution? Add two feet to the fuselage, decrease the size of the internal fuel tanks, and then get ready to add even more "bells and whistles." If in the process the plane lost the ability to fulfill its original mission, that too could be justified, since it now became a candidate for something like the AMRAAM (Advanced Medium Range Air-to-Air Missile), the latest in the thirty-year line of expensive and unsatisfactory "smart" weapons intended to identify and destroy enemies long before they come within visual range. This is gold plating taken to the limits of absurdity, since standoff weapons like AMRAAM make the original emphasis on air-to-air performance entirely irrelevant. The result is an ostensible fighter that is not allowed to fight, an air superiority plane performing a job that might just as easily have been handled by an F-15 or, for that matter, a modified executive jet.

The extent of the indignities to which the F-16 has been subjected can be seen by comparing the vital statistics for successive models, each a little more complex than the one before. The Fighter Mafia's prototype, the one that flew circles around Northrop's Cobra, matched a single Pratt & Whitney F-100 engine rated at 23,840 pounds of thrust to an airframe with a gross maximum takeoff weight of around 19,000 pounds. The initial production model, already burdened with the first of the innovations dreamed up by General Alton Slay and his Configuration Control Committee, saw the takeoff weight raised to 23,500 pounds without any corresponding increase in the size of the control surfaces or one additional pound of thrust from the engine. Even so, the first production models to enter service in 1978 made a lasting impression on the men who flew them. Briefing room banter was enlivened with talk of "bat turns" and other novel maneuvers, possible in no other plane. They were also impressed by little touches like the rear-sloping angle of the pilot's seat, which helped to maintain blood supply to the brain in "high-g" maneuvers.

Those who flew the next single-seater in the series could not help noticing that the magic had lost some of its sparkle. This was only to be expected, because the plane had acquired several thousand pounds of additional electronics, as well as a larger tail fin, which was introduced to compensate for a loss of stability incurred by the first round of modifications. "Compared to the 'A,' the 'C' was flat champagne," one pilot who has flown both airplanes lamented. Its successor, the 28,000-pound D-series F-16, has proved to be an even more sluggish performer, so much so that it now comes very close to matching the F-15 in acceleration, turning ability, and general agility. But the plane most distantly removed from the Fighter Mafia's ideal is something called the F-16E, "nothing but a little B-1," in the words of one disgusted critic. Complete with swept, full-delta wings and weighing almost 40,000 pounds, it represents the final, debased variation on the original concept. Although details of its performance are still largely classified, it is safe to assume the "E" will prove to be by far the most insipid performer of them all. How could it be otherwise? With a gross takeoff weight more than 100 percent greater than the prototype, and the same overtaxed engine to push it along, it must have limitations preordained by elementary mathematics.

More than malice or stupidity is required to explain a process that takes a perfectly good weapon and systematically strips it of the qualities that made it a success. Odd as it may seem, the real reason is that for all its alarums about the Russian threat, the consistent pattern of its actions indicates that the Pentagon views the actual business of war as an abstract notion rather than the sole justification for its continued consumption of public funds. As the gilding, and gelding, of the F-16 demonstrates, the chief preoccupation is not the development of weapons but the cultivation of bureaucracies to manage them. Only when viewed from this perspective does the wave of late enthusiasm for the F-16 begin to make sense. Slipped into the system without the benefit of a large supporting network of friends and patrons, it offered few bureaucratic incentives and little opportunity for empire-building. Today, its "further development" the bread and butter of literally thousands of Air Force personnel, it has suddenly become an invaluable defender of the nation's security.

The ironic consequences of the approach are enshrined in several decidedly odd, one-of-a-kind versions of the F-16, which underwent extensive testing throughout the early 1980s. One of these, the AFTI (Advanced Fighter Technology Integrator), features two vertical fins mounted on either side of the engine's air intake, while a second model sports an additional set of short, stubby wings astride the cockpit. According to an Air Force spokesman, who waxed lyrical about these and other Flash Gordon excesses, like a voice-activated fire control system under consideration for the next generation of fighters, future airplanes will be able to "outturn, outshoot, and outfly any other aircraft in the world today." So great was his enthusiasm, it seemed churlish to point out that the first, unadorned F-16 prototype could do all that and much more besides, at a fraction of the cost. And make no mistake, the next frontline Air Force fighter will cost enough to make even the F-15 look cheap. Rick Blucker, a civilian cost analyst attached to Wright-Patterson

Air Force Base, explained why. "We basically used off-the-shelf technology for the F-15," he said. "But the technology we expect to incorporate into the Advanced Technology Fighter is *beyond* the state of the art. The evaluation and demonstration stage could last four years, and will probably include some hardware development; consequently the contracts will be worth a lot more."

No description of the F-16's emasculation would be complete without mention of a curious press statement issued by the Air Force in March 1985, which managed to depict the plane's principal asset as a liability. The F-16 was *too* maneuverable, the release maintained, adding that the abnormally high rate of unexplained crashes was due to pilots blacking out during "high-g" turns and loops. Skeptics within the Department of Defense were more inclined to lay the blame with General Alton Slay and the panel that had come to be derisively known as the Configuration Add-On Committee. But first, a little technical background.

All aircraft require a backup power system, although the F-16's need is greater than most since the basic shape of its body is inherently unaerodynamic and can be made to stay in the air only so long as there is a working on-board computer to "sense" instability and make the immediate slight corrections to maintain pilot control. This so-called fly-by-wire system was one of the Fighter Mafia's original specifications, the advanced technology justified by the advantages to be had in eliminating weighty on-board hydraulic systems and the freedom the designers were given to turn just about every square foot of outer fuselage into a "lifting" surface. However they embraced the benefits of computerization, Sprey and the others were adamant about the need for a simple and reliable backup power system, like the banks of batteries incorporated into the first prototypes.

The Configuration Control Committee had its own ideas, replacing the batteries with what a critic described as "a bomb right under the pilot's seat." The bomb in question is something called a hydrozene generator—hydrozene was used to fuel Hitler's V-2 rockets—which has often proved reluctant to come to life. According to officials of the Washington-based Project on Military Procurement, a frequent critic of Pentagon policy, pilots have reported that the generator often takes thirty seconds or more to restore lost current. The consequences can be easily imagined: A pilot is climbing or turning when, quite suddenly, the jet engine experiences compressor stall, a not uncommon occurrence with the immensely sophisticated and high-stressed F-100 engine. "The pilot has two options," a former Air Force fighter jockey explained. "He can bail out and wave goodbye to $16 million worth of equipment or he can stay on board, trying to get the engine restarted." With no power, no control, and no communication with the ground, the pilot can only hope the hydrozene generator cuts in before it is too late. The reason the Air Force opted for a system with so many inherent disadvantages would, under normal circumstances, defy explanation. With something like the redesigned F-16, it is, however, just another sin of commission.

The F-16 had dropped into General Dynamics' lap, a gift bestowed by providence rather than good management. The company had followed the Fighter Mafia's instructions, defeated Northrop's ill-conceived Cobra, and then sat back when Defense Secretary Schlesinger thrashed out his bargain with the Air Force. Good fortune of this magnitude had no right to happen twice. All the same, it was about to do just that.

In January 1975, Schlesinger once again took the initiative and threw General Dynamics into the quest for what the European press was calling the "Sale of the Century." He could hardly have picked a less well-equipped champion to carry the flag. Northrop, Lockheed, and McDonnell Douglas prided themselves on being the unofficial diplomatic corps for the U.S. aerospace industry, their people acutely sensitive to the often strange needs and norms of the world beyond America's shores. By contrast, Fort Worth in general and General Dynamics in particular were burdened with a reputation for cultural insularity. With the exception of Canadair's Sabres, the company had had no extensive dealings with foreigners since the days of Basil Zaharoff and the "special transmissions" that Senator Nye had found so intriguing. There had been airliner sales, of course, but they dated back to the almost forgotten era when San Diego still reigned as one of the world's leading commercial airframe producers. The only recent foray outside the Lower Forty-eight had been the agreement with Australia to supply twenty-four F-111s, and that incident could hardly have been expected to leave a favorable impression on the four NATO nations currently searching for a new joint fighter.

Northrop's globe-trotters had always believed their international expertise would be one of the deciding factors in the Cobra's favor. Fort Worth's heritage was strictly bush league; it could never hold its own against the likes of Saab and Dassault. One of the early figures in the Fort Worth aviation industry, oil millionaire Amon Carter, was reputed to be such a hometown chauvinist that he took bagged lunches to business meetings in nearby Dallas rather than spend any money there. Admittedly Carter had no connection with the modern Fort Worth but the impression of a company run by straw-chewing hayseeds lingered on, confirmed in part by David Lewis' recent trip to Paris, where he unsuccessfully sought to involve the Dassault company in the coproduction of F-16s for the European and international markets. According to a somewhat condescending Dassault spokesman, Lewis arrived with a quantity of American cakes, which he presented to the

French firm's founder, Marcel Dassault, as a small token of his friendship and esteem. "The cakes, they were very hard to eat—not a good taste," the spokesman later explained, adding that the idea of a joint venture with General Dynamics dissolved with the first unpalatable mouthful. Since the trip took place before Northrop's defeat, its overconfident executives must have been laughing at the idea of a U.S. executive setting forth to negotiate a billion-dollar deal with no better ammunition than a firm handshake and a suitcase full of sticky confections. As their own international successes testified, given the choice between cake and cash, foreigners chose the money every time.

How Northrop might have represented its country must remain a matter for speculation, but it could hardly have adopted a more successful tactic than General Dynamics. Just as it had done at home, the company allowed James Schlesinger to become its chief salesman. As usual, the defense secretary's first priority was to stifle opposition from within the Pentagon, where people like General John Vogt, a former commander of NATO's air forces in Western Europe, were doggedly advancing the F-15 as the only airplane equal to the task. "I know it is popular now to talk about airplanes that are cheaper and airplanes that can do the job for less money, but the fact of the matter is that it is false economy," he lamented not long after the NATO decision was announced. "The F-16 is virtually limited to daylight, good weather conditions. . . . The airplane does not possess the capabilities that are so necessary. It is going to be limited to doing the job under eyeball conditions." Schlesinger responded by cutting General Vogt and his colleagues off at the pass. While Vogt and other F-15 advocates were given permission to fly both Saab's Viggen and Dassault's Mirage F-1E, all requests to test their own government's official contender were summarily turned down. What little enlightened guidance the Europeans received from the U.S. Air Force came almost exclusively via John Boyd, who was brought in to explain how the F-16 meshed with his air combat theories, and the relatively small group of development, liaison, and project officers the plane acquired when it left the prototype stage and entered full production. Rather than stressing the plane's unique qualities, Schlesinger's campaign remained from first to last an exercise in alliance politics.

Pressed into the fray, General Dynamics found itself in an alien, cutthroat environment, although not at a disadvantage. Dassault and Saab were both entitled to stress that their aircraft would be built entirely in Europe, and each enjoyed the benefit of neighborly proximity with at least one of the prospective purchasers. But those strengths were also part of their weakness. The Mirage was popular in Belgium, particularly among the French-speaking population in the south, but it was viewed with little enthusiasm in Northern Europe, where Denmark, and possibly

Norway, were seen as natural supporters of the Swedish Viggen. Belgium had bought both French and American planes in the past, Finland flew Starfighters, while Holland, which would have been Northrop territory under normal circumstances, was insisting that its final choice would be based on each contestant's willingness to better the others' proffered packages of jobs, technology transfers, and, perhaps, reciprocal trade agreements. Schlesinger's strategy played on the NATO allies' traditional suspicions. When the French spoke of the Mirage's potential as a base for a pan-European aerospace industry, all but the southern Belgians interpreted the remarks as an invitation to surrender technological control of their own industries to Dassault's technocrats. A spokesman for the French firm conceded that the victor would be in an enviable position, but qualified his remark with a geographic allusion. "France is a European country after all. I don't see what is the difference between American dominance and French dominance except that America is six thousand miles away." There was evidently little difference between the French and the American talent for prompting scandals. At about the same time that news of Lockheed's and Northrop's indiscretions were surfacing in Washington, a member of the Dutch parliament charged that he had been offered a large bribe to support the Dassault plane. Untainted by imputations of any sort, Saab realized that while it might sign up Norway, Denmark, and perhaps even Holland, it would take a near-miracle to sway the southern Belgians from their preference for all things French. The F-16 thus became the only universally acceptable alternative, albeit one that traded lower levels of European involvement in the manufacturing process for a quick and harmonious decision. But first, before the players could bring the drama to its almost inevitable conclusion, they had to enact the ritual moves that accompany any major and politically sensitive international arms deal.

Saab opened the bidding by setting a firm price of $9 million on its Viggen. This was a good deal more than the projected prices for its two competitors, but the Swedes were adamant that it still represented good value for money since their plane, like the F-15, had pretensions to being a true all-weather fighter, while the Mirage and the F-16 were limited by definition, if not by fact, to "some-weather" operations. Saab rounded out its package with promises of substantial percentages of coproduction and the usual argument that a decision in favor of the Viggen would spawn a series of high-tech industries and create perhaps forty thousand jobs.

The French package was not terribly different, although the Mirage did have the advantage of a lower price, $6.9 million, and the experience of both Dassault's own seasoned salesmen and the men in the international affairs section of the Délégation Ministérielle pour l'Armement, whose efforts had helped to establish their country as an exporter of weapons second only to the United States.

Finally, there was General Dynamics, with its own two-pronged offensive. Not long after the French announced their official price, the cost of each F-16 was set at $6.09 million, a figure that made it both the cheapest and the subject of much skepticism. The company and its advocates in Washington moved quickly to allay the suspicions, stressing that the quoted price was not some hollow paper promise but an ironclad "not-to-exceed" figure. Since all sales of U.S. weapons have to be conducted through the conduit of the government, the sum came stamped with the approval of the Ford administration and the implied promise that it would remain firm come what may. As subsequent events would demonstrate, even the best of friends should consult the fine print before striking a bargain.

European content, the one area where General Dynamics was at a pronounced disadvantage, demanded similar sacrifices. Should the four NATO nations adopt the F-16, they were promised the right to produce 40 percent of the component parts for their own planes as well as 10 percent of the total airframe and trappings for the U.S. Air Force models being built back in Texas. In itself, that part of the American offer was not terribly different from the offsets being touted by Saab and Dassault. There was, however, one additional and potentially lucrative extra. Besides taking part in their own and the United States' F-16 programs, the Europeans were promised the right to produce 15 percent of the components used in all planes sold to third-party nations, a figure General Dynamics' salesmen divined at somewhere between 4,000 and 5,000 aircraft. Once again, official efforts on the plane's behalf banished perceptions of hucksterism. If future defense secretaries worked as hard to sell the F-16 in other parts of the world, there was little reason to question the assertions that it would replace Northrop's F-5 as the world's most ubiquitous fighter. In terms of employment, the General Dynamics proposal amounted to some 25,000 jobs, with reason to hope the number might rise as high as 40,000. Even better, the work done in Europe would entail more than mere assembly work, since General Dynamics also pledged to share the secrets of its advanced design and construction technologies, even to the extent of helping companies with no previous aviation experience to establish themselves in the new field.

The second flank in the American assault on Europe was dominated by Schlesinger and his Defense Department diplomats. The NATO countries should standardize their equipment with that of the alliance's senior partner, innumerable visitors from Washington asserted. The argument was strictly logistic, although the implication was politically explicit: If you want American protection, then you had better make sure you buy American.

The combination of pressure and promises worked. In May 1975, after four months of increasingly frenzied lobbying by all three contenders and their respective governments, Holland, Denmark, and Norway released a joint statement that endorsed the F-16 as their plane of preference. The one holdout, Belgium, a last bastion of enthusiasm for the Mirage, was soon to be bludgeoned into submission by perhaps the most notable advocate General Dynamics has ever had, President Gerry Ford. After attending a meeting of NATO leaders in Brussels, Ford retired behind closed doors with his host, Prime Minister Leo Tindemans, and advanced the advantages of the F-16 in no uncertain terms. Dassault's officials professed repugnance at such unorthodox tactics. "Could you imagine us using our president as a salesman?" one French executive told British journalist Anthony Sampson. Though it may have scandalized the opposition, General Dynamics' use of Gerry Ford did it little immediate good. When Ford flew home, the Belgian parliament was no closer to a decision than before.

As seen from Washington, the situation appeared promising but perilous. Having recently committed themselves to the F-16, the Dutch were suddenly gripped by a last-minute outbreak of second thoughts. A number of Holland's trade union leaders had broken with the official line and embraced the Viggen as the plane that offered the most attractive financial offsets and incentives. Meanwhile, a left-leaning faction of parliamentarians and party spokesmen had taken to urging that all three of the original competitors be banished from contention in preference to a closer look at one or another of Northrop's products. Obviously, Belgium was the key. If the political leaders in Brussels could not be brought into line, there was a very strong chance that Holland's vacillations would set off a chain reaction in Norway and Denmark. Ford had failed to lock up the deal. Now the task fell entirely on Schlesinger's shoulders.

As usual, Schlesinger prosecuted his campaign on a purely political level. Inviting Belgian defense minister Paul Vanden Boeynants to Washington for a final round of bargaining, he repeated the logistic rationale for a common NATO fighter and sweetened his argument with a few little extras. If Belgium signed up with General Dynamics, Schlesinger promised that his government would purchase some $30 million worth of machine guns from Fabrique Nationale, one of the country's largest industrial concerns and a firm already slotted for a major role in the production and assembly of the European version of the F-16. It was a neat diplomatic solution that suited both parties' needs. The French-speaking south won the jobs while the north obtained the right to the plane it had always wanted. Early in June, after representatives of the four countries had met and drawn up a joint statement, the result was announced, to the mixed delight and dismay of the world's aerospace executives gathered at the Paris Air Show.

David Lewis was overjoyed by the announcement. "This is great for the Air Force, great for NATO, and great for our company," he said. "We are certainly glad this great little plane of ours has taken another giant step forward." Lewis was quite candid about his company's strategy and the benefits of innocence, informing journalist Andrew Cockburn that the result might have been different if the company had been more experienced in dealing with foreigners. "We didn't get ourselves involved in a number of things we would have regretted. We also didn't need to use any nonstandard sales techniques at all," he said. It required very little deductive reasoning to see his remarks as an oblique reference to the activities that had helped to land Northrop and Lockheed in so much trouble before Senator Frank Church's Subcommittee on the operation of multinational corporations. But then, bribes were hardly necessary with persistent and insistent advocates like Gerry Ford and Jim Schlesinger.

Perhaps the most remarkable thing about the F-16's early years is the way, time after time, the best thing that could have happened usually did. David Lewis and the new regime sat tight, did as they were told, and waited for others to make the pieces fall seamlessly into place. If Sprey, Boyd, and Riccioni had been less perceptive or determined men, there would have been no lightweight-fighter program. Nor would the plane have progressed beyond an ill-favored experiment had it not been for David Packard's near-indiscriminate enthusiasm for competitive prototyping or, later, Schlesinger's willingness to defend it at home while promoting it aggressively abroad. Fortune had not merely smiled; she had taken General Dynamics into her embrace like a loving and long-absent aunt.

"Three months before," David Lewis commented, "there was nothing except some warm feelings here and there." Now, quite suddenly, the company had a future. Initial orders for Europe and America came to a round one thousand aircraft—by far the largest production run since World War II—and it was likely to grow still further as the international sales campaign rolled through the Middle East, Asia, and South America. It was precisely what the company needed to launch a new era: a safe and assured financial mainstay that would eventually produce 40 percent of the company's operating profits. At Fort Worth, the sense of relief was palpable. Employment had fallen from a peak of 24,500 men and women at the height of the F-111 program to less than 10,000. Now it was rising again, growing in fits and starts as a new production line was set up to handle the first orders from Europe and the United States. On Wall Street, General Dynamics was beginning to look like a safer proposition. With the European contracts in the bag and the Air Force indicating that the initial order for 650 aircraft might soon be increased to perhaps as many as 1,700, the stock price began its long, steady rise toward the dizzy heights it would reach more than a decade later. In Europe, General Dynamics officials had recruited thirty-three principal subcontractors and some four hundred "third-tier" suppliers, while back at home, the family of associated firms brought into the U.S. F-16 construction program had risen to almost four thousand. It was, at a glance, a complete success story from start to finish. The Europeans were getting a great plane at a good price, the U.S. Air Force was, if not totally happy, at least reasonably satisfied with the deal it had struck with Schlesinger, while back in Chicago, Colonel Crown and his clique were seeing the first concrete indications that they would receive financial rewards commensurate with their substantial investment.

Alas, there are few certainties in the real world and even less in the kingdom of strategic abstractions where major weapons systems are conceived, developed, and deployed. As the Europeans would soon discover, the much-vaunted "not-to-exceed" price was so far removed from reality it brought to mind one of Senator Sam Ervin's stories from the early days of the F-111 investigation. It was a little-known fact, the McClellan panel's resident wit and philosopher informed his fellow senators, that George Washington had spent his childhood in Texas, and moved to Virginia only after taking an ax to his father's cherry tree. " 'Son, we have to pull up stakes and go somewhere else,' " was the way Ervin quoted the elder Washington, " 'because if you can't tell a lie, Texas is no place for us.' "

For all the effusive assurances that the F-16 was a fully equipped, battle-ready warplane, the Europeans soon learned that they had not bought a weapons system so much as the right to play a minor role in an ongoing evolutionary process. Air Force liaison officers representing each of the four European clients arrived in Washington to take their places on a development board composed of U.S. Air Force personnel from the various supervisory offices, General Dynamics' own managers, subcontractors, various civilian functionaries from the Office of the Secretary of Defense, and a floating collection of technical specialists brought in to add their own flecks of gold leaf to the gilt accumulating on the lily. Every addition forced the price a little higher, and there were many, many additions. Originally set at $6.09 million, the fly-away cost grew to exceed $8 million and was well on the way to $10 million within two years. Today, ten years after the decision, the value of an average European F-16 complete with retrofitted extras stands somewhere in the vicinity of $16 million. Inflation accounts for part of the increase, while the duplication of tools essential to raise the levels of European content is responsible for another million dollars or so. The rest has been lavished on all the high-tech extras that the plane was designed specifically to do without.

When European politicians grumbled about their increasingly expensive investments, they received a short lesson from the U.S. General Accounting Office on the folly of taking a sales spiel at face value. "It is not realistic to attempt to establish not-to-exceed prices at a point very early in the acquisition process," a report published in mid 1979 explained. The chairman of the multinational development board, General James Abrahamson, described the escalation in terms of an unfortunate misunderstanding between good friends. It was really no different from buying a new car and then paying a little extra to have a cassette player installed, he maintained. Working solely from the official documents, the GAO investigators concluded that while the Europeans might complain about it, the inflated price was due largely to their own requirements for a multirole warplane. There was a paper record to support this conclusion, but its pertinence to the facts was open to serious debate. While the Europeans endorsed some of the additions—the "active" radar, for example—they were either divided or actively opposed to many of the other, more rococo enhancements. In August 1978, more than three months before the first European F-16 made its maiden flight in Belgium, the U.S. Air Force had already invested the plane with the capacity to drop nuclear bombs, control lasar-guided munitions, and carry the latest variant on the Sparrow radar-guided, standoff missile that had acquitted itself so miserably over Vietnam.

Another element in the increase was summed up by a line from the annual Defense Department report of 1977, the Pentagon's equivalent of a letter to stockholders. "Our NATO allies," the report asserted in reference to the F-16, "will wish to depend more on U.S. aircraft design and development." It would have been at variance with the perpetual optimism of Defense Department publicists if the other side of the issue had been raised. According to the GAO report, the 10 percent content of European components in the first 650 U.S. Air Force F-16s saw the price increased by a total of $241 million. And the Europeans paid even more. The list of additional expenses began with $369 million to recompense the U.S. for prior development and research. Finally, they were forced to absorb the additional overheads incurred by running two, independent assembly plants—the first at Gosselies in Belgium and the second in Holland on the outskirts of Amsterdam—when one would have been entirely sufficient. General Dynamics and successive U.S. administrations have cited the coproduction agreements as an example of the NATO partners' ability to cooperate for the common good. Ten years after the bargains were struck complaints have subsided to an occasional murmur. Outside the alliance of coproducers, however, the F-16 program was reviewed more critically. "We in the United Kingdom do not regard production of F-16s in Europe as an outstanding example of project," the chairman of a special parliamentary report stated in 1978. "This particular program appears to be too much of a benefit game for the U.S. industry and economy without equal technological rewards for Europe and NATO."

The Europeans did do better with sales to third parties. Israel became the first and one of the most enthusiastic purchasers, ordering an initial fifty F-16 A's and B's before upping the consignment to two hundred. Egypt collected fifty for its cooperation in the Camp David peace talks, while other aircraft found their way into the service of Venezuela, Pakistan, Saudi Arabia, Greece, Turkey, and South Korea. China was nominated as a prospective purchaser by another of General Dynamics' more notable advocates, Henry Kissinger, who was also in favor of supplying 160 F-16s to the Shah's Iran. That sale, scrapped by the first revolutionary government, has proved to be General Dynamics' only disappointment. The eventual tally will probably never reach the five thousand overseas sales predicted by General Dynamics' spokesmen during the battle for Europe, but still, with the figure now approaching three thousand, there is little basis for recrimination.

12. Two Weapons

Before proceeding to the innumerable embarrassments and scandals, both major and minor, that beset General Dynamics through the late 1970s into the '80s, it is worth a moment's pause to take a brief look at two more of the company's military programs. Critics of the way the Pentagon buys its weapons have often asserted that the procurement process operates back to front. New weapons might well be inspired by the demands of the battlefield, but the pace of their evolution—not to mention their final form—is determined by the bureaucratic imperatives of those responsible for their development. The later versions of the F-16 certainly fall into that category, as do the M-1 Abrams main battle tank and the TAKX supply ships, prime components of the Rapid Deployment Force.

First, the M-1.

In March 1982, the office of the Pentagon charged with overseeing the development of the M-1 circulated a brief note for the information of the various congressional defense committees. By the often convoluted standards of military grammar, the note was simplicity itself: The Chrysler Corporation, America's chief supplier of tanks for more than forty years, was about to quit the field. In return for a badly needed cash payment of $336.1 million, the automaker would relinquish its stewardship of the two government-owned factories that constituted one of its last two profitable divisions. The circular bestowed the Army's blessing on the transaction and advised recipients to note the name of the new prime contractor, the General Dynamics Corporation of Saint Louis.

The M-1 had more than its share of problems, but General Dynamics officials were enthusiastic to the point of exuberance, promising that the technical difficulties would soon be ironed out and the skeptics silenced. The tanks, each worth roughly $3 million, would then roll off the production line in an uninterrupted stream until the Army had the seven thousand or so that its spokesmen insisted was the minimum number needed to face the Russians on equal terms.

While it may have seemed entirely natural for General Dynamics to begin catering to the Army, the one branch of the armed forces with which it had enjoyed scant dealings in the past, there were many who wondered if this time the company had not gone too far. Even David Lewis admitted the price could not have risen much higher, but he was equally insistent that his company "had not bought an Edsel at a Cadillac price." Lewis glossed over the problems that had hounded the tank since its inception, issuing a string of folksy assurances that everything would turn out for the best. However, for all his genial optimism, it was still difficult to disguise the fact that much needed to be done to get the project back on the rails.

The most obvious problem concerned the physical condition of the twin plants, the first in Warren, Michigan, and the second in Lima, Ohio. The Warren factory had been thrown together in 1942, when the first buildings were erected to house a production line for Sherman tanks, and while it boasted washrooms trimmed with lavish amounts of white marble, it could not point to anything resembling industrial efficiency. Pretax profits had shrunk to a miserly $73 million on sales of $820 million in 1981, tools and much of the heavy equipment were outdated, while an unreliable system of inventory controls led to acute production line bottlenecks. In its World War II heyday, the factory had turned out as many as 5,000 tanks in a single year. When Chrysler placed it on the auction block, the problems were such that only 230 tanks had been delivered in the previous thirty-four months.

By themselves, the physical problems may have dissuaded most potential buyers, but they paled into insignificance beside the tank's most obvious defect, the enmity of a vocal opposition both inside and outside Congress. As with the Viper antitank missile and the Trident submarine, two General Dynamics programs much criticized for runaway costs, the primary objection concerned a price increase of well over 100 percent. Those of an allied viewpoint were more specific, assailing the uniformed and civilian decisionmakers whose combined meddlings had substantially altered the original concept. There had been repeated calls to suspend funding, as well as a number of General Accounting Office investigations that cast serious doubts on the tank's engine and transmission systems. Yet, in the face of numerous hazards, General Dynamics pursued the Chrysler prize with single-minded determination. The final figure was some $30 million in excess of the generally accepted projections of the subsidiary's worth and stood as a tribute to Lewis' and his board's faith in their own judgment.

Lewis explained that the M-1 project was a "target of opportunity," a sound long-term investment that would be worth every penny. The story of the M-1 began long before the Pentagon found a name for it. In the mid-1960s, the Army announced that the new Soviet T-62 tank was a wonder weapon far superior to anything in use in the West. The fact of the matter was that the T-62 was little different from earlier Soviet tanks, of which it was a direct descendant and over which the M-48 and the newer M-60s had already demonstrated a marked superiority. However, the Army persisted in its Chicken Little campaign, not only heaping undeserved praise on the T-62 but also going to great lengths to point out that the Soviets were producing them in numbers much greater than the United States could hope to match within the constraints of a peacetime budget. The Pentagon's solution was to go the Russians one better by building an even more sophisticated tank that would combine all the latest Western technology in one awesome, unstoppable package. The lavish application of technology would, it was argued, redress the imbalance by making a single U.S. tank the equal of several Soviet ones. That much of the technology was either untested or, in some cases, entirely theoretical did not disturb the Pentagon pitchmen in the least.

The initial result was the MBT-70, a clanking behemoth whose complexity was equaled only by its inability to live up to its designers' expectations. Even as a prototype—the stage of development that usually sees costs kept to a bare minimum in order to make procurement of the finished product more attractive—the tank ran way over budget. The price was not the only thing wrong with the MBT-70. Many of its vaunted technical features proved to be disastrously ineffective, a much ballyhooed automatic cannon loader not least of all. The tank was so irredeemably flawed, Congress took a most unusual step and, early in 1972, canceled the entire project. Surprisingly, the Army received the news with a minimum degree of protest, although its reaction had less to do with the tank's deficiencies than with the fact that there was a strong chance many of its component parts might have been built in Germany. General Creighton Abrams, a former protégé of General George Patton, and the man after whom the M-1 is named, expressed the Army viewpoint with unusual candor when he stated that German tanks, regardless of their merits, would serve in the U.S. Army only "over my dead body."

When the MBT-70 fell by the wayside, the Army returned to its drawing board with instructions to produce an all-American, slightly simpler, and much cheaper version, which was soon christened the XM-803. Once again, another costly monster emerged from the design process, and for the second time, the project was scrapped.

After two ill-fated efforts to replace the tried and true M-60, the Army girded its loins for a third attempt. On this occasion, a competition was conducted, with a panel of judges asked to evaluate a design produced by General Motors and the M-1—then called the XM-1, the "X" standing for "experimental"—submitted by Chrysler Defense. The selection advisory board favored the General Motors contender, but its reasoning played little part in the eventual outcome. Defense Secretary Donald Rumsfeld announced in the first days of 1976 that the M-1 would be the new Main Battle Tank, ostensibly because it boasted a greater degree of commonality with NATO's Leopard tanks. The real reason, however, had more to do with the poor health of Chrysler's crumbling car and truck operations.

Under normal circumstances, the Army might have been expected to raise at least a discreet fuss over the way its professional evaluation had been ignored. However, whatever their complaints, the Pentagon's tankmasters kept their objections to themselves. One gathers that after seeing its two previous tank proposals summarily quashed, the Army was simply overjoyed by the prospect of finally getting the kind of weapon for which it had yearned so long.

In creating the M-1, the designers discarded many of the conventions embodied in past tanks and strode boldly into the future. Although originally conceived as a "simple" tank, the M-1 proved to be anything but. From the very beginning, the most astonishing innovation was the power plant, not a conventional diesel but a state-of-the-art gas turbine engine that is, in essence, little more than a jet engine wedded to a traditional tank gearbox and power train. The 1,500-horsepower turbine provides the tank with sufficient acceleration to spring from a dead stop to twenty miles per hour in only 5.8 seconds. The M-1 would not be embarrassed in the stop-start traffic of any major American city, although, at a hefty sixty-two tons, it would not be popular with road repair crews. The Pentagon hailed the acceleration and the high top speed of almost fifty miles per hour as examples of the sort of thing that would make the Russians think twice before setting a jackboot on the soil of Western Europe. Unfortunately, the mechanical systems are also cursed with a number of inherent weaknesses, in which the Soviets can take some considerable degree of consolation.

The turbine is indeed powerful, so much so that it rapidly "ages" the other transmission components to the point where they sometimes come unstuck under the enormous strain. This may well be the reason the Army has never staged an official test of the tank's ability to perform an extremely simple maneuver that remains one of the best methods of dodging enemy fire. This procedure, best likened to the way in which a matador waits until the last second before sidestepping a charging bull, involves nothing more complicated than accelerating at top speed for fifty feet or so before suddenly applying the brakes and coming to a dead stop. The "sprint and stop" tactic makes it that much more difficult for an enemy gunner to plot a target's course with any degree of certainty. In a typical tank battle of the kind that has been fought repeatedly in the Middle East over the past two decades, action has consisted of a series of free-for-alls which resemble nothing so much as the ultimate arcade video game brought to life. The range is usually no more than eight hundred meters and seldom less than three hundred. Death comes quickly to the unlucky, the inexperienced, and the ill-equipped. Tanks lurk behind the crests of hills or in holes dug for them by retinues of armored bulldozers; they fire "over-the-shoulder" shots while running backward at top speed. Not surprisingly, those that survive present the smallest possible target, on the least number of occasions, while making optimum use of their guns. In this context, the M-1's ability to withstand the punishment of repeated high-speed dashes and sudden short stops is almost as important as armor plate itself.

Despite what many might consider a potentially fatal flaw, the Army has continued to wrap itself in an air of haughty disdain, dismissing criticism with the blanket explanation: "Testing has not been done for the 50 foot dash since this is not an Army requirement."

The turbine's superheated stream of exhaust gases figures as another area of concern, the M-1's detractors pointing out that the blast of seven-hundred-degree air would serve as a beacon for heat-seeking missiles. This deficiency is supposedly negated by the tank's ability to leap nimbly over broken ground with all the grace of a pirouetting hippopotamus in Disney's *Fantasia.* Not surprisingly, the engine must be in working order if it is to achieve this feat, but thanks once again to the exhaust system, this is never entirely certain. Tests have shown that the M-1's ducted exhaust has a tendency to work loose, redirecting a jet of high-pressure gas onto the ground beneath the tank's hull and stirring up clouds of dust. When this happens, the turbine will likely choke to death—unable to suck air through its clogged air intake filters, it slows, shudders, and eventually stops dead in its tracks. If this ever happens in the smoke-and-dust-filled chaos of a close-range tank battle, the disabled M-1s will be nothing more than expensive and highly vulnerable pillboxes. They will also be death traps for their unfortunate crewmen.

If the grim day ever comes when M-1s receive their baptism of fire, those charged with getting them to the scene of the action will first be faced with a long list of logistical problems. Keeping just one battalion on the move requires twenty-six additional fuel tankers, since the M-1 guzzles gas at about twice the rate of the M-60. The process of transporting large numbers of M-1s with any degree of haste is also much complicated by the tank's unique bulk. The largest Air Force transport, the C-5, can carry two M-60s but only one M-1, and the same figures apply to railway flatcars—surely two of the most vital cogs in the logistical machinery optimistically envisioned as being capable of supplying frontline forces for the duration of hostilities.

Fuel truckers and maintenance crews are not the only members of the M-1's expanded retinue of essential camp followers. There are also support personnel who drive and maintain ACEs—an Army acronym that describes the world's first hot-rod bulldozers. Because it is slower, the M-60 can travel quite comfortably with a fleet of relatively conventional bulldozers. The fleet-footed M-1 unfortunately leaves old-fashioned bulldozers in its voluminous clouds of jet-propelled dust. So the Army came up with the ACE, a diesel-powered "super-dozer" that can scuttle over broken ground at a rattling thirty miles per hour. Predictably, each ACE costs the best part of one million dollars and the Army wants at least six hundred of them.

Finally, the M-1 provided those interested in the subtle manifestations of international diplomacy with a classic example of the way in which military considerations may be compromised for political ends. In this instance, the lesson is embodied in the tank's main weapon, its gun.

The original M-1 design specified a tried and tested 105 mm weapon with the same rifled barrel that was the standard weapon on all M-60 tanks. The gun was relatively simple, easy to operate and fire, and in comparison with the one that replaced it, cheap as well. In 1978, the situation changed quite suddenly.

In its efforts to encourage the West German government to purchase a package of military hardware that included AWACS reconnaissance planes as well as the M-1's own gas turbine engine, Washington agreed to buy German Rheinmetall tank cannons as part of a quid-pro-quo exchange. The decision reflected a desire to placate Bonn. When West Germany pulled out of the original MBT-70 project, Washington agreed to consider the German Leopard as its next Main Battle Tank. This proposal quickly ran afoul of Army policymakers like General Abrams, who drew up a long and biased list of performance specifications intended to cast all but American-made tanks in an unfavorable light. The Germans withdrew in a huff and were still irate when the decision to rearm the M-1 with the Rheinmetall gun was announced, a move that had everything to do with diplomacy and very little with defense.

Since the new gun is a smooth-bore weapon that requires substantially larger and longer shells than those fired from the old 105 mm, the M-1 can carry only forty rounds instead of the fifty-six that can be stowed in an M-60's ammunition bay. That amounts to a decrease of roughly 30 percent in the firepower a single tank can direct against an enemy before being obliged to withdraw from the engagement and travel several miles to the nearest forward supply depot.

The official rationale behind the new gun, turbine engine, and all the other gadgetry in the M-1 is that the old M-60 is no longer superior to the latest generations of Russian armor, the allegedly ferocious T-72 in particular. Superficially, the T-72 does appear to be a better, more formidable tank. It is lower to the ground, carries a thicker shield of frontal armor, and boasts an oversize smooth-bore gun that was once hailed by the Pentagon as "one of the most powerful guns in the world."

Recent experience in the Middle East, however, indicates that the T-72 is far from the invincible beast suggested by its squat, lethal silhouette and by innumerable Pentagon tributes to its Russian engineers and designers. For example, the T-72 boasts an automatic cannon loader that reportedly "eats" crewmen, stuffing arms and legs into the breach with indiscriminate "efficiency." The engine is based on a design originally intended for French dirigibles of the late 1930s; it breaks down regularly and appears to be underpowered.

Even if the T-72 were everything its Western admirers claim, there would still be little military justification for the move to the M-1. In the hands of the Israelis, American-made M-60s have proved themselves equal to the task of demolishing all current Soviet tanks up to and including the T-72s, which burned with no less ferocity than their predecessors during the Israeli Army's drive through southern Lebanon in the summer of 1982.

Any uneasiness General Dynamics may have felt about adding yet another much-criticized weapon to a stable that already included the increasingly expensive and controversial Trident submarine must have been mitigated by the official mood prevailing in Washington. The attitude was summed up the chairman of the Senate's Tactical Warfare Subcommittee, Barry Goldwater. At a hearing not long after the tank subsidiary changed hands, the venerable Arizona Republican began the proceedings by saying the M-1 would become a magnificent weapons system "if and when we get all of the little bugs ironed out." But bugs or no bugs, the senator made it clear the United States needed the M-1.

"We have followed the Soviets through the T-54/55, the T-62, T-64 and the T-72," he explained before going on to invoke the ritual justification for just about any new weapons program, ". . . and we believe they are now developing one even more capable than the T-72." Continuing, he said that the Red Army's tanks were the subject of regular evolutionary improvements and wondered why, when the U.S. Army attempted to improve its own M-1 in the same way, it was "pilloried in the press."

Goldwater's use of the word "evolutionary" is an example of weapons-think semantics. Russian tanks are certainly the result of an evolutionary process; remember, the T-72 still depends upon a variation of the same French dirigible engine that has powered its ancestors since before

World War II. However, if one sticks to the evolutionary analogy, it soon becomes clear that the M-1 has crawled from a very different kettle of ooze. With its gas turbine, laser range finder, computers, and new gun, it is the intended progenitor of a new species rather than another, slightly different link in a long chain. If the first amphibians had been similarly unready to succeed in their new form and environment, the sole evidence of their existence would be a scant fossil record of the species' short-lived existence.

Having spelled out his military version of Darwin, Goldwater made a point of extending a particularly warm greeting to Walton H. Sheley, Jr., an investigator from the GAO. In one of its most recent reviews of the M-1, the independent auditing body concluded a sometimes favorable, sometimes scathing appraisal with two pieces of advice. The first was a general observation that the tank did everything except run reliably, the second, a specific warning that Congress take the precaution of "conditioning further appropriations for large scale production of the M-1 on the power train meeting the Army's durability requirements."

The Army's apparently contradictory desire to take delivery of tanks that did not meet its own performance specifications was explained by Army Under Secretary James R. Ambrose, who asserted that the problems were not as serious as they might appear to the untutored eyes of the GAO's civilian investigators. As for the problems, the next witness did his best to dismiss them as troublesome but essentially unimportant defects in an otherwise perfect piece of machinery. General Glenn Otis, from the M-1 project office, said that an expert panel had been recently convened to suggest remedies that would then be applied to all future tanks and retrofitted to existing ones.

In his earlier testimony, the GAO's Mr. Sheley warned of the optimism that was to be found in abundance at the Pentagon. Two previous "blue ribbon panels" had failed to remedy the M-1's ills, he said, adding that if this third group proved no more successful, the U.S. government would be forced into paying billions of dollars for tanks "whose power train components may need frequent replacement."

The search for solutions was to take much longer than twelve months, and may never reach a satisfactory conclusion, at least in respect to the tank's most fundamental deficiencies. As David Lewis predicted, General Dynamics' imported managers did revitalize the production line and eliminate much of the shoddy workmanship that was symptomatic of Chrysler's chaotic decline. According to official statements, the Army now regards the tank's durability in the field as "entirely satisfactory" and has demonstrated its faith by assigning the first M-1s to the Fulda Gap in Germany, the spot considered most likely to witness a Soviet armored breakthrough into Western Europe. The other questions, the tank's most basic flaws, remain immune to retrofitted remedies.

It still lacks the range of the M-60 and still carries less ammunition. Tread life is around half the predicted level, apparently because the young men who drive the M-1 cannot resist the temptation to make maximum use of the fifteen hundred horsepower at their disposal. Nor does it fulfill another task traditionally assigned to tank groups, that of escorting advancing infantrymen. This is now out of the question, since the tank's average "walking" speed is a good deal greater than that of the average soldier, while the rush of its seven-hundred degree exhaust gases makes close proximity extremely painful.

The last objection to the M-1 is perhaps the most basic of them all: It is too expensive. At roughly $3 million each, the M-1 costs between two and three times more than the old M-60 while performing many of its tasks less ably. A better, more practical, and far cheaper solution might have seen the good points of the M-1—its laser range finder and fire control systems, for example—grafted onto the older tank at a fraction of the final price. That, however, would have defeated the point of the exercise, the defense not of the U.S. or NATO but of the armor establishment's undiminished share of the procurement budget.

Not every weapon possesses the glamour of a charging tank, nor are all obliged to endure intense congressional scrutiny. But noticed or not, the same rules that produced the M-1 tank also work in the more obscure corners of the Pentagon. Here begins the story of our second case study in bizarre procurement decisions, a vital but seldom mentioned component of the national defense, the TAKX ships of Military Sealift Command. This time the beneficiary of the system was not Land Systems but the Quincy shipyards.

In 1982, Quincy landed contracts to build five cargo vessels specially designed to support the Rapid Deployment Force. The $792 million coup was summed up by David Lewis in his chairman's preface to that year's annual report: "Quincy maintained a strong research and development effort which paid off late in the year when contracts were received to build and operate under charter five ships of the new TAKX class. These ships will provide equipment, fuel, ammunition and other matériel for this country's Rapid Deployment Force. General Dynamics will be responsible for obtaining equity and long-term financing of these ships which are to be chartered to Military Sealift Command for five years with options for up to 25 years."

It sounded like a wonderful deal, and indeed, for General Dynamics, its stockholders, and the consortium of banks and other investors lucky enough to have a slice of the action, it represented a potential gold mine.

Unfortunately, their gain was the taxpayers' loss. As before, a little background is required.

Any army on its way to war needs ships to carry its equipment, a job traditionally handled, in this country, by the Navy's own private cargo and tanker fleet, under the control of Military Sealift Command. In the early 1980s, in response to former President Carter's decision to establish a military force capable of giving battle in any future trouble spot within days of the decision to mobilize, the Navy asked Congress for an initial $2 billion to buy a number of medium-sized fuel tankers and cargo vessels, the latter equipped with their own heavy cranes, helicopter launching pads, ramp-loading rear doors, and broods of flat-bottomed landing craft that could be slung over the side to shuttle supplies up to the beachhead. Congress was sympathetic in principle to the aims of the Rapid Deployment Force, but it balked when asked to authorize several billion dollars for ships that would do little more than float about the Indian Ocean and the eastern Mediterranean waiting for trouble to erupt. Determined to get their share of the RDF bonanza, those in charge of Sealift Command began to wonder if there might not be some way to obtain the ships without running the gauntlet of congressional approvals. What followed remains a testament to the wiles of the bureaucratic mind, not to mention the Navy's cavalier disregard for orders and its contempt for the public purse.

Shortly before his sweeping victory in November 1980, Ronald Reagan issued a position paper entitled "A Program for the Development of an Effective Maritime Strategy." Among its other suggestions, the paper seized upon Sealift Command as an example of the advantages to be had in injecting the profit motive into government operations. "The Navy today is facing a critical shortage of trained personnel. With the commercial industry assuming increased responsibility for many auxiliary functions, substantial cost savings can be achieved and a large reserve of manpower released to provide crews for the growing naval fleet," the relevant section of the document began, before concluding on a note of hopelessly misguided optimism. "This is an example of the way we can increase defense mobility without adding burden to the taxpayers."

What the President had in mind was the near-dismemberment of Sealift Command. Under the Navy's control, support and supply vessels are manned by civil service sailors who collect their weekly paychecks whether the ships are at sea or tied up in their home ports. Commercial ship operators, on the other hand, take on full billets only when their ships are ready to put to sea. The Reagan plan promised distinct advantages for all but the Navy. The federal government would cut its expenses, while America's ship owners and operators saw the prospect of taking over Sealift Command's old function as a much-needed shot in the arm for an industry increasingly beset by cheaper overseas competition.

Safely in office, President Reagan instructed the Navy to put his theories into practice. The reactions at the upper levels of Sealift Command would not have been much different if the White House had ordered every state secret in Washington loaded onto one of its ships and consigned to Vladivostok. Handing over operations to private enterprise would mark the beginning of the end for the cargo fleet's bureaucracy. In 1970, when the Vietnam war was raging, there were 253 ships shuttling supplies around the world, their operations overseen by a single two-star admiral. Twelve years later, the war was over and the number of ships had fallen to 150, but the Washington headquarters' staff had grown to number a two- and a three-star admiral, with a work force of more than a thousand deskbound underlings that was soon to be bolstered by the addition of a further 450 civilian employees. With Sealift Command's cozy sinecures in danger of extinction, the Navy decided to fight back.

In May 1982, Navy Secretary John Lehman invited bids from companies interested in taking charge of three small oil tankers used to transport fuel to minor military installations in the Pacific. Nine companies responded to the invitation, or rather attempted to respond, because the lack of official information made it almost impossible for any of the tendering firms to beat the "inside" price being prepared by Sealift Command's own extremely imaginative estimators.

According to a lawsuit filed in July 1983 by the Joint Maritime Council, a trade organization representing U.S. ship owners and operators, the competition was conducted with the specific intention of proving the Navy's contention that civilian-operated ships would lead to greatly increased costs. It was, in the words of the complaint, "a sham competitive procurement . . . arbitrary, capricious, an abuse of discretion . . . without observance of procedure required by law."

The predetermined outcome of the competition was exemplified by the absence of critical information on the *Nordaway,* the oldest and by far the sorriest of the three T-1 class tankers. The rear hull sections were leaking so badly, bilge pumps had to be operated around the clock, while there was so much structural damage and faulty equipment above the waterline that the Navy had booked the ship into a South Korean shipyard for a bow-to-stern overhaul. The tendering firms knew nothing of the impending renovation, since the scant details laid out in the official Request for Proposal stated only that the successful bidder would be expected to maintain the tankers "at high standards of safety and repair." Understandably, each of the "outside" entrants included the cost of a massive refit in their estimates. Sealift Command, however, did not, calculating its price upon the lower maintenance and operating costs of a fully overhauled vessel. Since the cost of essential repairs ran into the millions of dollars, the Navy's sleight of hand would have been enough in itself to knock Sealift Command's erstwhile competitors out of contention. But the Navy was not prepared to leave well enough alone. Intent on demolishing the President's dangerous fancy once and for all, it denied the entrants access to repair and dockyard records, essential information without which no bid could be anything more than educated guesswork. The last and by far the most blatant indication of the Navy's real intent was its insistence that the bids be based on a firm figure of 330 sea days per year, even though, under its own control, the three tankers had averaged only 180 days at sea over the same period. This provision totally negated the rationale behind the presidential directive. If the tendering firms were obliged to keep their ships fully manned even when tied up in port, they were not being asked to better the Navy's record but to match it, inefficiency for inefficiency.

The result was announced late in 1982: Each of the tankers was to remain under the control of Sealift Command. The verdict would later be set aside, when the Joint Maritime Council sought and obtained an investigation by the Defense Department's inspector general, but in the meantime, the Navy brandished the result as the ultimate refutation of the President's preference for private enterprise. More important, it used the competition as a forward base from which to proceed with the demolition of the greater threat to its future, the possibility that the TAKX amphibious landing ships for the Rapid Deployment Force would also go to the private sector.

Having proved to its own satisfaction that the President's idea was fatally flawed, the Navy was really no better off, since it still had to contend with Congress' persistent refusal to grant the money it needed to buy the ships outright. The solution was found in yet another of the Reagan administration's initiatives to improve the lot of large corporate investors.

In mid-August 1982—shortly before General Dynamics landed its five construction contracts—the Navy took receipt of a consultant's report that advanced what purported to be a universally acceptable alternative. Instead of buying the ships, or allowing them to be manned by outsiders, the Navy could charter them from a civilian operator and man them with its own people. This would save at least 20 percent on the estimated cost of purchasing them outright, the report asserted. Later described by a General Accounting Office spokesman as "a mockery of common sense" in testimony before the Joint Committee on Taxation, this critique was repeated by an outside accounting firm, Coopers & Lybrand, and management consultants Booz Allen & Hamilton, which both maintained that the Navy's plan would increase costs by between 20 and 60 percent beyond the cost of simply buying the ships outright.

Implemented within weeks of their own internally prepared report, the Navy's scheme also promised numerous advantages to those companies lucky enough to get in early. The details read like an adventure playground for accountants:

General Dynamics would build the five TAKX ships at Quincy, financing their construction with money put up by a consortium of investors that counted Manufacturers Hanover and Morgan Guaranty banks among its most prominent members.* Borrowing the $1.9 billion required to begin work might have been difficult under normal circumstances, but here the risk was minimal since the venture was backed by the Navy's assurance that each of the boats would be put to work the moment it entered the water. Meanwhile, investors would be able to buy equity shares in the ships, repaying the initial bank loans while simultaneously turning each vessel into a particularly lucrative tax shelter. The investors would be entitled to immediate tax credits amounting to 10 percent of the estimated $178 million final price for each boat, as well as enough accelerated depreciation allowances to bring the total tax relief to around $1.3 billion over the first three years alone. "But don't call your broker and try to get a piece of this deal," *Forbes* magazine warned its readers. "Wall Streeters say that it is so good the equity positions will almost surely be snapped up by the construction lenders and shipbuilders."

There was only one potential pitfall to this cozy arrangement, a strong probability the Internal Revenue Service would refuse to accept the Navy's premise that the charter agreements were "service contracts" rather than conventional leases. If that proved to be the case, most of those lush tax deductions would melt away like morning dew. Determined to get its own way, the Navy moved to reassure the investor-builders by guaranteeing an after-tax return of 11.745 percent on invested capital. This meant that if the investors were stripped of their tax deductions, the Navy would return the difference together with all penalties, interest charges, and legal fees incurred in skirmishes with the IRS. The money, so General Dynamics and the others were assured, would be drawn from the Navy Industrial Reserve, a floating cash account usually devoted to capital improvements at naval dockyards. The inconsistency of the Navy's position is staggering: Whatever the IRS took with one hand, the Navy pledged to return with the other.

The happy result is recorded in both the files of Sealift Command's contract office and David Lewis' 1982 letter to his stockholders. General Dynamics received its contracts, the banks put up the cash, the Navy got

*Altogether thirteen ships were involved. The others, including large T-5 oil tankers, were to be built or converted by Maersk Line and Waterman Steamship Corp.

its ships, and all concerned had a good laugh at the expense of Congress, the IRS, and the quaint notion that presidential directives are meant to be taken seriously. The contempt for common sense remains the saddest feature of the whole affair. As it stands now, the total cost at the end of the twenty-five-year lease period is anyone's guess—and the Navy will still have to pay for its own repairs as well as the cost of full-time crews. The patent illogicality becomes even more pronounced when one considers the job for which the TAKX ships were conceived. If on some dark day in the future the Rapid Deployment Force actually goes into action, there is every chance one or more might suffer a fate similar to the *Atlantic Conveyor,* the press-ganged British container ship sunk by an Exocet missile during the Falkland war. If that should ever happen, the U.S. Navy's loss would be more than a ship; it would also have to recompense General Dynamics and the other owners for their lost investment.

13. One Last Scandal

Those who worked at Groton for P. Takis Veliotis saw them first, tiny cracks in the familiar arrogant facade. In someone less renowned for icy self-control, the symptoms of mild anxiety might have gone unnoticed. But Taki's was such a larger than life presence, the small moments of perplexed distraction assumed a prominence they would not otherwise have had. On his flying visits to corporate headquarters in Saint Louis, the man reputed to be General Dynamics' toughest executive grumbled uncharacteristically about vague ailments and recent visits to the doctor. At Electric Boat and other of the company's outposts, he made fleeting references to family problems that required immediate attention in his native Greece. There were even rumors that he was about to tender his resignation, strange behavior indeed for a man viewed by many as a future leader of the world's most diverse arms conglomerate.

It is easy to understand why David Lewis would have been loath to let the imperious Greek slip away. Thick-set, heavy-jowled, and all of six feet three, he was said to have stopped the chatter at cocktail parties merely by walking into the room. He had certainly stopped a congressional committee in its tracks; around the executive offices of Electric Boat, that particular performance had passed into the realm of legend.

Admiral Hyman Rickover, the head of the Atomic Reactor office and the man popularly identified as the Father of the Nuclear Navy, had declared war on Electric Boat in 1977, accusing the yard's executives of deliberate fraud, criminal deception, and shoddy workmanship during the construction of eighteen 688-class fast-attack submarines.

Having agreed to build the boats for around $1 billion, Groton lodged a series of supplementary claims for cost overruns that ultimately amounted to more than $800 million. The Navy's Claims Settlement Board balked at the increases and reduced the figure to a mere $125 million in a decision that pleased no one. General Dynamics sued the Navy, while Rickover, loudly proclaiming that the company did not deserve a penny, selected from the thick log of claims eighteen items worth $13 million, analyzed them, and forwarded his findings to the Justice Department with a demand that criminal proceedings be instituted without delay. The claims were a catalog of greed, he said, a compendium of distortions, fabrications, and fraudulent deceits. When there was no action, the admiral returned to his favorite forum, venting his anger before various Senate committees. Veliotis was the star witness for the defense.

The men were studies in contrast, both in physique and in their styles at the witness table. Where Rickover repeated his accusations to the senators with the colloquial candor of general-store gossip, Veliotis struck for the high ground. Rather than counter Rickover's persistent charge that General Dynamics and other suppliers did not care if they were "building ships or manufacturing horse turds," Veliotis wrapped himself in the mantle of an offended, and highly indignant, patrician. His most famous performance was pure theater.

Leaning back from the table, he produced a bag of jangling iron plates and scattered them before him. This was the raw material of the nation's defense, he said. It was hard, cold, unforgiving metal, not some drawing board fantasy that could be corrected with an eraser and a french curve. If the Navy would refrain from making last-minute changes to the plans it supplied his workers and stopped Admiral Rickover from treating the shipyard like a private playground, Veliotis promised, the overruns would end. Until those conditions were met, he refused to guarantee anything, not even that work would continue.

It had been a masterful performance, but it belonged to the past. The Veliotis who drifted about General Dynamics' far-flung dominions during the early months of 1982 evidently had much on his mind. Another of the rumors doing the rounds maintained that he was about to be hauled over the coals by a grand jury investigating charges that he had extracted millions of dollars' worth of kickbacks from subcontractors in return for promises of work. Many found the theory attractive, particularly those whom Veliotis had offended, or shouldered aside, in his spectacular rise. Few, however, found it credible, even allowing for General Dynamics' historic propensity for drifting blindly into scandal. Would the Pentagon's most diverse supplier jeopardize $13 billion worth of backlogged contracts by elevating to its board a man likely to end up behind bars? The idea was just too farfetched. If the grand jury really was the source of Taki's restiveness, then it had to be nothing more than a case of preappearance nerves.

David Lewis evidently believed in Veliotis' innocence, if indeed he thought much about it at all. There had been any number of warning bells, some dating back as far as five years, but they had all been ignored. The most recent sounded in mid-1981, when Navy Secretary John Lehman successfully demanded Veliotis' removal from the shipyard before agreeing to restore Electric Boat to his list of approved Navy shipbuilders. Lewis agreed, reluctantly. Taki could be abrasive, painfully so at times, but he was also effective. "He was a valued man, tough as hell to live with, but he had done a good job," Lewis admitted some time later. "He had alienated a few people here and there in the Navy and a few other people in our company. But still, good shipyard managers are not easy to find."

On his first day at Groton, Taki sacked three thousand people, and then responded to union complaints by eliminating four thousand more. A couple of months later, he solved the Rickover problem, the bane of all previous managers, by the simple expedient of refusing the admiral access to the plant. Profits and man-hour productivity were both up, overheads diminished, and quality control, at least according to Veliotis' own reports, was very much improved. Even more impressive was his toe-to-toe slugging match with the Navy. During one of the periodic outbreaks of rhetoric and litigation that marked the overrun dispute, he remarked that the U.S. government should order the next generation of nuclear submarines from Russia if it was not prepared to reimburse its weaponsmiths' fair and reasonable costs. The implied threat that Electric Boat might shut down rather than submit was fully appreciated in Washington. Quite apart from the fact that the submarine builder was the largest employer in both Connecticut and Rhode Island, there was only one other qualified submarine builder in the country, and that lacked the experience and the facilities to produce the new Trident submarines, the planned mainstay of the nation's seaborne nuclear deterrent into the next century. The confrontationist approach produced results. Although the Settlement Board had recommended only $125 in additional payments, Electric Boat managed to obtain all of $486 million.

As a reward, Lewis created a new division. Taki became a vice-president in charge of marine operations and international affairs. The title had an appropriate diplomatic ring, since Veliotis was to be ambassador at large and chief salesman to the world. Throughout its eighty years in the arms trade, the company had looked to just one customer, the U.S. government, for its livelihood; now that situation was supposed to change. The multilingual Veliotis would be dealing directly with civilian cabinets and military establishments, dispensing not just the maritime products in which he specialized but the entire General Dynamics arsenal of jet fighters, large and small missile systems, tanks, and robot gun systems. Around Saint Louis it was confidently predicted his first success would be in Greece, where the air force was casting about for a new air superiority fighter. That deal alone would be worth some $1.6 billion.

There may also have been another reason for Veliotis' appointment, one that had more to do with boardroom politics than with international affairs. In the estimation of Lewis and several other shapers of company policy, Taki was a little too ambitious and much too impatient. It was taken for granted that he considered himself the most suitable replacement for Lewis when the current CEO, then sixty-four, announced his anticipated retirement at some point over the next two or three years. A good deal more disconcerting, however, was a rumor that soon reached Henry Crown in Chicago. Acting at Veliotis' request, E. F. Hutton, the Wall Street financial house, was said to have prepared a computer analysis of a plan to seize control with an offer of fifty dollars a share for stock, then selling for around thirty-one dollars. Crown made little of the episode. "I thought it was ridiculous," he said later. Others who were about at the time remember it differently, pointing out that while the takeover scenario may have been overly ambitious, it proved that Veliotis needed watching, closely. "They thought that keeping him on the team meant that he was playing on their side," a former General Dynamics executive recalled. "What they failed to see until it was too late was that Taki never played for anyone but himself."

The breadth and depth of Taki's ambition made his decision to resign all the more astonishing. Late in April 1982, he asked to be relieved of his duties as soon as possible. He was in poor health, he said, and now wanted a more sedate life to settle his mother's complicated estate and take charge of the family shipping business in Athens. Lewis was taken aback and urged his associate to think things over. Two weeks later, Taki was back in Lewis' office to reaffirm his decision and demand that his resignation take effect within ten days. Once again, the request fell on deaf ears. On May 6, four days after receiving the second tender of resignation, Lewis and the other directors reconfirmed Taki's seat on the board, despite the fact that the man in question was at home in Milton, Connecticut, preparing for an extended overseas trip.

Much later, when all the talk was of "twenty-twenty hindsight," General Dynamics officers were apt to profess amazement at their own naïveté. It was certainly warranted in Lewis' case, because he, more than anyone else, had had every reason to doubt Veliotis' motives. Some months previously, Veliotis announced that he had been summoned to New York by a grand jury looking into the kickback allegations. There was nothing in it, he added, assuring Lewis that he was eager to take the stand and settle the matter once and for all. Yet only five days before his appointment in court, Veliotis wanted out. More suspicious minds might have seen something other than mere coincidence in his timing. Even though the chain of cause and effect that led to Taki's sudden departure was forged of circumstantial links so obvious they should have compelled action, Lewis retained total faith in his chief troubleshooter.

The first steps on the road to the impending, but still unrecognized, crisis were taken in 1973, when Lewis, not long at General Dynamics himself, went in search of someone to straighten out the Quincy shipyard. The quest led him to Quebec and the Davie Shipbuilding Company, where Veliotis, then forty-seven, ran the operation with the unchallenged authority of a feudal lord. He had been there since migrating from Greece in 1953, and had spent the last ten years in a succession of increasingly important jobs that provided ample opportunity to indulge his talent for aggressive bargaining. He beat down subcontractors, enforced penalty clauses, played hardball with the unions, and made a pile of money for the Canada Steamship Company, Davie's parent. "We built every ship for less than we built the one before," Richard Lowery, his old boss, fondly recalled. "Every year he ran the yard, we made money." Lowery offered a part ownership in the yard, but Veliotis rejected it outright. At Davie he was a big fish in a small pool; with General Dynamics, he would be part of a company large enough to match his ego and ambitions.

In mid-1973, Veliotis arrived at Quincy with a fearsome reputation and a loyal deputy, James Gilliland, a Scottish-born technical manager, whose career had prospered in his master's shadow. What they found was a situation not unlike that which confronted Fort Worth in the wake of the F-111. The fourteen-ship contract with the Navy that helped to precipitate Roger Lewis' downfall had finally run its course, concluding on an appropriately sour note when Henry Crown ordered a final $30 million worth of additional write-offs on top of the $240 million posted by the previous administration. Veliotis' first task, even before he went in search of fresh work, was to knock the yard into some semblance of order. The Navy ships had been marred by an initial outbreak of overconfidence that saw Quincy take on far too much work and at prices that left no margin for error or bad luck. Unfortunately, there were plenty of both. Faced with designing three distinctly different types of vessel, Quincy's naval architects and engineers soon realized they could not handle all the work on their own and began farming out subcontracts to consultants and sometime competitors. The organizational logistics proved to be an expensive nightmare. Plans arriving from different sources failed to overlap or simply did not turn up on time. It was the old familiar story, this time set on the banks of the Fore River rather than the flatlands of northern Texas. Overheads soared, the work force boomed, while profits and productivity dropped through the floor. In addition to the last batch of write-offs, General Dynamics had just sunk another $30 million into capital improvements; Veliotis was to be the man who made sure the company received a return on its investment. "If the job was not done the way he wanted, you could pack your bags and go because your career at Electric Boat was over. He reamed out the place from bow to stern," a former employee recalled.

Things began to look up almost straightaway. Veliotis began by fighting a war of attrition with the yard's construction unions, beating down their pay demands over the course of a four-month strike while simultaneously winning the repeal of several internal work rules that had contributed to the yard's excessive overheads. The folks in Saint Louis could not have been happier. Quincy's unions were tough nuts, but the new man whipped them into sullen compliance. Just as Lewis had hoped, Veliotis was every bit the iron fist in the iron glove.

He was also a good deal more than a hatchet man, as was manifested that same year when he struck a $700 million deal with the British-based Burmah oil company's tanker and transport subsidiary. As the nature and background to the contracts of five large liquefied natural gas tankers made abundantly clear, he was both an astute commercial negotiator and an appreciative student in the ways of the federal government, without whose help the Burmah deal would never have been possible.

A little background is required. Under the letter of the law, the U.S. Maritime Administration is empowered to grant loan financing guarantees to American shipowners engaged in local and international trade. Burmah's problem was simple: It was not a corporate citizen of the United States, and without federal assistance the ships would never be built, at least not at Quincy. The scheme worked out by Veliotis and the head of Burmah's shipping operations, fellow Greek Elias Kulukundis, succeeded in changing that situation, although not without a good many preliminary problems and embarrassments.

Five corporations were set up, one to sponsor each of the ships. Incorporated in Delaware and known as the Cherokee group, the five shelf firms were to "buy" the tankers and lease them back to Burmah, which would then be free to operate the fleet as its own in all but name. General Dynamics' involvement was everywhere apparent in the groundwork leading up to the applications. When approval for the loan guarantees was finally granted, three years following the application's filing, the decision was made only after General Dynamics acquired an equity interest in several of the corporate entities. That happened in January 1977, by which time Burmah's relationship with one of Washington's more notorious figures had thrown the deal into the spotlight of congressional scrutiny.

Burmah's first problem, soon to become General Dynamics' as well, was a payment of some $3 million to Korean businessman Tongsun Park, the sometime operative for the Korean Central Intelligence Agency whose covert contributions to U.S. and international political leaders led to the so-called Koreagate hearings during the Ford administration. Burmah certainly needed a friend wise to the ways of the inscrutable East. Shortly before he parted company with Burmah in 1976, Kulukundis had entered into an agreement with the Japan Line shipping company to lease two supertankers at a cost of $40 million per year. The Greek had visions of his subsidiary growing into one of the world's largest oil and gas shippers, and he had spent billions, much of it borrowed, to lease, buy, and build tankers. The trouble was that there was neither enough oil to keep his fleet fully occupied, nor sufficient incoming revenues to service the debt. Driven to the edge of bankruptcy by its subsidiary's ill-advised investments, Burmah humbled itself before the Bank of England and begged for the financial assistance needed to ride out the crisis. As a concession, and an indication of its willingness to turn over a new leaf, the company agreed to the bank's demand that Kulukundis be shown the door. Though he was removed, the mess he created remained to be cleaned up.

That was where Park entered the picture. Kulukundis' replacement, American businessman and naval architect John J. McMullen, asked his corporate counsel about the chances of abrogating the supertanker leases. The legal opinions were not very optimistic. Japan Lines had agreed to take back the boats only if they were accompanied by a payment of $120 million. It was far too much for Burmah to countenance in its weakened state, so the matter was turned over to Tongsun Park, who claimed to be a close personal friend of Japan Lines' senior managing director, Misashi Matsunaga. The Korean proved to be as good as his word, and Japan Lines soon reduced its claim to $56 million. And that was not Park's only success. In addition to slicing Burmah's liability in half, he obtained an undertaking from the Japanese to pay 37 percent more for the Indonesian natural gas being shipped to Yokohama aboard Burmah's other ships. Park accepted the specified $3 million fee for their part in the settlement and divided the money among two corporate accounts held with a Bermuda bank.

In 1975, when congressional investigators began scrutinizing Park's records, the Bermuda bank accounts became an immediate subject for speculation and conjecture. The pattern of large deposits followed by equally impressive withdrawals and transfers to banks in just about every corner of the world suggested a well-oiled machine for the collection and distribution of bribes on a global scale. When he first set up the deal to supply Japan, Kulukundis established an intimate relationship with General Ibnu Sutowo, a senior officer in the Indonesian army and the head of the state-run Pertamina oil company which was to supply the liquid natural gas. So warm was the relationship that by 1976, the Indonesian government, hardly a model of propriety at the best of times, had charged

Sutowo with corrupt practices and placed him under house arrest. Further, the Indonesia junta of General Suharto also threatened to cancel the LNG supply agreements, arguing that the deals were "not the product of arm's length negotiations."

General Dynamics' position was particularly difficult. It had taken on the original LNG ship contracts in the expectation that loan guarantees would be easily obtained, but by January 1977, the ongoing investigations of Tongsun Park's relationship with Burmah, an SEC inquiry, and the lurid stories emerging from Indonesia had placed the future of the deal in limbo. Quincy had five partially completed ships on its slipways, five more ready to take their place, and a customer that would not be able to pay unless the guarantees were approved. Fortunately, it also had a friend at court, the Ford administration's secretary of commerce, Elliot Richardson.

The onetime ambassador to the Court of St. James's—a perennial Republican party appointee to high office—was in a good position to know a good deal about the mechanics of international bribery. Only a few months earlier, Richardson had served as chairman of a blue-ribbon presidential panel charged with formulating government policy on overseas graft by U.S. corporations, a body spawned by the earlier revelations of Northrop's and Lockheed's activities in Europe and Japan.

On January 20, 1977, his last day in office before making way for the new Carter administration's appointee, Richardson announced that the long-awaited guarantees had been finally approved. The commerce secretary justified his last-minute decision with a twelve-page press release, which stressed that the loan guarantees would safeguard five thousand jobs in his home state of Massachusetts. As for the accusations raised in Congress that a large part of Park's $3 million lobbying fee had been distributed as bribes, Richardson said that he was satisfied the amount did not relate to the loan guarantees in any way. Presidents normally make a point of appointing gifted men to their cabinets, but judging by the confident tone of his comments, Richardson's talents were truly unique. Along with his loyalty, intelligence, and facility with words, the commerce secretary was evidently something of a clairvoyant, because the Justice Department was still actively investigating Park's Bermuda bank account, while Burmah itself remained the subject of undiminished interest by the Securities and Exchange Commission.

For that matter, so too were David Lewis and three other senior General Dynamics officers. On December 19, 1976, at a time when the loan subsidies and Quincy's future hung in the balance, Lewis disposed of 100,000 shares and pocketed an even one million dollars in pretax profits. Henry Crown sprang quickly to his chairman's defense, dismissing as "ridiculous" the suspicion that Lewis and the others had sought to benefit from inside information. He and other members of the Crown family had bought 80,000 of the shares, he explained, adding that the remaining 20,000 had been divided evenly between two other members of the board, Robert Reneker and Albert Jenner. Just as it would do with several subsequent investigations of General Dynamics executives' trading activities, the SEC declined to cite any of the officials for breaches of the law.

On paper, if not on the high seas, Veliotis' reign of terror appeared to work wonders. The ships rolled down the slipways ahead of schedule and under cost, much to the delight of Saint Louis' executives, who were greatly impressed by the way the new recruit handled the many different problems. He fought and won another, prolonged strike in 1977, and somehow managed to keep the yard in the black despite significant increases in suppliers' prices. A little more than three years later, he was rewarded with a directorship.

Perhaps it was the combination of Veliotis' personality and his performance that led Lewis and the others to ignore some of the more unsettling aspects of his style. Burmah, for example, later complained that penny-pinching construction methods had tripled the tankers' annual maintenance costs. But what should have been the most disturbing evidence of Taki's rapacious appetite surfaced in 1975, when a small and relatively insignificant subcontractor, Ogden Engineering of Indiana, filed suit for payment of $342,235, money it claimed was owing on prefabricated fittings delivered up to a year earlier. Although it was not mentioned in the court papers, Ogden's executives were incensed by what they described as Veliotis' stand-over tactics. He had ordered the money withheld as a lever in an attempt to prize out a 20 percent share of the company for himself. The case was settled out of court not long after General Dynamics countersued for $9.4 million, almost ten times Ogden's worth in sales assets.

Another of Quincy's numerous subcontractors was more receptive to Taki's unorthodox suggestions. Gerald E. Lee, the chairman and CEO of Frigitemp Insulation, was a former sailor and high-school dropout who had taken charge of his family's small Brooklyn refrigeration business in the early 1960s and turned it into an $85 million a year success story. By 1973, the year Lee heard that Quincy's current insulation subcontractor, East Coast Cold Storage, might be in difficulties, Frigitemp was already working for Litton Industries' shipbuilding division and was at the point of exchanging contracts with Lockheed's marine subsidiary. Lee set out to add the Quincy contract to his collection. The plan required careful preparation and a new employee, George G. Davis, who joined the team after being offered the position of vice-president in charge of marine operations.

Davis, a large and beefy man whose down-home accent testifies to a Deep South childhood, was far from idle at the time he was taken on. The owner and operator of a number of small but successful companies supplying marine fittings and specialist labor to shipbuilders in the U.S. and Canada, he traveled enough to justify a private plane, which he flew himself. At Lee's suggestion, he canvassed his associates in Quebec and Montreal, hunting for someone who knew Veliotis well enough to help Frigitemp land the insulation and joining contracts for the first three ships. He found the first candidate very close to home. Peter Brading was the president of a Montreal marine supply and equipment leasing firm in which Davis had purchased a majority holding some two years earlier. Brading announced that he was on a first-name basis with Veliotis, adding that he had dined with him on the evening before his departure for Quincy. In return for a verbal promise of commissions amounting to 2.5 percent on future work, Brading undertook to approach his friend and arrange a meeting. "It was what I call a dog and pony show," Davis drawled from the witness box during his 1984 trial in New York. "We showed our slides and sketches of the insulation system we wanted to install on the LNG tankers. When we were finished, Mr. Gilliland told us they would let us know."

The answer, when it came, was most disappointing. Veliotis, who arrived late and left before the presentation was over, indicated that he would direct the insulation contracts for the giant spherical tanks to an Ohio firm, Pittsburgh-Des Moines. The setback only strengthened Lee's desire to land Quincy as a customer. Once again Davis returned to Canada, to sound out another prospect who claimed he could deliver Veliotis' signature on the dotted line. The new hope was Sukhame "Shook" Bose, Brading's occasional business partner and a native of Calcutta, who operated a fluctuating number of Asian shipping ventures from his Montreal home. He, too, demanded a commission; this time it had doubled, to 5 percent. Lee replied that Bose could name his own price if he proved as good as his word.

At this point, the details of the backroom negotiations become rather clouded, all too often buried by a mass of conflicting courtroom testimony that saw Bose, Davis, and Lee each accuse the other two of proffering the first bribe. Regardless of its origin, the promise of substantial under-the-counter payments worked wonders. At Veliotis' instruction and in the face of subdued opposition from his own technical advisers, Quincy withdrew from its negotiations with Pittsburgh-Des Moines and awarded the insulation contract to Frigitemp. Several weeks later, Lee pocketed a second contract, for joiner services. According to trial testimony, in return for his cooperation, Veliotis received an initial payment of fifty thousand dollars, which reached his numbered account with Crédit Suisse in Lucerne after being routed through a number of Davis' shelf companies in Canada and the Cayman Islands. The system of transfers must have met with Taki's approval, because it was not long before Gilliland opened his own numbered account, apparently on his friend's recommendation. Anyone in Quincy's accounting department who might have questioned why the big man was so keen to favor a company with minimal LNG experience and cost projections substantially higher than its competitors' evidently decided not to pursue the matter. All knew that Veliotis' word was law, to be challenged only at one's peril.

It is a little harder to understand how the public watchdogs in the Maritime Administration came to overlook what now appear blatant signs of foul play. Since all of the ships were destined to receive a federal subsidy of one sort or another, the shipyard was required to submit detailed estimates of the construction costs to a Washington review board. A senior Maritime Administration official and adviser to its Subsidy Board, John McGowan, later recalled noticing something about the figures, although not enough to warrant an investigation. Along with all its previous applications, Quincy had enclosed insulation estimates from several companies other than the eventual winner, East Coast Insulation. On the LNG ships, however, only Frigitemp's quotes were sent to Washington. "We looked askance at what was going on," McGowan said. Looked but didn't touch. Despite the insulation anomaly, all ten vessels qualified for subsidized mortgage insurance, while one received a direct grant of $30 million. One gathers that the idea of fraud did not occur to the technical-minded investigators. McGowan could recall only one instance of bribery in his twenty-seven years with the department, and that particular case required very little sleuthing since it was listed as a "finder's fee" in a statement of the applicant's incidental overheads.

The Frigitemp conduit to Lucerne was equally efficient, directing the payments through a series of dummy companies and bogus bank accounts in Canada and the Cayman Islands. Whenever a payment was due, either Lee or Davis rang Montreal and instructed Bose to prepare a fictitious invoice on the letterhead of any one of several bogus companies. Frigitemp was billed for insulation molds, consultants' services, and raw materials that were never made, rendered, or delivered. When the payments arrived in Canada, Bose transferred them to a local branch of the Bank of Nova Scotia, which passed them on to its office on Grand Cayman. There it was likely to remain up to a month, usually resting in an account belonging to one of Bose's Asian shipping companies, before being split up and transferred to its final destinations.

If the parties to the arrangement had restrained themselves, there is every reason to believe the pattern of contracts and kickbacks would be operating to this day. The Maritime Administration was happy. Quincy's ships were roughly 40 percent more expensive than vessels of comparable size and type then under construction in Asia, but judged against the inflated standards of the United States, they were considered cheap enough. David Lewis was happy, so much so that he resolved to place Veliotis in charge of the submarine operation as well as the civil shipbuilding division. As for Lee and the Frigitemp crew, their prospects could not have been more cheerful. The Quincy contracts added some $45 million worth of new business to the books, enough to guarantee a healthy profit and still leave room for the $1.8 million assigned to Veliotis and Gilliland.

Unfortunately, self-control was in very short supply at Frigitemp's Brooklyn headquarters. Since Lee, Davis, and Mel Silver, Frigitemp's president, were already diverting money to outsiders, not just at Quincy but also at Litton Industries' and Lockheed's shipyards, the temptation to skim a few extra dollars into their own offshore bank accounts must have been irresistible. Lee, a man with a well-deserved reputation for conspicuous consumption, was perhaps the worst offender. An eagerly awaited visitor to the casino islands of the Caribbean, he settled one particularly large debt in Haiti with money the corporate records listed as payments for raw materials. On another occasion, Frigitemp paid for renovations to his Park Avenue penthouse that included one charge of $150,000 for an Italian marble floor. Mel Silver bought houses, while Davis invested in a new plane, a motor yacht, and several shipping companies, including one of Bose's operations, Mamoni. Much later, when the former friends were at each other's throats in court, Bose claimed that Davis' investment was a long-term, low-interest loan in recognition of services rendered. Davis told a very different story. While he was milking Frigitemp—something he still denies—Bose was busy milking him, siphoning money, profits, and the proceeds of an insurance settlement on one of Mamoni's ships into a family company of his own. Still, Davis cannot have found the losses too painful. When he was charged with violations of the racketeering and corruption act in 1983, he produced $5 million in bail without so much as a murmur.

Joseph Blackman, a former Frigitemp executive, who quit in disgust before the final collapse, described his former bosses as "pioneers in the area of fraud." "They moved from the nickel and dime rip-off to the million dollar and the multimillion-dollar rip-off," he said.

Concealing the evidence presented a major challenge. If anyone had taken the trouble to compare the records of incoming supplies with the actual warehouse inventories, they would have noticed some astonishing discrepancies. In one instance, the books showed a number of substantial payments for timber needed to complete the decks of Navy ships under construction at Litton's shipyard in Pascagoula, Mississippi. However, a subsequent investigation revealed that the wood had not been bought; indeed, as Silver explained after being granted immunity as a government witness, "the trees were still growing in the forest." Litton's willingness to accept the situation without complaint was hardly surprising, he said, since at least one Pascagoula executive was also on the take. When delivery could not be put off any longer, Lee and Silver reverted to the last resort of embezzlers at the point of being discovered: they robbed someone else to make up the first deficiency. A federal investigator who was later called upon to unravel the maze of false receipts and conflicting documents recalled his difficulties. "It was just about impossible to figure who owned what and whose money had paid for it. Sometimes we'd find an item like wood or insulation materials that had been paid for three times by three different customers."

By late 1976, the conspirators must have known they were pushing their luck too far. But instead of backing off, Lee and Silver made one last effort to postpone the inevitable. Together, they hired a new financial officer, Emmett O'Sullivan, who soon realized that his chief qualification was innocence. "I guess I was the Mr. Clean they wanted to interface with the banks," he said. O'Sullivan spent the first half of his four-month stay at Frigitemp trying to make sense of the books and the second demanding answers that were not forthcoming. By the first week of April 1977, Lee and Silver were beginning to think they had made a major mistake. Terrified the new man might run to the authorities, they debated two possible solutions.

First came the soft approach. Lee invited O'Sullivan to stop by his Park Avenue home for a quiet drink and a chat with him and Silver about business. The weekend invitation carried the unmistakable promise of a bribe about to be offered, but O'Sullivan elected to attend anyway. After four months of frustrations and unsolved mysteries, he was owed an explanation, not to mention an apology. Had he known the extent of his host's mounting anxiety, he might have chosen to stay at home.

As Lee must have known, O'Sullivan proved immune to a variety of enticements. After rejecting a six-figure salary and a place on the board, the accountant confirmed Lee's worst fears by once again refusing to sign the company's quarterly financial statements. Lee's affable, low-key speech about the peculiar norms of the marine construction business did nothing to change O'Sullivan's mind. As he was leaving, O'Sullivan announced that he was going home to write out his resignation. Lee and Silver had been nervous at the start of the meeting, but they were panic-stricken by the time it was over.

Even by the standards of previous misconduct, the second plan to silence O'Sullivan was unprecedented. Silver and Lee had already made use of a certain New Jersey businessman and onetime company official, Michael Ferrarie, to deliver payoffs to various shipyard executives. In their panic they wondered if Ferrarie might not also be able to arrange a murder. The idea was discussed in depth at a meeting in a Newark sheet metal factory a few days later. In return for fifty thousand dollars, thirty-eight-year-old Michael Ferrarie agreed to have O'Sullivan plucked off the street and executed. Prompted perhaps by a morbid imagination or the first stirrings of a troubled conscience, Lee asked where the body would end up. That was easy, Ferrarie said, displaying a professional's eagerness to talk shop. O'Sullivan might disappear into a car crusher or he could end up in a meat grinder. The last was too much, even for Lee. Although Ferrarie was originally instructed to begin his preparations, the order was soon canceled when Lee, as he later admitted, "went hysterical" at the idea that O'Sullivan might finish up as a string of sausages.

While Lee and Silver were taking their problems to gangsters, Davis was preparing his own escape route. During the ten months that passed between O'Sullivan's departure and Frigitemp's final collapse, Davis stayed in constant touch with Veliotis, keeping him informed about the latest developments and preparing what he hoped would be a painless solution to their mutual problem. Veliotis realized that Frigitemp's impending bankruptcy would jeopardize his ability to bring in the last of the LNG ships on time and under budget. All of Frigitemp's equipment, tools, and supplies would be placed under bond until the receivers sorted out who owned what. Davis, too, had much to lose. His position at Frigitemp and his carefully cultivated friendship with Veliotis had made him a rich man, and he was loath to let it all slip away. The solution seemed obvious. Less than a month before Frigitemp went under, Davis incorporated a new company, IDT Engineering. A few days later, it was charged Veliotis fulfilled his part of the bargain by booting Frigitemp out of Quincy and reassigning the contracts to Davis and IDT. As an additional service, he pressed his legal department to drop a planned suit against Frigitemp, arguing that it would serve no useful purpose. A new contractor had already been found, he said, and the essential plans, tools, and molds placed under lock and key in General Dynamics' own warehouses. In return for his help, Veliotis allegedly walked away with additional payments of $1.35 million, which he split more or less evenly with Gilliland. It was money well spent, as far as Davis was concerned. IDT now held contracts worth more than $23 million, not bad for a company that had been in existence for less than a month, possessed little equipment of its own, and employed a work force of less than ten people.

And so the scam continued in another form. Davis took on workers, many of them former Frigitemp employees, and grappled with a monumental mess in the warehouses where General Dynamics had dumped Frigitemp's equipment. The job had been done hastily and poorly on the day Frigitemp's bankruptcy was announced. Delicate Formica molds were chipped or broken in transit, while several tons of full-scale plans were found to be so jumbled Davis had to hire a new warehouse, spread each out on the floor, and recatalog them like pieces in a giant jigsaw. As if that was not bad enough, he soon found himself fighting a legal battle to recover damages from General Dynamics, something Veliotis did nothing to stop.

Given his problems, Davis must have been in a poor frame of mind when Lee and Silver threatened to deliver a boxload of sensitive documents to the FBI unless he agreed to pay each man $100,000 a year. Mel Silver was particularly incensed when Davis refused. It had been IDT's seizure of the General Dynamics contracts that precipitated Frigitemp's ultimate collapse, and Silver seemed to think he had a moral right to share in Davis' good fortune. And there was another, more pressing reason, which Silver tried to explain during a heated telephone conversation. Now that Frigitemp had gone under, he and Lee were both under siege by lawyers representing the insulation company's creditors. "It wouldn't look good," Davis recalled Silver saying, "if we were going around spending money so soon after going Chapter Eleven." Davis was unmoved. He gambled that Lee and Silver would not want the U.S. government to join the ranks of their inquisitors, and dared them to do their worst. Their bluff called, the would-be blackmailers backed off, muttering curses and what, at the time, were hollow threats.

Davis' conviction that he could avoid detection was curious, to say the least, since it was based on the dubious assumption that Lee and Silver would emerge from the bankruptcy investigation unscathed and unindicted. The logic seems to have hung upon the fact that his scams had been somewhat more discreet than those of his former partners, who could not finger him without further implicating themselves. By late 1979, however, that hope was long gone. Having dug through layer after layer of deceptive bookkeeping, the investigators were beginning to see the first hard evidence of massive fraud. With their own survival now at stake, Lee and Silver began looking for someone to share the blame, and Davis, the man who had spurned their request for "a little walking-around money," was an obvious candidate. As Davis slid into the scandal's vortex, Veliotis must have known he could not be far behind.

The hammer fell with a muffled blow in February 1980, when Veliotis and Gilliland were summoned to Boston for deposition interviews with Lawson Bernstein, Frigitemp's receiver in bankruptcy. David Lewis was understandably concerned when he heard the news, but not enough to question Veliotis' assurance that he was the innocent victim of a smear campaign mounted by the real villains. The deposition session might even be for the best, he said. A relatively informal affair, the perfect place to quash the rumors before they surfaced in the public forum of a full-blown court case. Lewis had no doubts about Veliotis' ability to handle himself under oath, but just to be sure, it was decided to send two General Dynamics lawyers along to look out for the company's, and the witnesses', best interests.

Just as Taki predicted, the interviews went off without any major embarrassment. The General Dynamics executives both denied taking bribes and maintained they had little personal contact with any of Frigitemp's principals. Some of the questions about the transfer of contracts to IDT were a little harder to field, but Veliotis managed to scoff at suggestions that he had been party to a conspiracy. The decision, he said, was the result of an understandable desire to avoid disruption in the production schedule, and nothing more. The ninety-minute session ended with what must have seemed a reasonable request when the General Dynamics lawyers repeated their demand that Veliotis' and Gilliland's testimony be sealed, hopefully precluding its use in any and all future court cases. Much later, when the Veliotis affair was well beyond Lewis' control, that request was to be the source of considerable embarrassment in Saint Louis.

The civil investigation bubbled along for two more years, gradually accumulating a volume of circumstantial and documentary evidence that attracted the attention of federal prosecutors. A grand jury was convened to examine Frigitemp's collapse and the question of who made off with how much. Predictably, one of the first demands was for the sealed records of interview compiled by Lawson Bernstein during the Boston deposition session. The reaction in Saint Louis was one of utter dismay. General Dynamics and Veliotis each mounted independent court actions to restrict the grand jury's access to the files, a strange position since both plaintiffs were adamant they had nothing to hide. When the decision went against them, Veliotis made what was to be his last, vain attempt to postpone the inevitable. If he could not stop the grand jury, he realized it would be only a matter of time before he was summoned to the witness box. So, even though General Dynamics had declined to lodge an immediate appeal, Veliotis mounted one of his own, buying a little extra time to consider what few options were left.

That Veliotis managed to contain what must have been a growing sense of panic until almost the last minute was entirely typical. Up to those final days in April and early May, when he began to grumble about his health and talk wistfully of retirement, he was the picture of aggressive executive efficiency. He continued to pursue an antagonist's approach in his dealings with the Navy right up till December 1981, when David Lewis bowed to Navy Secretary Lehman's demand and promoted his toughest executive to the position of vice-president in charge of international operations. There was no hint of admonition in the appointment, nor was it taken that way in the wider world outside Saint Louis. On Wall Street and in the financial press, the move was generally seen as an indication that General Dynamics was going after the international arms market with the same enthusiasm that had made it the Pentagon's leading supplier, an impression reinforced by Veliotis' exuberant predictions about his new job.

The recent promotion was one of the things that made the events of May 1982 so hard to accept, particularly for David Lewis, who continued to invest his total faith in Veliotis until the bitter end. After tendering his second letter of resignation—the one the board chose to ignore when it elected company officeholders at the annual meeting on May 6—Veliotis left for Greece, flying out of New York less than nine days prior to his appointment with the grand jury. The apparent coincidence troubled Lewis not at all. He still clung to the hope that Veliotis would reconsider his resignation, and firmly believed that the grand jury would not pose any major problems. The last thing he expected was the terse message to the legal department that arrived from Athens less than two days before Veliotis' scheduled appearance in the witness box. He would not be coming home, it said, nor would he cooperate with the grand jury. As Lewis later recalled, the news that Veliotis was taking the Fifth Amendment was a "devastating blow." There was, however, much worse to come.

In his haste to flee America, Taki was obliged to leave some $4.6 million worth of personal property behind. His Connecticut estate, his motor yacht, parcels of shares, an interest in a group of Massachusetts banks, and other assets were frozen by General Dynamics, which now faced the unpleasant possibility of having to reimburse Frigitemp's creditors for the money Veliotis and Gilliland had received in kickbacks. The seizure was like waving a red rag at a bull.

Although technically a fugitive, Veliotis maintained anything but a low profile. Quite apart from suing General Dynamics and, later, *Newsweek* magazine over a story that claimed he was a security risk, he began to tell a growing throng of visiting U.S. newsmen that they were missing the real story. Why bother with small fish like him when sharks like General Dynamics were left unmolested. In comparison with the misdeeds of his former employer, any indiscretion of his own amounted to no more than small change. General Dynamics dismissed the allegations as the desperate lies of a rogue and embezzler but the denials were not entirely convincing. What made the assertions of scandal and fraud so much harder to refute was the fact that Veliotis had packed an insurance policy before fleeing to Greece. Upon arriving at Electric Boat after his stint at Quincy in 1977, Veliotis discovered that one of his predecessors had installed a speaker phone system in the general manager's office. One of those petty symbols of corporate status, it had seen some occasional use when lesser executives were gathered before Veliotis for conference calls with Saint Louis. Veliotis found an entirely different use for it: recording his conversations with Lewis and other company officials on a growing library of cassette tapes. It was their words, he said, rather than any unsupported accusations of his own that presented the final and undeniable proof of his assertions. It was not long before carefully selected transcriptions of the conversations began to appear in the press.

The seeds for this last and latest General Dynamics scandal had taken root. By the time the investigations had run their course, David Lewis had been forced to resign, a grand jury was investigating a former assistant Navy secretary, George A. Sawyer, and many of the company's most embarrassing secrets had become part of the congressional record. The old joke had come true once again. In its brightest and most prosperous hour, General Dynamics was once again about to be struck by a bus.

14. Messing Around with Boats

Back in the days of Jay Hopkins, it was not uncommon to hear General Dynamics executives referring to their fiefdoms as close and happy families. The analogy might have been convenient for the purposes of public relations, but it came nowhere near describing the actual situation. As in any large and wealthy clan, particularly one divided by both geography and individual inclinations, there have always been favorite sons and black sheep, outstanding success stories and sources of shame. During the fifties, Electric Boat and Canadair had been the family's paragons, while Convair's lumbering B-36s, its accident-prone Hustlers, and the financially disastrous jetliner program figured as its chief disappointments. Twenty years later, thanks to the F-16 and Groton's inability to keep its house in order, the situation had been neatly reversed.

As success piled on success, Fort Worth's execution of the F-16 contracts became a fixture in the speeches of congressmen and Pentagon critics, who cited it repeatedly as an example of the way a military procurement program ought to be run. Under the plant's general manager, Herb Rogers, the airplanes rolled out on time and more or less in accordance with the promised delivery prices. Even when the schedule was threatened by factors outside Fort Worth's control, Rogers and his team always managed to get it back on the tracks. At one stage in the late 1970s, for example, the plant's engineers learned that large quantities of "soft" aluminum were causing problems at construction sites and car assembly plants across the country. "We damn near turned those planes inside out to find out if any of the stuff had found its way into our supplies," Jim Ashton, a former Fort Worth executive who would figure prominently in Electric Boat's later troubles, recalled in 1985. None was found, nor were the production schedules disrupted to any great extent.

Thanks to P. Takis Veliotis, things were also much improved at Quincy. The yard still had its problems—quality control was not what it should have been and general, unanticipated overheads continued to figure in financial statements—but at least it was not losing money. By the early months of 1977, thanks to the LNG ship contracts and Commerce Secretary Richardson's timely decision to approve the loan guarantees, the cash flow was sufficient to cushion the yard's worst inclinations to industrial anarchy. Electric Boat, by comparison, was an unredeemed mess, with a long list of troubles that grew more acute with every passing week. Alone among the company's corps of senior executives, Veliotis boasted the record and the drive to return the enterprise to an even keel.

Veliotis was well aware of the peculiar perils of the submarine business. Although it now poured its energies into the construction of surface ships, Quincy's history as a builder of submarines stretched back to the days of Isaac Rice, Charles Schwab, and the deal to supply the British with a fleet of covertly constructed boats that had raised so many eyebrows and produced such lucrative profits in 1917. Its submarine successes were, however, ancient history. The last of Quincy's submarines had been launched less than five years before Veliotis took charge and the memories they evoked were not pleasant. Like just about every other project the yard handled during the sixties, the two boats were neither profitable nor free from mishaps. "If you screwed up on a surface ship they would send you to a submarine to work on," one Electric Boat technician complained to naval writers Norman Polmar and Thomas Allen. "The tradesmen—the craftsmen—did not like to work on submarines; they didn't like to work in tight corners."

The seeds of Electric Boat's then and later misfortunes were sown in 1968, when Admiral Rickover began lobbying Congress for the funds to build a new generation of nuclear attack boats. As usual, they were to be larger, faster, and deeper-diving than any of their predecessors. The vast weight of Navy opinion was directly opposed to the scheme, and the reasons were not solely the result of the surface fleet's desire to protect its own budget and prestige, the self-serving explanation Rickover so often presented to his political allies on the Hill.

Until 1970, when the admiral pulled out all stops to get the program off the ground, the most advanced nuclear attack submarines were relatively modest affairs like the SSN *Sturgeon,* the same class that had caused Quincy so many problems during the mid-sixties. Rickover was not opposed to the older boats; indeed, he was hardly in a position to say anything against them, since he had worked assiduously to win congressional funding for them some ten years earlier. It was just that by the start of the new decade, he felt something "more capable" was needed, and that, in the Rickover lexicon, meant boats with larger and more powerful reactors. No less prepared than the surface admirals to invoke the Russian threat when it suited his purposes, Rickover maintained that the Soviet Navy was at the point of introducing some new boats that would make short work of their U.S. counterparts. Sturgeon-class boats, he maintained, were too slow to elude a Russian dragnet and too lightly armed to defend themselves when cornered. As always, the key to his counterstrategy was represented by preliminary plans for an enormous reactor capable of generating 30,000 horsepower, twice the output of any previous submarine power plant.

The surface admirals had never taken kindly to Rickover or his tactics, and this time their opposition mixed technical arguments with deep personal animosity. In 1959, for example, Rear Admiral Elmo Zumwalt, who would later become the head of the Navy's systems analysis section, had been subjected to one of Rickover's infamous interviews. Zumwalt was being considered for the command of a nuclear-powered guided missile frigate, and he realized that his future in the "nuclear navy" depended upon creating a favorable impression with its founder. Just as he had done with hundreds of other aspiring "nucs," Rickover gave the candidate very little opportunity to present his best face. The admiral began the interview by lambasting Zumwalt for being a "stupid jerk" and progressed through a series of vituperative questions that had absolutely nothing to do with warships, nuclear power, or the candidate's fitness to command a ship of the line. Zumwalt was quizzed about Plato's *Republic,* Clarence Darrow's philosophy as a defense counselor, and the importance of English and history in the Naval Academy curriculum. Periodically, and without warning, Rickover would break off the interview and order Zumwalt to the Tank, a bare and tiny office where he was obliged to sit silently at a small desk in front of a blank wall while he waited for his inquisitor to resume the ordeal. Rickover concluded the session by upbraiding the candidate for being no better than a "greasy aide," and sent Zumwalt on his way with a terse "Get out of my sight."

Much to Zumwalt's surprise, he was offered the job as commander of the nuclear-powered *Bainbridge,* but decided to turn it down in favor of a similar though conventionally powered ship that would allow him to remain outside Rickover's jurisdiction. Zumwalt might have felt his interrogation was harsh, demeaning, and perhaps irrational, but the fact of the matter was that he escaped lightly. Other, less senior candidates had been forced to sit in silence while their families were criticized for perceived genetic defects or, in the case of one unfortunate officer, obliged to wear a woman's wig while the "gnome-like figure behind the desk" dished out his litany of ritual humiliation.

In 1967, when Rickover unveiled his plans for the new generation of boats to be built over the course of the following decade, Zumwalt was both skeptical and annoyed. It was "sprung upon me as a *fait accompli* when I was director of the Division of Systems Analysis, the office where all new concepts for weapons were supposed to be worked up first," he said later. "Somehow Admiral Rickover had gotten the work done elsewhere and without my knowledge."

The case against the proposed new class of boats, already christened SSN-688, was not far removed from the Fighter Mafia's insights into the consequences of the high-tech sophistry that had loaded the F-15 with counterproductive electronic systems. Zumwalt and most of the other admirals reasoned that a 30,000-horsepower reactor would demand a submerged displacement of around 7,000 tons, a wider hull, and an overall increase in bow-to-stern length of perhaps sixty feet or more. Yet even though they were to be half as large again as the Sturgeon-class boats,* the 688s promised only a marginal increase in speed, since much of the new reactor's additional output would be devoted solely to overcoming vastly increased water resistance. Speed, crush depth, and most other details of submarines' performance are highly classified, but the informed body of opinion within the Navy, both then and now, rates the 688s as being no more than five to eight knots faster than their immediate predecessors. Moreover, as the Navy pointed out, the new submarines would be no less vulnerable to Russian attack, since any slight advantage imparted by the higher speeds would be negated by a corresponding increase in the levels of hull and engine noise, every modern submarine's two most lethal adversaries. The Defense Department, still very much dominated by McNamara's emphasis on cost efficiency, found itself in near-full agreement and bolstered the Navy's technical argument with economic objections. Dr. John Foster, the man who succeeded Harold Brown as the director of Research and Engineering, attempted to point out that while Rickover had a clear concept of the nuclear reactor design, the proposed boat's other characteristics "are not fully defined." It would be better, he and other civilian officials asserted in appearances on the Hill, if the Navy acquired additional Sturgeon-class boats, which, since they were cheaper, could be purchased in larger quantities.

Rickover had faced similar obstacles in the past but had always managed to circumvent them by taking his case to his friends, admirers, and supporters in Congress. In March 1968, he did so again, landing Dr. Foster before a panel of Rickover loyalists from the Armed Services Committee. Foster tried to point out that the Vietnam war was costing U.S. taxpayers roughly $80 million a day and he stressed that the nation could not afford to develop a new submarine merely to provide Rickover with a vehicle for his latest reactor. It was to no avail. Committee members cited Rickover as "the world's leading authority on nuclear submarines" and credited him with a near-oracular command of the strategies

*The Sturgeon boats are small only in comparison with their successors. They displace about 4,500 tons. The 688s are 6,900.

and issues involved in underwater warfare. Even the argument about the proposed boat's enormous and still undisclosed cost fell by the wayside. If the U.S. government was pouring its resources into an indecisive and increasingly unpopular Asian land war, surely it could afford a few additional pennies for some much-needed naval equipment.

Rickover, ever invincible on the Hill, carried the day yet again, and in 1971 the Navy entered into an initial agreement with the Newport News shipyard of Virginia. Electric Boat's only surviving rival was to produce the first 688 boat, the *Los Angeles,* after which the class was named, before turning over its plans and authorized revisions to Electric Boat, which would build seven of the next ten, for which Congress had approved funds.

General Dynamics was delighted to take on the work. David Lewis' aims for Electric Boat were similar to his as yet unrealized ambitions for Fort Worth: a stable, consistently profitable contract that would guarantee the yard's long-term stability and provide breathing space in which to enact some badly needed internal reforms. But first, before it could fix a firm price to each boat, the yard had to know something about the nature of the task it was taking on, and in those early days the final specifications were still in a state of flux. Hardly twelve months had passed since Rickover reluctantly abandoned plans to have the hulls constructed out of a new "miracle" steel, which, it was hoped, would lead to substantial improvement in deep diving capabilities. Unfortunately, the metallurgical researchers could not complete their studies in time and conventional steel was chosen for most areas of the boats.

At one time or another, everyone commits some reckless, unthinking, or just plain stupid act that is no sooner done than regretted. The courts are full of such hapless folk, and by mid-1972, so too were the executive suites in Electric Boat's administrative headquarters. In taking on the job, the yard had assured the Navy that each boat would be built for no more than $69.1 million, a figured derived from the $59.1 million figure presented to Congress as the "official target" price plus an additional, and unmentioned, $10 million to compensate for inflation and inevitable errors as Electric Boat's work teams learned their new jobs. The problem —and the root cause of the controversies that were to dog General Dynamics for the next fifteen years—was that it was simply not enough to cover costs, particularly those of a shipyard with so many existing and impending problems.

As of early October 1971, C. B. Haines, the deputy program manager for the 688 boats, advised his superiors that trouble was looming. "Electric Boat Division has not developed the necessary planning tools required to properly schedule conflicts and overload conditions until it is too late to manage them successfully. The problem is compounded by a lack of visibility in the manufacturing schedules and by a lack of realism in manufacturing intervals," he wrote in a confidential internal memo. Roughly translated, Haines was predicting bottlenecks and lots of them, unless immediate action was taken to set the yard's machinery on something resembling an efficient footing. What made his observations all the more acute was the fact that at the time he put pen to paper, the yard was actually running ahead of schedule on the early and relatively simple job of laying keels and arranging the first hull sections in their correct sequence. When the easy part of the contract came to an end, Electric Boat's house of cards would collapse in chaotic disorder. And that, thanks to Rickover's characteristic insistence on contracts with a low tolerance for noncompliance, was certain to inspire vast losses.

The first cracks began to appear in mid-1972, when the shipyard's construction unions and management negotiators reached an impasse in contract deliberations. In the years to come, General Dynamics would be hit by a succession of sometimes ugly scenes outside its divisions' strikebound gates, and this episode, even though the workers remained on their jobs, set the tone for the bitter wrangles to come. Over the four months of meetings, Electric Boat's negotiating position produced a costly victory. Determined to hold its hard line, the company took on no new workers and restricted overtime to the bare minimum. Just as C. B. Haines predicted, the result was the total collapse of the original construction timetables, which was aggravated in late fall when the Navy sent three older submarines to Groton for extensive refits and repairs. When the contract talks finally came to an end in October, there was, as an internal corporate report asserted, "low worker morale and lower-than-normal productivity."

Documents later obtained by Senator Proxmire's Subcommittee on International Trade, Finance and Security Economics indicate that when the yard began to rehire, it did so with little thought and less forward planning. In some areas of the yard there were so many unskilled and semiskilled workers performing the same job that a visiting Navy inspector dashed off a terse note in which he criticized Electric Boat's then general manager, J. D. Pierce, for allowing "idleness" and the lack of "an aggressive effort to reduce it." Yet while the yard was overburdened with unskilled workers, it was also suffering from a severe shortage of welders, work gang supervisors, and quality control assessors. Pierce replied to the critique with what would become the formulaic response to all Navy queries: Conditions at the yard were not as bad as they seemed and would soon respond to recently instituted corrective measures.

Despite the cheerful assurances, nothing of the kind took place. A deputy general manager was appointed to iron out the production problems, but by that stage it was a lost cause, as another internal memo makes clear: "During the three month period 12/72 to 3/73 SSN 690 has slipped six weeks behind schedule. . . . The rate of slippage has not decreased. We are continuing to lose approximately one week progress for every two weeks of calendar time." Simple solutions were discussed at length but never implemented. An in-house education program to train new workers in the complexities of submarine construction might have been started or there could have been a serious attempt to regulate the flow of work teams and materials between one part of the yard and the next. Once again, nothing of the kind was done. Work areas remained cluttered with unnecessary equipment, fork-lift trucks, and the other symbols of a sloppy shop. Workmen who had labored on a specific task on one boat were transferred to the next in the series, only to find that it was so far behind schedule they were not yet required. Machining essential parts, work that should have been done in Electric Boat's own shops, had to be farmed out to subcontractors, something that exacerbated the already strained relations with the yard's unions.

Many of the problems were in the grand tradition of General Dynamics' divisions at their worst, the visible evidence of a management system divided by its executives' distinctly different imperatives. In order to alleviate the overcrowding, the people in charge of engineering and day-to-day operations had been pushing for a large undercover construction shed. This ran counter to the ambitions of the comptroller's staff, whose prime objective was a clamp on the rapidly escalating costs. Finally, when a new work area and several large cranes were installed, it was only after much preliminary procrastination and, as another memo asserted, "with great reluctance." Much the same attitude blighted all attempts to rationalize the incoming flow of materials from suppliers. The production people wanted to continue ordering supplies, even though the lagging schedules meant they could not be used right away. The accountants consistently delayed approving the purchase forms, ignoring the engineers' argument that their suppliers would be swamped by the rash of last-minute orders that would have to be submitted eight months or a year down the road. "We are still better in putting out fires than preventing," a report to director of planning N. D. Victor stated in August 1973, adding that there was "no advance planning function" worth the description.

By the start of 1974, the situation was becoming desperate. Remedies were not being applied and fresh obstacles continued to arise. Schedule slippage that had been running at two weeks per month were approaching three weeks, while the original estimate that it would take roughly four years to finish the first boat had risen to a little less than six.

General Dynamics could have done any number of things to ease its position with Electric Boat. It might have 'fessed up to the Navy, explained the problems, and asked to have the contracts renegotiated. In this case, honesty was not the policy of choice for two very good reasons. The Navy had visions of operating as many as fifty 688-class submarines and had already lined up the appropriations to fund the next dozen or so in the series. What chance had Electric Boat of picking up follow-on contracts to the first batch of 688 boats—known as the Flight One series —if it admitted what a botch was being made of the job? Even more important was the second need to present an efficient facade. While he was busy negotiating the SALT treaty with the Russians, President Nixon had also been bargaining with his own Pentagon chiefs of staff. If the arms limitation pact was to survive its passage through Congress, the services would have to be bribed into acquiescence. The Navy's part of the payoff was the Trident ballistic missile submarine, for which Congress had just allocated a $908 million down payment. General Dynamics weighed the situation and opted for the only possible alternative. It decided to continue cooking its books.

As with every major defense project, the Flight One boats were the subject of voluminous progress reports compiled by General Dynamics' own people and the two hundred or so Navy observers and overseers stationed at Groton. The progress reports, complete with estimates of the number of man-hours needed to finish each boat, were of major importance to Electric Boat's financial health. Progress payments were made in accordance with the amount of work done, or rather the amount of work the company claimed to have done, since the resident Navy people appear to have been uncritical overseers. In the first quarter of 1974, for example, Electric Boat's status reports estimated that it would take a further 32.2 million man-hours to finish the first seven boats. The statements submitted to the Navy, however, put the figure at only 26.8 million man-hours. One year later, when the company was blaming design changes for the increased costs, the internal estimate had risen to 43.1 million man-hours, while the Navy was being informed that it would take but 39.8 million man-hours.

Late in October 1972, the Navy lived up to the defense industry's expectations by informing its chief suppliers that an additional 11 Los Angeles-class boats would be ordered the following year. Electric Boat received its official Request for Proposal on February 1, 1973, by which time its executives were already hard at work drafting a reply. The resulting document was more a work of fiction than a product of experience. Even though N. D. Victor had received a memo in January that summed up most of the major problems delaying the Flight One boats and concluded with the observation that there were serious questions "about our ability to realistically bid anything but the bare minimum number of boats in the current RFP," the eventual proposal asserted that the yard could deliver each of the next eleven boats at roughly the same price it had specified for the first seven.

Documents obtained by congressional investigators indicate that when the yard was preparing the draft of the final Flight Two proposal, there was some pretense at candor. Like the Flight One bids, those for the second series were based on production experience with the earlier and smaller Sturgeon-class boats, a convenient base for negotiations since it obviated the need to mention the foul-ups already encountered. However, a draft proposal contained in a larger, general report on Electric Boat's fortunes, which was circulated through Groton and Saint Louis in March 1973, estimated the amount of labor involved to be almost 20 percent higher than the figure submitted in the official bid later that year. More important, the draft went on to describe the estimate as a product of "construction experience to date on current 688 work."

The disclosure did not survive very long. Early in April, a two-man delegation arrived in Saint Louis from Groton—comptroller Art Barton and 688 program manager Henry Hyman—for a meeting with David Lewis and other head-office potentates. In the course of the conclave and the three days of reflection that followed, the documents were stripped of all reference to "current 688 work" and the estimated time required to complete each boat was trimmed by 300,000 man-hours. General Dynamics would later admit that the pruning had been done at the instruction of no less a person than David Lewis himself.

In their later correspondence with the Department of Justice, the company's lawyers denied any attempt at fraud, contending that the Navy was aware of the reductions at the time they were made. Hadn't Rickover lashed the yard for indolence and incompetence, upbraided it for overmanning, and demanded that the total be cut by a further 100,000 man-hours per boat? While both these explanations had their points, they skirted the real issue: Did Electric Boat "buy in" to the Flight Two contracts, promising a low initial price with the intention of recouping future losses through bogus, overinflated claims?

Here again the company's own records undermine its later denials. At one point, J. D. Pierce and Henry Hyman were discussing how best to handle the weekly progress reports demanded by Rickover. "As regards Item Two, which relates to the manning of the SSN 688 class ships, there is probably no good time, but it is better for us to stop reporting on this item as soon as possible," Hyman wrote. "We'll never be able to make a claim hold up if we are reporting inadequate manning." It should be noted that this comment was made long before the first claim was filed. Some months later, Hyman was moved to ponder the shipyard's future. "We must recognize that we are fighting for our existence," he informed Pierce and deputy general manager Mel Curtis. One of his suggestions demanded a concerted effort to keep track of late design changes, which were then arriving at a rate of about 27,000 per year. "Additionally, a plan should be developed to ensure that the proper groundwork is established to support any 'claim' action that may be appropriate for Electric Boat division to initiate. I have an outline for such a plan that I intend to discuss with you in the next week or so," he wrote. His timing was remarkable, ominously so. Only two months had passed since the division's receipt of a contract commitment for the next eleven attack submarines.

From time to time in their interminable exchange of memos, various Electric Boat officials bemoaned what they described as the yard's loss of its "can-do spirit." The situation at Groton had degenerated dramatically since the early days, when Rickover handed Jay Hopkins a nuclear monopoly by awarding Electric Boat the contracts for the *Nautilus* and the *Sea Wolf.* It had been one big, happy family back then, a band of friends struggling to advance the frontiers of science. Now, although the contracts kept on coming, everything was an uphill battle. Nothing better illustrated the transformation than the preliminary wrangling that culminated in Electric Boat's being selected as the contractor for the gargantuan Trident submarines.

There were any number of reasons for Electric Boat to have chosen to let this cup pass it by. The 688 boats were mired in problems, there was a chronic shortage of work space, and the supply of skilled craftsmen was never enough to meet existing contract demands. If the yard further burdened itself with one of these monstrous ballistic missile carriers, past experience indicated that its already lengthy list of woes would grow exponentially.

In fact, the yard did refuse the Navy's initial offer, although not for the reasons prudence might have dictated. In September 1973, at about the same time Electric Boat obtained the contracts for the Flight Two 688 boats, Congress voted to proceed with the first Trident, even though the Navy still had only a vague idea what the finished vessel would look like. This was not Electric Boat's qualm. Happy to compound its existing problems, General Dynamics at first refused to take on the job until the Navy agreed to sign an agreement that was later described as "a marvelously inventive rubber document." Electric Boat was to build the hull for $253 million, a figure euphemistically labeled as the "target cost." With the possible exception of the congressmen who approved the deal, no one entertained even fleetingly the notion that the job could actually be done for this figure. So in order to allay Groton's understandable reluctance to commit itself to a project that would almost certainly run into the red, the Navy added a string of soothing qualifications. The admirals agreed to pay 95 percent of all overruns up to $26.6 million, as well as 85 percent of the next $106.2 million in excess charges. And the Navy agreed to absorb the full cost of inflation, revising the figures on a regular basis in step with rises in the industrial equivalent of the consumer price index. Later, when the Trident program was being held up as an example of the way not to buy a submarine, a former Electric Boat executive expressed some sympathy for his old employer. "The contract was kind of like putting an alcoholic in charge of a liquor store. You can't really complain about the mess."

Like the F-111, the Trident originated in one of the officially sanctioned excursions into pure theory. Although there was no immediate need to a find a successor to the Polaris and Poseidon ballistic missile submarines, the Navy began to speculate in the mid-1960s about the kind of vessel that would replace them. Their ruminations followed the classic pattern. Conceived as a simple, slow underwater launching platform and tentatively dubbed the Strat-X (Strategic Submarine Experimental), the mooted boat was to be an insurance policy against the possibility that Congress might be tempted to spend more on land-based missiles and bombers than on the Navy's leg of the nuclear deterrent triad. At that early stage, ideas about the boat's final cost, size, and shape were necessarily vague. There were some innocent souls who pictured it as being powered by a combination of diesel and electric engines—a sensible approach since there was no need for a high top speed, but one doomed from the beginning by the Navy's institutional reluctance to abandon its nuclear fixation. Others wondered if the cost might not be kept down by equipping the boat with external missile tubes, an innovation that would have reduced the price considerably.

It was not to be. When President Nixon unveiled his bribes to win the services' support for the SALT agreement, the Navy was forced to make its own, internal peace treaties. Admiral Zumwalt wanted a new boat, but before he could present his plan to Congress, he had to win Rickover's support and that meant accepting whatever reactor the aged admiral put forward as an "ideal" power plant. Rickover outdid himself, insisting on a 60,000 horsepower nuclear reactor that he had originally conceived as the foundation for a submarine equipped with cruise missiles. As with the 688 boats, the size of the power plant determined the dimensions of the ship that was to be built around it.* Once again, an inexpensive and effective concept had been scuttled, not because it was unsuitable but out

*The Trident was the biggest sub in the world until the Russians beat it with an even larger, titanium-hulled monster, but it is still an awesome piece of engineering. At 560 feet and 18,000 tons, it is large enough to hold five or so repicas of the Nautilus.

of the need to process military goals through the twisted conduit of procurement politics.

As might have been expected, the new Trident work load further compounded all of Electric Boat's existing problems. As 1974 drew to a close, the first of the 688 boats, originally set for delivery in June of the following year, was running more than eighteen months behind schedule, while those following it on the production line were, at the current rate of slippage, likely to be handed over up to four years late. The yard was conducting its business amid a mounting barrage of short-tempered, acrimonious recriminations. Some two years later, the mood was captured in a memo compiled by Bruce M. Prouty, a visiting examiner from Arthur Andersen, one of Electric Boat's auditors. Prouty recounted his discussion with Art Barton, the division's comptroller, and recorded the latter's objections to the administrative style of Mel Curtis, the division's deputy general manager, who had come across three years earlier from Convair. "It was very evident that the operations people and Curtis had the prevailing hand and that Art, in essence, was told to stay out of trying to project how many hours it was going to take to build the ships. Art said that the twenty-five million hours, in his estimation, was unreasonable. However, when they started talking about thirty-one million hours last August, he felt there was a possibility of achieving that at that time. Now even that thirty-one million hours is unrealistic, and they are going to have to come up with a new number since they have not made the improvements they anticipated."

Prouty went on to record his observations about the shipyard's inability to reach a consensus: "Curtis had been running the shipyard like a dictator and whatever hours he said it was going to take to build the boats were the hours which nearly everyone in the Division supported," he began. "How much can we rely on the estimates of the Division at the current time if there is someone forcing a number down the throat of the Operations personnel, and they are afraid to come out of the trench and tell their true feelings about the hours required to complete the contract?"

Finally, after remarking that "the credibility of nearly everyone at the division has been dealt a severe blow," he noted that the insistence on unrealistic man-hour estimates placed Electric Boat in a precarious legal position. "There are apparently people at the Division at a high level who have raised questions about the reasonableness of the numbers and, under oath, some very damaging testimony could probably be obtained." Prouty concluded optimistically, suggesting future talks with Electric Boat's new acting general manager, Gorden MacDonald, General Dynamics' vice-president for finance, who had been recently dispatched to

Groton with instructions to clean up the mess. He was to receive little guidance from Curtis, who, struck down by a cerebral hemorrhage at the end of April, was rushed to the hospital and then into premature retirement.

MacDonald was to be the second prominent victim of Electric Boat's resistance to change, but he was certainly not the last. While conditions in the yard remained the same or, in some cases, grew steadily worse, the occupants of executive row changed places with the frequency of ducks in a shooting gallery. Of the top three hundred administrative positions, more than half saw fresh incumbents come and go over the course of a single year. The Navy continued to receive doctored progress reports, while old problems, on the rare occasions when they were actually solved, were replaced by fresh ones.

David Lewis passed through Groton in late January 1977, and recorded his observations in a note to MacDonald that is almost poignant in its exasperated disappointment. "The short visit we made to the yard on 26 January was very revealing and extremely painful," he lamented, adding that while manning levels had risen by as much as 100 percent in some cases, productivity had not improved at all.

"I doubt that most of our people really want to loaf and the majority will work if they know what to do and how to do it. . . . There was almost an air of arrogance about these 'stand-around' people. They made absolutely no effort to appear busy when officers or supervisors of the division came around. They continued their conversations without embarrassment and certainly without reaction to the presence of the top people in the division."

Everywhere he turned, there were appalling spectacles. The new construction building, opened only a few months before at a cost of many millions of dollars, was "the most deplorable of any operation I have seen in my life." It was, he said, "just another piece of real estate in which to operate in the traditional way of Electric Boat workers." Lewis found so much amiss, the sheer volume of complaints sometimes led him to dwell on trivialities like the work habits of fork-lift drivers. "It is almost laughable to see those expensive transportation vehicles carrying one or two tiny little parts on a great big wooden pallet from one end of the yard to the other and then returning for one more little part on one more big pallet."

MacDonald managed to rectify none of the major problems, although he did scrap some of the management systems Mel Curtis had imported from the West Coast, attempting to reestablish direct lines of communication within the division. It did little good—and some of the things that occurred during his administration made the existing situation a good deal worse. By early March, the yard's quality control supervisors had begun to notice that large quantities of "soft" steel had been used in some of the boats' most vital structural components. Since it was already welded into place, the yard had no choice but to detail specialist "deconstruction" gangs to cut it out and replace it with sound metal acceptable to the Navy.

Lewis was soon looking for a new man to take the weight off MacDonald's overtaxed shoulders. In March, at about the same time that the steel problem erupted, a rumor swept the yard that the third general manager in as many years would soon take charge of operations. One executive was excited enough to make a note of the whisper in his desk diary, recording the information that the new man would be a shipyard specialist with prior experience in submarine construction.

The candidate was Takis Veliotis, who might have been brought down from Quincy immediately had it not been for one major problem. While he held a Greek passport and a Canadian one, Veliotis was not an American citizen and did not possess the top classification security clearance required to supervise work on the secrecy-shrouded Trident and 688 projects. The clearance was not given until October, and even then David Lewis had to enlist Rickover's assistance in having the process accelerated. In the light of later events, it would seem that the investigation of Veliotis' background was conducted with perhaps a little too much haste. When he moved from Canada to Quincy in 1973, Veliotis supplied a résumé that described him as the recipient of an engineering degree from a Canadian college. Not only the degree was a figment of the Greek's imagination; so, too, was the college that bestowed it. This was not the only time General Dynamics and the custodians of the nation's secrets slipped up on their jobs. Three years before, Henry Crown's son, Lester, was granted a top security clearance even though he was under investigation for his involvement in an attempt to bribe Illinois state legislators. As David Lewis would later admit before a House subcommittee, General Dynamics "forgot" to mention the probe when it endorsed the younger Crown's application.

Veliotis' talents were but part of his appeal. With delays mounting and costs soaring—the first 688 boat was delivered almost twelve months late and $56 million over budget—there was a pressing need for a point man who could present the best possible case for the yard's claims against the Navy. The first, submitted in 1975, had been settled the following year for $97 million, less than half the amount for which the company had asked. Now there was a much larger claim, for $544 million, in the pipeline, and the supporting evidence was certainly no better than that which had backed up the first. In the first demand for out-of-pocket expenses, Electric Boat had been able to depict itself as the victim of its partner in the 688 program, Newport News. Under the terms of the contract, the Virginia yard would build the first boat and then deliver all its plans to Groton, where they would be used to construct seven of the next twelve in the series. Unfortunately, many of the documents arrived late or, more frequently, became the subject of postdated revisions. Design changes are a ubiquitous hazard in the shipbuilding industry and they can be extraordinarily troublesome, particularly when they involve alterations to work already done. But that was seldom the case at Electric Boat, at least according to the testimony of congressional investigators and General Accounting Office officials who appeared before the numerous congressional committees that raked over the coals of the claims affair almost one decade later.

The rest is history. A special Navy panel set up to arbitrate builders' financial grievances approved what its members considered a final settlement of $125 million. General Dynamics eagerly accepted the money and then went on to pocket a lot more besides. While negotiating on one front with Admiral Vincent Manganaro's Navy Claims Settlement Board and fending off Rickover's accusations of criminal fraud on another, David Lewis and Gorden MacDonald were drawing up their own agreement with the Carter administration's Navy secretary, Edward Hidalgo. By the time the result of these independent negotiations was announced, on June 9, 1978, the total cost to the Navy had risen to $648 million. General Dynamics, which would soon submit yet another claim, for a further $100 million, grumbled that it was still being asked to swallow a $300 million–plus loss, but its bleatings drew little sympathy. They were to attract even less when subsequent investigations revealed that upon leaving government service, Hidalgo had gone to work for General Dynamics as a free-lance consultant and adviser in its latest campaign to sell tanks and F-16 fighters to Spain. And that was to be but one small squall in the storm of investigations, accusations, and embarrassing disclosures that would lead one U.S. congressman to suggest a new corporate motto: "Catch us if you can."

15. "Catch Us if You Can"

Late in February 1985, Iowa Senator Charles Grassley addressed a Washington convention of the American Conservative Union, an assembly whose views span the breadth of right-wing passions. Grassley was on home turf from the moment he walked into the auditorium; once rather uncharitably described as "one of the troglodyte Republicans who came to the Senate on President Reagan's coattails in 1980," he was a natural mouthpiece for the most deeply cherished tenets of his audience's faith. He shared their repugnance for abortion, saw nothing wrong with prayer in schools, and accepted the necessity of deficit budgeting with the reluctance of a minister obliged to conduct services in a house of ill repute. Needless to say, the senator has always been staunch for a strong defense. It was, in fact, the subject of his speech.

A few years earlier, as Grassley freely admits, his views on what constitutes a strong military would have been no less wrapped in tradition than his statements on most other subjects, a rule-of-thumb equation that has always calculated security in terms of the Pentagon's total budget. On that Saturday afternoon, however, he took a very different line. "The nation's largest entitlement program has nursed a new generation of welfare queens," he said. "It is the defense industry." Just as they would do on four other occasions during the course of his address, the crowd clambered to its feet and rewarded the speaker with a prolonged standing ovation. As Grassley would later explain during an interview in his Capitol Hill office, his conversion had been a direct consequence of several ongoing investigations into the activities of General Dynamics, inquiries that at times "make me feel like I'm going to be sick."

As with so many of the other slings and arrows that came crashing down upon General Dynamics during the late winter and spring of 1985, Grassley's disgust at the antics of the nation's largest weapons supplier originated in the events of eight years before, when Navy Secretary

Hidalgo stepped in to help the company survive the mess it had made of the 688 attack-class submarines.

On August 4, 1977, General Dynamics' board met for a council of war. The Navy Claims Settlement Board, which had been considering the company's request for an additional $544 million, had made it clear that only a fraction of that amount would be paid. This was simply not good enough as far as Lewis, MacDonald, and the rest of the board were concerned, and by the end of the month they had adopted a very hard line, informing Hidalgo, at that stage an assistant Navy secretary, that it might well become necessary to stop work on the submarines unless a more equitable settlement could be reached. This was nothing if not blackmail. For years, the Navy had been depicting the 688 boats as vital to the national defense, an indispensable counter to the ever-growing Soviet underwater fleet. If Electric Boat downed tools, as Lewis claimed it would, the program would be delayed still further—and the lost time might extend to years if the Navy decided to fight fire with fire by moving the partly completed hulls to another yard.*

Whether or not Lewis actually planned to carry out the threat remains a subject for speculation. While he left Hidalgo with no doubt that he was serious, precedent would tend to suggest that the announcement was a high-stakes bluff. Two years earlier, Newport News had threatened to stop work on the nuclear-powered aircraft carrier *Carl Vinson,* and was proceeding with another boat, a light nuclear-powered cruiser, only because a judge had ordered it to do so. The Virginia yard did win a victory of sorts. By the time Edward Hidalgo set out to end the war between the Navy and its contractors, the *Vinson* was being built under an ad hoc agreement that saw Newport News collect its costs plus a premium of about 7 percent. A similar agreement would do Electric Boat nicely until a better, renegotiated deal could be arranged a year or so down the road. The company was certainly pushing hard in that direction. In the two months after the August meetings of the board and the executive committee, the company discussed a four-point settlement with then Navy secretary W. Graham Claytor, indicating that it would be prepared to continue work if it received a written assurance that renegotiations would continue through the following year.

In December, the two most crucial events in the chronology of General Dynamics' later embarrassment took place. On December 1, Hidalgo informed the Navy Claims Settlement Board that while its findings would be welcome, he was taking charge of the matter "to achieve organizational objectives of singleness of authority." What that meant was any-

*This was never a serious possibility but, in keeping with the tone of the dispute, Navy officials refused to rule out any option no matter how ridiculous.

body's guess, because the board was then within four weeks of handing down its final decision and the assistant secretary's intrusion into the affair merely added one more team of players to the field.

Hidalgo met several times with Lewis and another Electric Boat official, Max Golden, his attitude one of eager conciliation. Like the supplicants, the assistant secretary harbored no doubts that Electric Boat deserved a large infusion of cash. The only point of disagreement, or so it would seem from the documents, was how this might best be packaged. For his part, Hidalgo favored a settlement similar to the ones that had saved Lockheed and Grumman, a government-funded bailout based on the contention that large defense companies are so vital to the nation's health and safety, they must be preserved at all costs. It would have been the perfect solution but for one major problem: While Electric Boat was profligate in its losses, the parent company was doing respectably well and could not be considered a candidate for imminent disaster while Fort Worth and the West Coast operations were prospering with their F-16s, missile systems, and cruise missile research.

Lewis was quite open in admitting that General Dynamics could survive losses at Electric Boat of $700 million or more, and suggested instead an alternate strategy based on the original log of claims submitted back in 1976—postdated design changes, unprecedented inflation, and Newport News' failure to supply the plans for the lead ship on time and in the finished format. Hidalgo was adamant that the approach would lead to a further worsening of the company's relationship with the Navy, especially with "you know who" and all the other "top blue suits." The only way Electric Boat would get any money was if it went before the Congress with a plea based on Public Law 85-804, the legislation enacted to save Lockheed and Grumman. Despite General Dynamics' pronounced reservations, Hidalgo made it promise to open its books for an independent review by an outside auditing firm, Coopers & Lybrand. Lewis and Golden agreed only after establishing their right to withhold what they considered proprietary information. In other words, the people being audited were to be allowed to exert a major influence on the outcome of the final report.

The second major event of that 1977 Christmas season was to figure as even more important in General Dynamics' ill-starred future. On December 10, two days after the last conference between Gorden MacDonald, Max Golden, and Hidalgo, Admiral Rickover concluded his own uninvited investigation of the Electric Boat claims and filed a detailed report with his superiors, demanding that criminal proceedings be brought against Electric Boat and some of its executives. There were at least eighteen examples of falsified claims and deliberate fraud, and doubtless a good many more which could not fail to come to light during a full-scale investigation by the appropriate authorities, he wrote. Hidalgo made a point of ignoring Rickover as much as possible, reportedly filing the admiral's unsolicited advice and instructions in his wastepaper basket. On this occasion, he did take some action. The Navy's chief counsel was instructed to refer the matter to the Justice Department. In the meantime, Hidalgo continued with the negotiations as if nothing had happened, faltering in neither intent nor pace, even though Admiral Manganaro had recently brought a number of other "questionable claim demands" to the attention of Togo West, the Navy's legal counsel.

Defending himself later before an unashamedly critical Senate panel, Hidalgo explained that he was motivated by nothing more than a desire to cleanse the "hatreds, the animosities and the poisoned wells." The Navy's relations with its suppliers were in a deplorable state. There was the *Carl Vinson* being built down in Virginia without a concrete contract; there were the charges, the claims, and the threats to stop work at Electric Boat. And of course, there was Rickover, forever running to the Hill to denounce all and sundry for indolence and incompetence, treason and fraud. He could have left the matter in the hands of Admiral Manganaro, but it would have done nothing to remedy the worst aspects of the situation. "The terms of those 1971 and 1973 contracts were an outrage," Hidalgo said. "There was no way this new class of boats could be built with a margin for error of only four or five percent." That General Dynamics brought the predicament upon itself by submitting an unrealistically low bid bothered him not then nor later, when he was officially informed that the Navy had been supplied with what the company's own records indicate were intentionally doctored progress reports. Had he known, he said, he would have stopped the settlement in its tracks. Steadfast in his ignorance, however, he was not disturbed by Rickover's and Manganaro's suspicions, and continued to wrangle with Lewis about the best method to restore Electric Boat to full health.

Lewis still had serious reservations about a plea of financial impairment—not least because it would mean the widespread distribution of the Coopers & Lybrand investigation—and he was unimpressed by the settlement figures being advanced by the Navy. In March 1978, assistant secretaries Claytor and Hidalgo were informed for the second time that Electric Boat would stop work unless the talks took a more constructive turn. A mediator was brought in. Democratic Senator Abraham Ribicoff of Connecticut summoned Lewis, Claytor, and Assistant Defense Secretary Charles Duncan into his Capitol Hill office in an attempt to establish some common ground and win an extension on Electric Boat's thirty-day deadline. The conference failed to produce an offer that matched the company's demands, but it did see the day of reckoning postponed for a further two months. Lewis returned to Saint Louis with the news that the future, while still far from settled, was looking a little more cheerful. Hidalgo, the assistant Navy secretary, would shortly sound out his contacts in Congress and ascertain what sort of additional expense the members would agree to authorize. Even better, he had indicated that any assistance General Dynamics cared to provide in drafting the contract amendments to go before Congress would be gratefully appreciated.

The Coopers & Lybrand audit of Electric Boat's and General Dynamics' respective financial situations was vital to this endeavor, since it was to provide Congress with a ready reckoner of the cost to complete the remaining submarines and the likely effects of a major loss. Like the Navy's progress reports, the final document was remarkable more for its omissions than for its summation of the facts.

Of course, that was not the way Hidalgo presented it to Congress when he testified before the Senate Armed Services Committee in August 1978. It was, he said, "a comprehensive review with full access to General Dynamics' financial records." Strictly speaking, that may have been true of the original report, though certainly not of the précised version contained in the actual memorandum of decision that accompanied the proposed legislation.

In a draft of the report supplied simultaneously to Gorden MacDonald and the Navy in late March, the investigators estimated that General Dynamics could survive a loss of up to $1.14 billion, substantially more than the $744 million in past and future losses they envisioned under the existing contracts. If the company could afford to lose almost three quarters of a billion dollars without going to the wall, how on earth could there be any justification for a plea of pending financial collapse?

From the day the Coopers & Lybrand team walked into the shipyard, Electric Boat officials had tried to keep a tight rein on the flow of information. The auditors were consistently denied permission to interview the company's in-house analysts, Bruce Prouty and the other members of the Arthur Andersen accounting firm, and they were also refused access to Electric Boat's bankers, even though the division's credit rating would have been of prime importance in determining the division's ability to absorb and survive future losses. Faced with the fruits of the Coopers & Lybrand investigation, MacDonald dashed off a letter to the Navy reminding Hidalgo and Claytor of their earlier promise to allow General Dynamics to review the report "for factual accuracy."

The company did conduct a review of sorts, although the articulation of a catholic perspective on the yard's problems would seem to have been low on its list of priorities, since the document finally submitted to Hidalgo was fully one-third shorter than the original, with an estimated sustainable loss of not $1.14 billion but $744 million, the exact amount predicted as the total write-off for the 688 program.

The reduction was substantial; a loss-absorbing cushion of $366 million was simply willed into unmentioned oblivion, and it was not to be the last. When the final Navy version was presented to Congress, with Hidalgo's assurance that it represented a full and complete picture of General Dynamics' financial affairs, the actual wording did not mention the $1.14 billion or $744 million figures at all, but said instead: "General Dynamics would remain a viable corporate entity if it absorbed a fixed loss in the order of magnitude ultimately agreed to—$359 million." Congressional investigators would later make much of the Navy's language, suggesting that it was meant to suggest that the company could take a $359 million loss but no more, a contention clearly at odds with Coopers & Lybrand's first reports and the internal appraisals generated by Arthur Andersen.

When Senator Grassley quizzed Hidalgo during a hearing conducted by the Subcommittee on International Trade, Finance and Security Economics in April 1985, the former assistant Navy secretary's richly cadenced tones acquired a note of offended innocence. Congress was not informed of the larger sustainable loss estimates because "they were irrelevant," Hidalgo explained. Grassley's fellow committeeman Senator William Proxmire had already speculated in the press that the deletions and amendments might have been intended to mislead Congress, and now Grassley took up the point. What right had the Navy secretary to decide what information Congress should and should not review? Congressmen are among the busiest of folk, he continued, but they are still capable of digesting information and arriving at their own conclusions. Hidalgo's counter to this and other charges that he had been more than fair with General Dynamics seldom varied. The final agreement must have been fair, he insisted, because David Lewis was still complaining about it long after the renegotiated contract had gone into effect.

It is a little hard to imagine what Lewis could have found to grumble about. Electric Boat had originally demanded $544 million, but by June 1978, when the settlement was announced, the total package amounted to a little more than $640 million. Hidalgo had begun by taking the Manganaro board's recommended settlement of $125 million and using it not as a final figure but as a starting point for future negotiations. Next, the parties agreed to split the difference between the Manganaro figure and the Navy auditors' latest estimate of past and future losses, which had recently leaped by another hundred million dollars or so, to $843 million. This brought the running total to $484 million, with several additional contract revisions yet to be taken into account. The first of these was an allowance for future inflation. Electric Boat was to pick up the first 7 percent of increased labor charges and 6 percent of material cost increases; the Navy paid the rest, eventually handing over a further $108 million. The last and smallest ingredient in the package was something called Future Cost Growth, which was to contribute a further $50 million to Groton's incoming revenues before the last of the eighteen 688 boats put to sea late in 1984, four years late and $89 million over budget.

At the time, and in his subsequent appearances before Grassley and Proxmire, Hidalgo made much of the settlement's impact on General Dynamics. The $359 million write-off that the company had agreed to declare as its part of the settlement—half of the spread between Manganaro's estimate and the figure drawn from the Coopers & Lybrand investigation—was larger than any previous loss declared by a U.S. shipbuilder, Hidalgo said, adding that it was more than Electric Boat had earned since the *Nautilus'* launching in 1955.

Even by Pentagon standards, $359 million is a tidy sum, but as usual, it is doubtful if the consequences of the enforced loss were as severe as Hidalgo and Lewis insisted. As Senator Proxmire's chief investigator, Richard Kaufman, pointed out in a background paper, "such business losses are tax deductible and, according to General Dynamics' annual reports, the after-tax loss was reduced to $187 million." Kaufman noted another of the settlement's less obvious sweeteners. Since the Navy agreed to pay General Dynamics with an initial, lump-sum installment of $300 million, the company was able to save many additional millions of dollars in interest on funds it would otherwise have been required to borrow. According to Kaufman, "General Dynamics estimated the savings in interest cost to range from $150 million to $200 million over the years 1978–1984."

Hidalgo went on to replace Claytor as Navy secretary not long after the 688 deal was finalized, and remained in office until the end of the Carter term. He was proud of his record and believed himself to be the model of a good citizen, forsaking the rewards of a prosperous practice in international law to answer his government's call for talented men. That was, in fact, what bothered him most about the allegations of impropriety that first surfaced within months of the settlement and continued to reemerge over the next seven years. Testifying before Proxmire and Grassley in mid-1985, he began by denouncing the *Washington Post* and the ABC television news program *20/20,* which had recently made Hidalgo's subsequent employment as a General Dynamics lobbyist and consultant in Spain a feature of a scathing report on the company's many scandals.

The report was "irresponsible character assassination," Hidalgo charged, "a slimy specimen of yellow journalism" that had caused his family much distress and embarrassment. However, it was not the public insults but the damage such criticism did to the country as a whole that was the greatest sin. Throughout his entire public life he had worked tirelessly to serve his country, first as a general counsel to the U.S. Information Agency, and later in the two top Navy posts. What would the young people of America conclude, he asked the panelists waiting patiently to begin the interrogation, if the reward for years of service was a cheap smear job by the sensation-hungry vultures of the media? It was outrageous, he continued; enough to dissuade any bright young man or woman from entering government service.

The diatribe wound down to the accompaniment of barely suppressed guffaws from the crowded press tables and a total absence of sympathy from the waiting senators. After pressing for an explanation of his motives in approving a settlement four times larger than Admiral Manganaro had suggested, the committee moved rapidly to examine Hidalgo's brief but lucrative career as a pitchman for General Dynamics' tanks and jet fighters.

They learned that the job offer had come from P. Takis Veliotis himself, in the fall of 1981. General Dynamics was running neck and neck with McDonnell Douglas in the race to sell Spain a hundred or so new jet fighters, when Veliotis, recently appointed to head up the new international sales division, asked Hidalgo to visit Madrid and chat with some old friends. Veliotis wanted someone who spoke Spanish and "knew a great many people in Spain," qualities that made a man with Hidalgo's thirty years' experience in international law an ideal candidate for the job. "What I did with the F-16 does not worry me at all," Hidalgo stated. Proxmire was deeply skeptical about the witness' lack of sensitivity. Surely Hidalgo must have heard of the infamous revolving door? Even if his motives and conscience were clear, as a self-proclaimed duty-bound former public servant wasn't he even a little wary of going to work for a company with which he had recently negotiated a large and controversial deal? After all, there were some uncharitable and suspicious souls who might be inclined to look upon the job as a payoff.

Hidalgo refused even to admit that his actions could be interpreted as questionable. During his stint as an architect of naval policy, he had dealt solely with Electric Boat, he said; Fort Worth's F-16s were an entirely different matter. That his fees came from a common pocket bothered him not at all.

General Dynamics' onetime adversary at the negotiating table was certainly well paid for his efforts to swing the Spaniards toward the company's products. Even though he admitted to spending only three weeks as a pitchman for the F-16, his Virginia law firm received $47,500, excluding incidental expenses. And that was not the end of it. In October, several months after the Spanish Air Force decided to buy the rival F-18, Hidalgo returned to Madrid to sing the praises of the M-1 tank. No more successful than his first round of lobbying, this latest expedition earned him a further $18,500, to bring his total earnings up to $66,000 for what amounted to less than four weeks' work.

This was not the first time General Dynamics recognized a talented recruit among the ranks of the senior Navy officials with whom it dealt. Nor would it be the last time a secretary who had helped to settle some delicate dispute ended up on the company payroll.

Hidalgo may have thought he had solved the 688 claims issue once and for all, but by the last months of 1981 it was obvious Groton was back to its old tricks. Lewis and various other Electric Boat officials were shuttling in and out of Washington in quest of an additional $100 million to recompense the shipyard for the most recent batch of cost overruns on the 688s. The request and the imaginative legal argument on which it was based drew some venomous comments from Hidalgo's replacement, Reagan appointee Dr. John Lehman, who even raised the possibility of excluding Electric Boat from all future contract competitions.

All the same, the matter was settled more or less to Electric Boat's satisfaction. It did not get the money, but it did win contracts for two additional boats, which helped to make the loss a little less painful. This time, Lewis dealt almost exclusively with the new man in charge of logistics and contract negotiations, Assistant Secretary George A. Sawyer. Eighteen months later, when he left the Navy, Sawyer reappeared in Saint Louis as the new executive vice-president of the Land Systems division.

While much more was to be made of Sawyer's career decision in subsequent probes mounted by Congress and a Virginia grand jury, it proved to be but one of the scandals that deluged Lewis and his colleagues through the grim spring of 1985. The accusations that had been mounting since Veliotis' flight to Greece reached a self-sustaining critical mass in late February 1985, when Democratic representative John Dingell's Subcommittee on Oversight and Investigations placed Lewis and MacDonald under oath and forced them to admit that the company had billed the government for everything from Henry Crown's new mattress to dog boarding charges, baby-sitting fees, and in several instances, legal expenses incurred in fending off other official investigations.

This was ripe stuff and it was highly embarrassing. For close to four hours, Lewis and MacDonald did their best to present a brave front before banks of television cameras and reporters jammed into the overflowing hearing room. The mood made their task hopeless from the beginning. As Dingell and his fellow panelists walked to their seats, some unknown wit called, "All rise" and several hundred people duly shuffled to their feet. The chairman hid his amusement behind an impromptu civics lesson. One rose to show deference to the majesty of the law as personified by a judge, Dingell explained; since he was but a servant of the people pursuing his duty, there was no need for those in attendance to do anything but sit quietly and observe the democratic process in action. All the same, the courtroom atmosphere lingered, considerably enhanced by the abject demeanor of the two chief witnesses.

Lewis began by reading into the record a long, sonorous statement in which he made frequent reference to his company's honorable history and its "sterling record." General Dynamics had been "badly maligned by forces beyond our control," he said, welcoming the opportunity to present an accurate picture of the company's activities.

Plowing on through his forty-six-page statement, Lewis tried to head off some of the most embarrassing issues before they arose. It had been alleged in the press, he said, that he and other company officials used corporate jets to attend tugboat launchings or to fly home from Saint Louis on weekends. It had further been stated that this had cost the taxpayers some $20 million and that records had been altered or destroyed so that the flights could be billed against Defense Department contracts. He denounced the charge as a vicious distortion of the facts. It was true that the company did operate a fleet of corporate aircraft and that most weekends he did use one of them to fly into the private landing strip on his South Carolina estate. However, it was most unfair to characterize any of these trips as an illegitimate expense, since they were but simple and essential precautions against attacks by terrorists. Alluding to the latent enmity of Syrians, Iraqis, and other assorted Arabs who had been smitten mighty blows by Israel's F-16s, he concluded by pointing to another security threat on the home front. "Over the past several years," he said, "our company has been plagued with overt actions against it by various groups representing points of view different from our Government's, including those who are belligerently antinuclear and antidefense."

The speech made not the slightest impression on the waiting congressional inquisitors. In the opening remarks, one of them commented that the evidence about to be displayed constituted "a textbook case of how to fleece the American taxpayer," while another spoke of the public's mounting disgust at stories of four-hundred-dollar hammers and seventeen-hundred-dollar coffeepots. "There has to be a day of reckoning," he said. Evidently the sins of the entire defense industry were about to be visited upon General Dynamics.

What followed was an unrestrained mugging, as the two white-haired and elderly executives were confronted with the itemized details of their company's misconduct. The pattern of charge and limp denial seldom varied. Were either of the witnesses aware of someone, or something, called Thurston, and the reason the company billed the Navy $87.25 for his four-day stay at an establishment called the Silver Maple Farm on the outskirts of Saint Louis? Lewis and MacDonald professed ignorance, so the item was explained to them. Thurston was an executive's dog and the Silver Maple Farm was a boarding kennel where the animal was kept whenever his master attended General Dynamics executive conferences —another subject in which the panelists evinced great interest. What was discussed at these meetings in the Carolinas and Florida, and why were they always conducted at luxury resorts, usually locations in proximity to at least one championship golf course? The bill for one such session came to almost $100,000, which, as usual, the company passed on to the Defense Department.

Lewis could do no more than roll with the punches, for MacDonald had already demonstrated the folly of resistance. Once or twice, nettled into brief outbursts of testy sarcasm, the thickset and gravel-voiced chief financial officer had been put quickly back in his place by committeemen whose theatrical indignation undoubtedly gained something from the presence of so many television cameras. Lewis never tried to talk back. With what was to become a much-repeated line, he conceded that Thurston should never have been made a temporary ward of the public purse and apologized for this and other "occasional slip-ups."

As an exercise in damage control, Lewis' testimony proved a total failure. The next day, an official spokesman for the Pentagon described his testimony as "nauseating," while congressmen and editorialists strove to outdo each other in the rhetorical sweep of their denunciations. "A War Machine Mired in Sleaze," thundered the *New York Times* several days after Lewis and MacDonald made their second appearance in less than a month before Dingell's subcommittee. Arms companies like General Dynamics were "the kept creatures of the Pentagon bureaucracy, only so faithless because so robbed of dignity." There was even criticism from other defense firms, who found their own billing practices being subjected to unexpected and increasingly unrestrained scrutiny. In the months leading up to the Dingell hearings, National Semiconductor had already admitted its guilt in supplying the Air Force and other branches of the armed forces with untested and unusable silicon chips, while McDonnell Douglas had demanded that the Navy pay for some $25 million worth of legal expenses it had incurred in a dispute with Northrop. Like General Dynamics, Rockwell was found to have billed the government for its executives' country club fees, United Technologies for political contributions, and Sperry for work on MX electric systems that was never performed. The latest General Dynamics disclosures saw the screws twisted yet again. Defense Secretary Caspar Weinberger announced that half a dozen other weapons firms would be investigated to see if the abuses encountered in Saint Louis and Groton were commonplace. An official of another defense contractor commented after the first

Dingell hearing that General Dynamics' transgressions were like waving a red flag at a bull. "They've put a big sign on the industry's back saying: 'Kick Us Hard,' " he said.

It was of course inevitable that General Dynamics would bear the brunt of the assault. Lewis had gone into the first hearing prepared to surrender several hundred thousand dollars in disputed payments. At his next appearance, one month later, the tendered peace offering had increased to $23 million, which the panelists still considered woefully inadequate. By early April, the Pentagon announced that it would take steps to recoup all of $244 million in what were officially described as "contract overcharges," withholding the money due on progress payments for other weapons systems until the full amount was recovered. On Wall Street, General Dynamics stock, which had been selling for only a few pennies short of $86 at the time of the first hearing, lost around $15 over the next two months.

Spokesmen were consistent in citing the victory over General Dynamics and the similar, enforced extraction of $250 million from General Electric as evidence of the administration's ongoing commitment to root out waste and fraud.

It would have been churlish for the administration's critics to decry the successes. The sums involved were spectacular—and the money was being returned. It was also the first time in recent memory that a congressman had called on the chief executive officer of a major defense contractor to resign, a suggestion put to Lewis by Democrat Jim Slattery of Kansas. Still, there were some, including Dingell, Proxmire, and Grassley, who wanted to go much further. As with all General Dynamics' other troubles, the Dingell hearings were rooted in the settlement negotiated almost eight years before by Hidalgo and Lewis. The imputations of impropriety had never quite gone away, and when Veliotis began releasing his tape recordings, they flared back to life with a vengeance. The disclosures of dog boarding fees and country club memberships being billed to the taxpayers were an outgrowth of the renewed interest, but that was all. The chief issues remained to be resolved.

In addition to the Dingell hearings and the joint Proxmire-Grassley investigation, there were almost a dozen other inquiries in progress. Some were quiet, low-key affairs like the Defense Department's review of Lester Crown's right to hold a top-security clearance, while one or two others managed to maintain an attitude of benevolent neutrality. Navy Secretary Lehman, for example, managed to tell a joint committee on seapower and naval policy that he was "entirely satisfied" with the way Electric Boat had reformed and rehabilitated its operations. Such apologists were, however, firmly in the minority.

Another of Grassley's bailiwicks, the Senate Subcommittee on Judicial Oversight, was grappling with the Justice Department for access to files that might help to explain why federal prosecutors had consistently declined to follow the advice of both their own subordinates and FBI agents who had urged that the company and at least two of its most senior executives be charged with fraud and conspiracy. The senator's determination to get to the bottom of the mystery was evidenced by his strained relationship with the White House, which, when he first demanded the Justice Department files late in 1984, indicated it would oppose the bid, perhaps even asserting executive privilege to keep them under wraps. In what can only be considered an act of extreme political bravery, Grassley responded by taking steps to subpoena Attorney General William French Smith for contempt of Congress a mere two weeks before the November presidential elections. The phone lines linking Pennsylvania Avenue and the senator's Capitol Hill office ran hot over the next few weeks, but the senator refused to budge. Seven months after repeated assurances that the disputed documents would soon be released, the Iowa farmer was still waiting, the tide of his anger mounting with every passing week.

The other branches of the armed forces were also taking a closer look at their favorite weaponsmith. The Navy wanted to know how Veliotis had come to be in possession of sensitive, classified photographs depicting technical details of the reactors in its latest submarines. He had acquired the pictures, so the Navy learned to its immense displeasure, several months after fleeing the country as a fugitive. The Air Force was curious about apparent contract anomalies concerning Fort Worth's F-16s, and it shared with the Army an interest in discovering how many of its officers had enjoyed General Dynamics' generous hospitality.

The Navy had already learned of just such a disturbing case, involving none other than Hyman Rickover, a discovery that must have provided the recently retired admiral's old enemies with a moment or two of quiet satisfaction. The public scourge of naval shipbuilders in general and Electric Boat in particular had accepted at least twelve hundred dollars' worth of personal gifts and trinkets that included a pair of diamond ear studs and a jeweled pendant for his wife.* For reasons neither MacDonald nor Lewis was able to explain, Electric Boat's internal records listed the gift as "one dozen watches for retiring employees." Ever critical, Rickover was reported to have told MacDonald that the jewelry was fine; it was the box that was no good. With a few dollars drawn from petty cash, a member of Groton's publicity staff returned to the store to obtain a plain white presentation case of the kind the admiral favored.

**The total value of the gifts to Rickover was officially estimated at $67,600. The admiral insists that most were distributed to congressmen and others whose support might be valuable.

Across the Potomac from the Pentagon, at the headquarters of the Internal Revenue Service, agents were sifting through fifteen years' tax returns and other documents, reviewing the accountants' logic that had allowed the company to enjoy one of the most prosperous periods in its history without paying a penny in taxes. This had been made possible through the use of something called "complete contract accounting," a bookkeeping system more commonly employed in the construction industry, which allowed taxes on progress payments to be deferred until the project was complete. At that point, and with all the money in hand, gross income could be listed in counterpoint to a portion of the $3.1 billion in "loss carry-forwards" that the company had at its disposal as of late 1983. Far from portraying a loss, the impression the figures might present to an uninformed eye, the procedure allowed the company to chalk up some $930 million in profits between 1981 and 1983.

A good deal more intrusive was an IRS examination of a $1.3 million floating expense account used to recompense executives for their out-of-pocket entertainment charges. Among the other disclosures to emerge from the Dingell inquiry was the information that the fund had once financed a thirteen-hundred-dollar night on the town in Tijuana, attended by several middle-level General Dynamics executives and two congressmen, Democrat Don Fuqua of Florida and Republican Bill Lowery of California. Closer to home, it also paid for a testimonial dinner honoring Senator Lowell Weicker of Connecticut. Those individuals were perhaps unlucky to be caught up in the investigation because, as the IRS soon noticed, few of the expense vouchers contained any mention of the person being entertained or the reason why. This was not an oversight but company policy.

The Securities and Exchange Commission was also interested in General Dynamics' bookkeeping methods, although for different reasons. Once before, its investigators had looked into the charge that Electric Boat's books had been manipulated to conceal setbacks that might have precipitated a severe drop in the stock price. That probe began in 1978 and continued for three years, amassing more than twenty thousand pages of documents before fading away late in 1981. At the time and in their later denials of wrongdoing, General Dynamics spokesmen made frequent reference to the investigation and the fact that no charges were brought or reprimands issued. "They looked at everything and they talked to everybody who had anything to do with it," John Stirk, a lawyer in General Dynamics' Washington office, informed the *St. Louis Post-Dispatch.* Once again, the assertion of innocence skated lightly over some inconvenient facts.

When Dingell's indefatigable researcher, Peter Stockton, began looking into the circumstances of the inquiry, he discovered that it had never progressed beyond the planning stage. Boxes of documents surrendered to the SEC were still waiting to be opened. And even if they had been examined, it is doubtful whether anything incriminating would have been found, since the contents had been thoroughly screened, once by General Dynamics' own lawyers and later by Electric Boat's outside counsel. Nor did the case officers assigned to the investigation conduct any interviews, evidently because they lacked the time, manpower, and perhaps the encouragement to do so. A former SEC lawyer who worked on the probe told the *Post-Dispatch* that it was like looking for a needle in a haystack. "It was a big investigation and we weren't throwing enough people at it," he said.

All the same, the investigation that did take place hardly constituted the blanket acquittal to which David Lewis alluded in an earlier interview with *Business Week.* The SEC and various grand juries, he said, "had dug through these things in spades. There's nothing there." In yet another congressional briefing paper presented to Senator Proxmire's committee by researcher Richard Kaufman, Stockton's friendly rival in what had become an unofficial race to see who could dig up the most dirt on General Dynamics, the conclusion to the SEC report was quoted at some length. "The staff's analysis of the documents did not discover evidence that General Dynamics' claims were fraudulent, but did indicate possible disclosure violations in 1976," the author explained, adding in a later section that the passage of time militated against an indictment, "as the possible violations relate to a period of six years ago and as further investigation and any possible litigation will require the allocation of manpower that is currently unavailable and that could be better spent on other, more current cases."

Kaufman also pointed out that while the obstacles listed in the SEC document were undoubtedly valid, they were also at odds with the agency's conduct in regard to another, almost identical case, involving Litton Industries. Like Electric Boat, Litton had filed a series of large claims against the Navy, which were settled after negotiations with Edward Hidalgo in 1978. The sin in the eyes of the SEC was the company's failure to notify its shareholders and the stock-buying public that there would be substantial losses if the Navy failed to approve the sums involved. And there was a second common thread running through both cases. Litton's auditors, Touche Ross, were persuaded to present the company's line in their own, purportedly independent appraisal of the overall financial position, accepting the argument that the books could be squared at some unspecified future date when the Navy put aside its animosity and agreed to be reasonable.

Litton's problems were settled by an agreement unlike the one negotiated with Electric Boat only in that the sums involved were markedly less, $200 million instead of Groton's $359 million. Both firms were obliged to declare a loss in 1978 in order to qualify for consideration under PL 85-804 bailout legislation and each had its current contracts redrawn. The only difference was that the SEC had gone after Litton for failing to announce a loss, while leaving Electric Boat in relative peace.

As the Kaufman report explained, the SEC distinguished the cases by arguing that Litton was well aware it would not recover the full amount in dispute and had even informed its bankers that a large loss might soon be encountered. In fact, General Dynamics had been every bit as frank with its own bankers, advising them of impending losses on the 688 and Trident programs at a series of meetings that might have come to light if the SEC had taken the trouble to open a few more boxes. A memo obtained from the files of a vice-president at Chase Manhattan, Electric Boat's principal bank, quotes Gorden MacDonald as saying Henry Crown, Lewis, and every other member of the board were aware as early as June 1975 that large losses were inevitable. "Needless to say," the Chase Manhattan executive recorded, with the relish of a gossip columnist, "this shook up Colonel Crown and other members of the Board of Directors and there was much recrimination and discussion."

If the banks, Arthur Andersen, and the company's chief stockholder all knew that Electric Boat was about to take a bath, where then was the strength in the SEC's explanation that General Dynamics was unworthy of prosecution because it had been sincere in the mistaken belief that the Navy would eventually hand over the full amount in question? "Had the precedents in the Litton case been followed," Kaufman's report concluded, "the SEC might have proceeded against General Dynamics and its outside auditors, Arthur Andersen & Co."

Now, armed with Veliotis' tape recordings and thousands of documents turned up by congressional investigators, the SEC was reexamining not only its conduct of the first inquiry but a slew of other, more recent charges as well. Foremost among these were questions posed by one of the Veliotis tapes, which indicate that David Lewis and other company executives may have taken steps to conceal another loss in 1980, this one worth approximately $100 million. "If this silly bastard starts popping off," Lewis said in reference to James Ashton, an Electric Boat executive and a Fort Worth veteran, who had peppered the Saint Louis office with memos and telephone calls criticizing Groton's internal accounting procedures, ". . . then we bring Arthur Andersen in immediately, saying, 'Well, look now, if this is the true story we've got to record a loss, you know!' We didn't want that issue to even arise."

The way Electric Boat extricated itself from that mess prompted yet another of the later investigations, this one by the Justice Department. In the tape recording, Lewis informed Veliotis that the situation would be much improved if the auditors learned of the loss after the company had landed contracts to build two more 688-class submarines. "If we can get the letter contract before this announcement is made, we would have Arthur Andersen in a position where they would be much more relieved and pay little or no attention to this," Lewis said. And lo, that was precisely what happened.

In late August 1981, David Lewis and Edward J. LeFevre, the company's top Washington lobbyist and a vice-president in charge of "government relations," called at the Navy Department for a tense meeting with Navy secretary Lehman and his assistant secretary for logistics and contract negotiations, George A. Sawyer. By all accounts, the meeting itself was not a success. The glad tidings came later, when Sawyer went bounding out of the building to catch up with the two General Dynamics executives and hitch a ride across town.

Some minutes earlier, Lehman had rejected Electric Boat's imaginative plea to recoup its $100 million overrun by filing an insurance claim. Fire, natural disaster, or the depredations of an invading enemy might have qualified the company for relief under the terms of the legislation being invoked, but the real cause was none of these. As James Ashton would tell Congressman John Dingell's Energy and Commerce Subcommittee on Oversight and Investigations, Electric Boat's latest losses were, as usual, almost entirely due to the workers' collective incompetence and their executives' neglect. Late in 1979, a check on one of the Flight Two boats turned up the distressing news that many vital welds in the hull and internal plumbing systems were either missing, poorly executed, or bereft of the paperwork required before the Navy would agree to take delivery. "It became necessary to tear apart much of the finished boats and do them all over again," Ashton said, placing the blame squarely on Veliotis, who continually demanded greater speed but failed to replace many of the supervisors, quality inspectors, and experienced leading hands laid off during the first few tumultuous months of his reign. As with the earlier overruns, no mention of an impending write-off found its way into the annual report. The only hint of bad news was a fleeting reference to "minor but troublesome welding defects" and a few brief words about the need "to locate and replace large amounts" of substandard steel. "With the problems largely behind us," it continued, "the production log-jam appears to be broken and we foresee much improved progress in 1981."

Lewis had entered the meeting with no reason to expect anything better than a dressing down. One week earlier, Lehman had made some acid-edged remarks about naval shipbuilders, homing in on General Dynamics and its disaster insurance claim, which he described as "preposterous." He had rejected the notion that "the government always pays" and warned that he could be pushed so far and no farther. If General Dynamics continued to play legalistic games with the taxpayers, he indicated he would consider suing the company to recover damages for the interminable delays and late deliveries, at that point running some three years behind schedule. The speech attracted a good measure of attention in the national press, but it was the force of the comments rather than their novelty that won headlines. Lehman had taken the company to task for its impudence on several previous occasions, but this was by far the most pugnacious display to date. Curiously, it was also the last. One month after their August 25 meeting in Washington and to the astonishment of all but the participants, the Navy and Electric Boat fell once again into each other's warm embrace. As with lovers reconciled, the end to the estrangement was soon being marked by an exchange of gifts.

In return for withdrawing its insurance claim, General Dynamics received the delayed Trident contract, an order for one more 688-class boat with an option for a second, and the security of knowing its name had been restored to the Navy's list of approved contractors. This was precisely the kind of settlement Lewis wanted when he said the division needed fresh work if it was to get the auditors off its back. Moreover, it had turned up just when it was needed most. Lehman rejected all suggestions that a deal had been struck and presented the agreement as a major victory. As with the Hidalgo truce negotiated three years earlier, this latest settlement also invites closer inspection.

Lehman and the Navy had made a lot of noise about incompetence, pointing to the decision to withhold an option for a ninth Trident as proof of a new, hard-line approach. It was, they said, a preview of a "get-tough" policy that would lead to an unprecedented outbreak of contractor efficiency. While the Navy secretary's words may have gone over well on the stump, the substance of his assertions cannot survive even a cursory examination of the relevant figures. Far from punishing Electric Boat by allowing the option on the ninth Trident to expire, Lehman was actually doing General Dynamics a $94 million favor. When the original option expired, the nominated target price for the ninth Trident stood at roughly $350 million. However, the contract that was awarded four months after the public reconciliation put the price at $444 million and allowed Electric Boat to demand even larger sums for the next two boats in the series, on which Lehman had also requested option prices.

Like a sleepwalker in a Saturday morning cartoon show, General Dynamics sailed through encounters with government agencies and agents, blissfully immune to the mortal perils crashing to earth around it. An observer could have been forgiven for concluding there was some potent but unseen force steering the company away from danger. The Securities and Exchange Commission, the FBI, the Navy, and a succession of prominent public officials had all taken an interest at one time or another, but none had ruffled a hair on the corporate head. One Navy secretary demanded the declaration of a $359 million loss, which was immediately compensated with $642 million worth of renegotiated contracts; another maintained that he was encouraging industrial virtue at Groton by parceling out a package of new work that would generate at least $200 million in additional profits. Even the Internal Revenue Service, Al Capone's nemesis and the bane of millions of businesses and average taxpayers, could not lay a finger on General Dynamics.

Prophets and critics from the Eisenhower era to the present have warned of the mutually beneficial relationship that binds military suppliers to their political patrons and uniformed customers. Lehman's masters, like all incumbents of the White House, will tell you with nary a trace of embarrassment that while previous administrations may have tinkered with defense issues for base political ends, their own conduct will always remain above reproach. Speeches like those Lehman was wont to give in the months leading up to the unexpected settlement of the insurance claims dispute are typical of the approach. And while they inevitably go over well with voters—after all, no one likes to pay three- and four-figure sums for toilet seats and plastic caps for the legs of stools—they are less than convincing. All too often, the bravest public assertions are undone by the results of decisions made in private. The background to Lehman's change of heart was read by some to provide an excellent case in point.

On August 7, 1981, some three weeks before their encounter with the Navy secretary, Lewis and LeFevre called at the White House to chat with presidential counselor Ed Meese about the peculiar perils of naval shipbuilding. We know about the meeting because Senator Proxmire asked Meese for a list of the occasions on which he or members of the White House staff met with General Dynamics officials, insisting that the information be tabled before the future attorney general was confirmed in his new job. Meese could recall only that his guests were troubled by difficulties enountered in relation to the Trident contracts and that Lewis wanted to discuss "problems with the Defense Department and General Dynamics." There was nothing unusual about this, he continued in a written reply to a second round of questions from Proxmire. "My practice . . . is to meet with individuals who believe they have significant information, grievances or views to offer representatives of the President." Meese could not recall any important details of the fifteen-minute meeting, although he did concede he may have made some bland promise to contact Defense Secretary Weinberger to ask for "a meeting between the parties." However, he had no recollection and could find no written records to indicate he had actually done so. In any case, he was firmly of the opinion that the White House should not become involved, and insistent that he had made no effort to "lend any support whatsoever to General Dynamics' position."

LeFevre must have thought the White House was worth a second try, because on August 27, two days after his ride across Washington with Sawyer, he placed a telephone call to Robert Garrick, Meese's deputy. The attorney general–designate was unable to ascertain the exact reason for the call, because he could inform Proxmire only that "the purpose appears to be a follow-up to the 7 August, 1981, meeting." Two and a half weeks later, LeFevre passed again through the White House gates, leaving his name, and the only record of his visit, with the security detail on duty at the time. Garrick and LeFevre conferred for an unspecified period, although once again, Meese was uncertain what they found to discuss, advising Proxmire that "the subject was apparently the same." Why, after a personal visit and a telephone call, LeFevre felt moved to make his third contact with the White House in as many weeks, presumably to be told he was barking up the wrong tree, has yet to be adequately explained. Whatever the answer, his problems were almost over. On October 22, exactly eight days after the meeting with Garrick, Navy Secretary Lehman and David Lewis hosted their press conference at the Pentagon to announce that all was to be forgiven and forgotten.

Lehman, Meese, and Lewis would all later insist that the reconciliation was an entirely natural event, prefaced by no untoward tactics or the exercise of inappropriate influence. This may be every bit as true as the participants assert, but that is certainly very hard to believe. It becomes even harder when the three men's accounts of the background to the settlement are laid out for comparison.

First, Meese. At several points in his confirmation hearings, the future attorney general denied ever having taken steps to influence Lehman's handling of the insurance claims affair. Lehman's story was different; not by much, to be sure, but enough to cast Meese's recollections into question.

In response to another set of questions submitted by Senator Proxmire, Lehman said that while he had not spoken directly to Meese, he had discussed the case with a member of his staff. "Although the specific date is unknown, my best recollection is that a message was conveyed during that period from a member of Meese's staff to the effect that the Navy should continue to do what it was doing and that Mr. Meese did not wish to become involved in the matter," he wrote.

The third and most unsettling account of the White House's part in ending the dispute comes from Lewis via another of Veliotis' tapes. As with all the recordings, Lewis contests their accuracy and maintains that they may have been doctored to present both him and his company in the worst possible light. Congressional investigators agree that there are indeed breaks and moments marred by electronic static, but they are equally insistent that nothing new has been added or overdubbed. In the relevant tape, recorded not long after the meeting with Lehman at the Navy Department, Lewis informs Veliotis of the secretary's mood.

"Lehman got mad with us because we went to see Meese," Lewis said, adding that he and LeFevre stood their ground.

"You bet we went to see Meese," Lewis stated. "We were not getting any satisfaction. We had to go there. We had to go so we didn't back down one iota and neither did they." The relevant section of the tape concluded with a final comment from Lewis that was to intrigue many members of Congress: "It is obvious to me that someone told him to settle."

Perhaps it was Assistant Secretary Sawyer's eagerness to be of assistance that led Lewis to surmise the fix was in. Having endured what LeFevre would later describe as a very heated meeting, he and Lewis retired to the men's room, then paused briefly in the lobby to exchange pleasantries with a mutual acquaintance before proceeding to their waiting car. The two brief stops gave Sawyer just enough time to collect his things and catch up with them after what was presumably a spirited dash through the corridors of the Navy Department. This was too much for Minnesota Democrat Gerry Sikorski, one of Dingell's fellow committee members, who used the incident as the basis for a scathing rhetorical question.

"What small contractor could suffer the wrath of the Secretary of the Navy, go to the White House and meet with Mr. Meese, then have a pleasant meeting with the Secretary of the Navy that results in an Assistant Secretary running out to your car like a puppy dog to assure you that the Navy will take care of you?" he asked, pausing for dramatic effect before continuing with his attack. "And where else can an Assistant Secretary get hired eighteen months later as an Executive Vice-President?"

For a record of what transpired in the back seat of the General Dynamics limousine as it rolled across town, we must turn to another of Veliotis' tapes. As with them all, Lewis disputes its accuracy and its admissibility as evidence. But still, as Dingell pointed out when Lewis voiced his objections at a committee hearing, the questions raised far outweigh any legal niceties.

"'Look,' he said, 'we've got to find a solution,'" was the way Lewis recounted the conversation to Veliotis the following day. "'This is just between us and we've got to figure out a way to sit down and negotiate some contracts, give you some stuff that maybe we can do to find a solution.'"

Lewis responded that he would be delighted to take on fresh construction contracts, but he was adamant that General Dynamics would not abandon its insurance claims "without something in return." Sawyer was happy to oblige, assuring Lewis that he could write a contract for another 688 boat "right now" and promising to remove any and all obstacles holding up the award of the ninth Trident, the same ship Lehman had ordered kept in limbo until Electric Boat gave some indication of its willingness to reform.

Lewis raised the problem of the $355 million option price on the Trident, stating that "goddamned Sea Command" would hold the company to the figure it had nominated eighteen months before. "We couldn't start there, and we're not interested," he said. As before, he was assured that all problems could be made to disappear.

"You've got to trust me," Sawyer supposedly said. "I will see that that does not happen. . . . I want to have terms that get you out of this problem. That's a quote." The rest is history. General Dynamics withdrew its insurance claim not long after, while Sawyer, acting in close concert with Lehman, authorized the new contracts and options. None of the anticipated problems were allowed to arise. Sea Command did not enforce the option price and Lehman's belligerence mellowed in the display of jolly optimism in evidence at his joint Pentagon press conference on October 22. As one commentator later remarked: "If no untoward deal was struck, a remarkable set of coincidences occurred."

It was not to be the last time Sawyer would strike a bargain with General Dynamics. Over the further two years he remained in his job, Sawyer played a major in negotiating the TAKX deal for the supply ships to service the Rapid Deployment Force. And finally, not long after that deal had assured Quincy's financial security into the 1990s, there was the telephone call he received from Lewis in early March 1983.

General Dynamics' chief executive had heard that Sawyer was considering leaving government service, and was keen to offer him a job. The prospective recruit proved most receptive, so Lewis said that he would discuss the topic with his colleagues in Saint Louis and report back in a few days. He did just that on March 20, asking if Sawyer could rearrange his calendar of appointments to squeeze in a trip to Saint Louis in five days. When Sawyer agreed, Lewis told him he would have someone from the company's Washington office bring over a first-class round-trip airline ticket.

It would be nice to report Sawyer's version of events prior to his resignation and the start of a new career at Land Systems, but unfortunately, that was reserved for an Alexandria grand jury convened to look into possible violations of the federal conflict of interest laws. All that can be said is that there was nothing unusual about Lewis' overture. From the early fifties, General Dynamics has consistently led its fellow contractors in recruiting former military officials and retiring public servants, Secretary Hidalgo being the last notable addition to the payroll. The difference this time was all a matter of timing. Sawyer remained in office through the two-month course of the interviews and job negotiations, continuing to deal with Electric Boat as of old.

During a second round of testimony before Dingell's subcommittee in early March 1985, Lewis was at pains to dispute allegations of illegality and impropriety. The handsome, square-jawed former assistant secretary was an honest, capable, and well-qualified man, an asset to the team. Before entering government, he had worked for the Bechtel Corporation, and later, for J. J. McMullen, a firm of marine architects and engineers that acquired several lucrative consulting contracts during the Reagan years. Under Lewis' interpretation of the law, everything was aboveboard because the first telephone calls did not involve job discussions at all. They were, he explained from behind the cover provided by a thicket of split hairs, discussions about *later* job discussions. Continuing in the same vein, he described Sawyer's free trip to Saint Louis in March as part of an "exploratory" process based on the expectation that the future division chief might choose to come aboard at some unspecified date when he was free from the restrictions imposed by the federal government's code of employee ethics. Lewis did admit that as early as April, more than one month before Secretary Lehman was officially informed that his subordinate was considering a job with General Dynamics, he had suggested that Sawyer might eventually be moved to Electric Boat after getting a feel for the company at Land Systems. However, he characterized this not as a concrete indication of what Sawyer could expect when the negotiations began in earnest, but as another exploratory talking point.

Lewis might have been comfortable with his notion of what was, and was not, a breach of the law, but his inquisitors clearly were not. "The statute we are talking about has nothing to do with your narrow and self-serving definition of what negotiation means," committeeman Sikorski told him.

For something that was not a job interview, the March trip to Saint Louis certainly looked like one. Sawyer was picked up at the airport and driven to the company's headquarters, where he conferred with Lewis and various senior executives before being hooked into a conference call with several other directors and division heads that lasted for well over an hour. Both parties appear to have been impressed by what they saw, because Lewis soon invited Sawyer to come back for a second all-expenses-paid visit, in May. Sawyer agreed, dashing off a short note in which he thanked Lewis for the guided tour and the "fulfilling and informative" opportunity to see how the contractor organized its operations.

The second round of talks took place not only in Saint Louis but at a variety of locations scattered about the country. Armed with what must have been a thick sheaf of airline tickets, each booked and paid for by General Dynamics, Sawyer first visited New York to talk with Henry Crown's old friend and fellow member of the executive committee, Nate Cummings. The next day he touched down in Chicago for further talks, this time with Lester Crown, Milton Falkoff, and several other prominent members of the hierarchy. Finally, several days later and after an interim return to Washington, the assistant Navy secretary met with Henry Crown before returning to Saint Louis via Chicago for one last meeting with another member of the executive committee, Elliot H. Stein.

At some point during his comings and goings across the United States, Sawyer must have decided that he should make some effort to inform the appropriate authorities of his plans for future employment. On May 20, two months after Lewis first broached the subject, Lehman received a note stating that his assistant secretary wished to be removed from all responsibilities related to Navy shipbuilding. As his reason, he mentioned ongoing employment negotiations with General Dynamics, which he maintained had begun only three days earlier. This information was passed on to the Navy's acting general counsel, Hugh O'Neill, who conducted a brief review of Sawyer's recent decisions as a government official and concluded there were no obvious ethical breaches. Checking the records, he found Sawyer had last participated in a matter concerning General Dynamics on May 5, when he approved Electric Boat as a bidder for two more of the 688-class submarines. Since this took place three weeks before the date on which he claimed to have received his first job offer, O'Neill could see no reason why Sawyer should not go ahead with his plans. A few days later, Lehman and Deputy Secretary of Defense Paul V. Thayer met with Sawyer to discuss Lewis' proposal and the Navy counsel's opinion of its propriety. They, too, accepted the assurance that Lewis had not raised the subject until six days before and agreed that there was no apparent conflict of interest.

It was at this point, when all the legal niceties had been observed, that Lewis said he felt free to commence the "official" stage of the job negotiations. As might be expected, they were over almost before they began and Sawyer was soon off to Saint Louis for a short orientation period before being transferred to Michigan, where he replaced Oliver Boileau, General Dynamics' president and the first chief of the Land Systems division following its acquisition from Chrysler.

In one way or another, the Justice Department had been involved in the investigation of the fraud allegations against General Dynamics since late in 1977, when Admiral Rickover itemized what he claimed were eighteen cases of clear-cut fraud in Electric Boat's claim for an additional $544 million to cover overruns on the first few 688-class submarines. His findings were to remain the subject of an investigation for a full four years before being officially abandoned late in 1981. The inquiry's demise seemed to mark the start of a bright new era for Electric Boat, since both the accuser and the accusations were swept from the picture at roughly the same time. Like the case itself, the end to the admiral's sixty-three-year career was to provide a talking point long after he was gone.

Rickover had never enjoyed a reputation for conventional behavior, and his wars with all contractors, not just Electric Boat, were legendary. But of late, the tales were even odder and the behavior often quite startling. Stories had been popping up in the press that suggested the eighty-one-year-old sailor was no longer fit to command a submarine. In January, for example, he had taken out the nuclear submarine *Jacksonville,* exercising his jealously guarded prerogative to personally put each new boat through its paces. As usual, his bunk was laid out with the freshly laundered khakis and the basket of seedless white grapes he demanded at the outset of every maiden voyage. According to an Electric Boat report filed with the Navy and later leaked to the press, the *Jacksonville* was placed in mortal jeopardy several times when Rickover set out to test the boat's response to a radical maneuver known as "crashback," a sudden, short stop underwater achieved by throwing the engines into full speed astern. Rather than merely stopping the boat, Rickover was said to have allowed it to accelerate backward until it was bucking and dipping beneath the surface, a highly dangerous maneuver, according to the Electric Boat report. Later that year, the admiral gave a repeat performance with another submarine, the *La Jolla,* this time prompting a stern-first crash dive that took the boat 240 feet beneath the maximum depth planned for that particular sea trial.

There was some suggestion Rickover merely forgot to order the engines returned to forward power, but that hardly seems likely, since the *Jacksonville* performed the crashback test five times before he was content to move on to something else. The Electric Boat report warned of potentially fatal consequences unless the Navy found someone new to test its submarines: "It appears [the officers] will not override Rickover unless they believe the safety of the ship is in jeopardy. However, by the time they make such determination, the time available for recovery may be insufficient." Rickover denied the charges of recklessness and senility by arguing that the Navy had every right to know if its ships could perform a simple, safe, and essential evasionary maneuver. Far from endangering the boats and their crews, he was actually making sure the taxpayers received their dollars' worth of defense from the same crooked contractors who were now maligning his competence to command.

Rickover had fought off previous attempts to force him into retirement, but this time there was no escape. In November, Secretary Lehman announced that the father of the nuclear navy was finally stepping down. He was to be offered a post as a special adviser to the White House, but it was obvious this was but a courtesy; the troublemaker had been shackled and ejected with great ceremonial deference to his experience, reputation, and rank.

The charge that General Dynamics played no small part in Rickover's ouster became a question of largely academic interest. Now that he was gone, the situation was radically different. Electric Boat and the Navy put their quarrels behind them and were soon back on the best of terms. The insurance dispute was settled with dispatch and fresh Trident orders tumbled in, one after the other. Incredibly, Groton even won an award for being one of the Navy's most improved suppliers. The honeymoon ended when Veliotis fled to Greece and started the avalanche that would later crash down upon David Lewis and the rest of his colleagues. In mid-1984, just when questions were again being asked about the Hidalgo settlement and the validity of Electric Boat's $544 million claim, the *Providence* (R.I.) *Journal-Bulletin* unearthed a document that lent credence to the old story that Rickover had been retired in response to the contractor's demands.

The new evidence was a memorandum written by an aide to Rhode Island senator John Chafee, which indicated that a deal had been struck: Electric Boat would remove Veliotis from Groton if the Navy ditched Rickover. Written in March 1982, it recounted the conversation at a lunch attended by the Republican senator, Assistant Secretary Sawyer, and the memo's author, Scott Harris. Sawyer was said to have complained that while General Dynamics had promoted Veliotis to the post of vice-president in charge of international affairs, he had also been left in overall command of the Marine Division. "In passing," Harris wrote, "it was noted that the Navy had lived up to their part of the bargain (i.e., getting rid of their troublemaker), but that the ghost of Veliotis seems to have a great deal of substance still." The *Journal-Bulletin* asked Senator Chafee why Sawyer had chosen to raise the matter in his presence. A staff member explained, beginning his reply by reminding the questioner that Electric Boat was Rhode Island's largest single employer. The senator had often served as a broker in disputes between General Dynamics and the Navy, but would not have been involved in anything of this kind, he said. All the same, the spokesman did agree that it would have been in both parties' interests to order the two iron-willed antagonists from the field at the same time.

If no deal was arranged, this was another of those coincidences that always seem to crop up in General Dynamics' dealings with the government. Less than ten days after Rickover's retirement was announced, the

Department of Justice wound up its four-year investigation. In the light of the findings, and the prevailing mood of the attorneys and investigators who worked on the case, the decision was baffling. On the morning of November 12, the same day that the head of the Justice Department's Fraud Section rang Rickover to tell him no further action would be taken, his supervisor, Assistant Attorney General D. Lowell Jensen, attended a presentation at which the case officers summed up their evidence and urged the indictment of General Dynamics and two individuals. Nor were they alone in believing a case could be made against the corporation. A little over one week earlier, the FBI had made the same recommendation, a most unusual step since the agency usually leaves the interpretation of gathered evidence to the Justice Department's prosecutors. In this case, or so one gathers, the FBI man believed the department needed a little prodding.

It did little good. Assistant Attorney General Jensen explained that he could not proceed, for reasons that are intriguing in their selectivity. Both the Justice Department's own people and the FBI had indicated that a strong case could be made against two individuals, yet the letters Jensen wrote to General Dynamics and the Navy stated that it would be impossible to establish the requisite criminal intent. His second objection was that while "the claim adopted certain theories which were considered to overstate Electric Boat's claim position, the Navy was well aware of what was going on and thus, could not be said to have been deceived." The third and final objection was a universal cover-all. The case would involve "a difficult theory of prosecution" as well as "technical issues requiring the review and opinion of experts." Of them all, this was by far the most inconsistent. If the same logic were applied equally to homicide investigations, no case involving forensic evidence or the testimony of psychiatrists would ever come to court.

Before arriving in Washington with the California Mafia that accompanied the new President, Jensen was considered one of the most capable prosecutors in the history of Alameda County. He had been there since 1959, working first for Ed Meese as an assistant district attorney before stepping up to the top spot when his friend and mentor moved on to a new job in the governor's office in Sacramento. He was little known in the East when he arrived, but his credentials were sound and he had scored some notable courtroom victories. He prosecuted Patty Hearst's kidnappers, William and Emily Harris, and would have sent Black Panther Huey Newton to jail for murder if the conviction had not been overturned on appeal. A keen supporter of capital punishment and a professed "law and order Democrat," he admitted to relishing the cut and thrust of courtroom debate. "It is easier to convince a jury than a subcommittee," he said not long after taking control of the Justice Department's criminal prosecutions.

Jensen was gone by the time Senator Grassley began demanding access to the files of the four-year investigation, but his successor, Assistant Attorney General Stephen Trott, saw little reason to doubt the truth of the observation. Grassley proved to be a very hard man to convince indeed.

Grassley began looking into the Department of Justice's handling of the shipbuilding investigations early in 1984, when the Veliotis tapes had just started their chain reaction. Proxmire had already sent his chief researcher, Richard Kaufman, off to Greece and he had come straight back to immerse himself in vast amounts of Electric Boat's internal correspondence. The Justice Department's actions—or rather its inaction—became an issue straightaway and Grassley soon joined the hunt at the head of his own small Judicial Subcommittee of Practice and Procedure.

Grassley's staff admit with a note of perverse pride that people chronically underestimate their boss. "When he first came in Washington, the joke was that the Republican, Roger Jepsen, was Tweedledum and Grassley was Tweedledumber," one aide recounts with a smile. And it was easy to dismiss him as a hayseed, with his large, farmer's hands, broad midwestern twang, and tendency to punctuate exclamatory sentences with words like "gosh" and "golly." Still, whatever else Washington's old hands may have thought of him, the rough edges concealed a core of carborundum.

"It wasn't a question of bravery but a duty," Grassley said in explaining how he had come to cite Attorney General William French Smith for contempt of Congress just one week before the presidential elections. "I have a job to do and duties to uphold, and one of those is making sure the Department of Justice does its job. At this point, it does not look like they were doing what they are paid to do." What made the contempt citation even more unusual was the fact that Republican Strom Thurmond refused to convene the full Judiciary Committee in order to approve the appropriate documents. Officially, Thurmond blamed the pressures of his reelection campaign, but he still devoted most of his written reply to cautioning Grassley about becoming entangled in a subject "fraught with constitutional and legal problems." As a further indication of his disapproval, Thurmond refused to order a meeting between Grassley and Justice Department officials. The Iowa farmer was undeterred. Having previously discussed his problems with Senator Proxmire, the Republican had his citation issued through the Joint Economic Committee of his Democrat ally.

It was not the first time Attorney General Smith had run into flak over his own and the department's handling of the General Dynamics investigation. Once again, it was a question of consistency. One year earlier, he had refused to disqualify himself from an inquiry into how the 1980 Reagan campaign obtained briefing papers from the Carter White House, the so-called Debategate scandal. On that occasion, he professed to see no conflict of interest in allowing his Justice Department underlings to clear his campaign colleagues of any wrongdoing while he remained on the job to check their work. But when the submarine investigation resurfaced as an active issue in August 1984, Smith turned his back immediately. He explained that it would be improper for him to become involved in the investigation because his wife had recently accepted an invitation to visit Groton and christen one of Electric Boat's new submarines. Smith also raised the point that his law firm had once represented another of the company's divisions, something he asserted would call his impartiality into question. As before, the issue was consistency. If Smith considered it inappropriate to deal with one division of a company after having worked for another, why then did he and his administration colleagues in the Defense Department not take action action against Edward Hidalgo, who was off hawking F-16s and M-1 tanks to the Spaniards? Even more relevant, why was this same code of conduct not forced upon Assistant Navy Secretary Sawyer, who had been dealing with Electric Boat a mere one month before going to work for Land Systems?

The issue of what the *New York Times* called "The Ceremonial Attorney General" was, however, a passing sideshow in Grassley's continuing attempts to get his hands on the Justice Department's files. Assistant Attorney General Stephen Trott had resisted all previous demands for the papers, insisting that he was in no position to comply because a new grand jury had recently been convened to reexamine the old allegations. What followed was a legal game of cat and mouse that grew from two participants to three when the White House stepped in to side with Trott. Grassley's staff dug up precedents that indicated information before a grand jury could and had been released to a congressional committee on at least one previous occasion. The Justice Department countered with its own precedents, while White House chief of staff James Baker argued that though Trott was probably right, it would be better to settle the dispute after the elections. Baker was also diplomatically insistent that the White House should be given the opportunity to examine the files first if they were made available, while the President's counsel, Fred Fielding, raised the possibility that Mr. Reagan would choose to exercise executive privilege if the Justice Department was forced to yield.

What fueled Grassley's determination were the snippets of information his small staff had bludgeoned out of the criminal investigation. The allocation of manpower, for example, appeared to demonstrate a haphazard attitude at best. Investigating attorneys entered and left the case with the frequency of commuters passing through a large railway station. The Navy assigned a solitary lawyer to the case in January 1979, pulled him out four months later, and did not bother to send anyone else in his place. There was sporadic activity on the part of the FBI, which conducted interviews with the Navy people involved in negotiating the Hidalgo settlement, as well as several often heated encounters with the Coopers & Lybrand auditors who had been brought in by the Navy to determine Electric Boat's true financial position.* As for the Justice Department's own attorneys—the same people Lewis said "had dug through these things in spades"—they were few, far between, and much interrupted by frequent transfers to other assignments. Throughout the entire four-year course of the investigation, there was only one six-month period when two full-time Justice Department attorneys were on the case at the same time, as well as several periods when it was being handled solely by part-timers. As a subsequent aide to Proxmire pointed out, this "could not have strengthened the investigation."

Grassley's team, now drawing increasing support from the Proxmire committee's larger staff, also learned something of the vehemence with which the Justice Department's attorneys and the FBI agents had pressed for prosecution. One informed his superior that the department would be derelict in its duties if no further action was taken: "I don't see how our collective performance . . . can be viewed as anything but dropping the ball." Yet despite this and other similarly insistent pleas, Assistant Attorney General Trott was still prepared to go on the record during the summer of 1985 with the comment that "there was not one scintilla of evidence" to justify an indictment. Grassley replied that where he came from, that "scintilla" looked more like a truckload.

Yet still the documents remained under lock and key. At one point Grassley thought he had obtained them when he made their surrender a precondition of voting to confirm Ed Meese as the new attorney general. On a number of occasions prior to the final vote in March 1985, Grassley found himself taking part in three- and sometimes four-way telephone conferences with Trott, Meese, White House counselors, and his own staff members. The last of these produced what his aides regarded at the time as a firm commitment to hand over at least some of the documents, the sensitive sections relevant to the current grand jury investigation being either deleted, edited, or subject to antidisclosure provisions. Grassley duly voted for Meese's confirmation, only to receive a further disappointment. One year after striking what he believed was a firm deal with Meese, his aides were still waiting for the package to arrive.

*Although originally brought into the case to make an independent appraisal of Electric Boat's future prospects, the Coopers & Lybrand auditors were quick to defend the subject of their investigation. When an FBI agent quizzed one of the accountants about Rickover's fraud allegations, auditor Joe Kehoe shot back with a question of his own: "If I were to say that Rickover was a homosexual would I, as a private citizen, be able to demand an investigation?" According to a summation of a telephone call between Kehoe and Gorden MacDonald, "the agent immediately changed the subject."

EPILOGUE: Too Big to Whip

By the end of 1985, General Dynamics was operating like one of its own tanks, deflecting each latest barrage with the armor of its own sheer bulk and then racing forward under fire to secure the next objective. Despite all the hearings and the indignant editorials, even perennial critics like Senator Proxmire found their former outrage being replaced by despair. Simply put, the company was just too big to whip. "Defense contractors like General Dynamics have so much leverage against the government they can flout the laws that govern smaller companies and individuals," the senator wrote late in the year. The events of the previous few months made his lament particularly persuasive.

Not long after the revelations of dog-boarding fees and executive mattresses, David Lewis fulfilled a long-standing prediction and announced that he would resign before the year was out. His replacement was to be former TRW chief Stanley C. Pace, another defense veteran and a man whose appointment would seem to indicate that General Dynamics has no immediate plans to pursue the civilian markets that Henry Crown and others have spoken of so wistfully. For Lewis, there was no other choice. Despite avowals of innocence and denials that his departure was in any way related to the recent mauling in Washington, he remained a marked man. Back in May, the Pentagon's inspector general, Joseph H. Sherick, had said that he had been sickened by Lewis' testimony and urged that the chief executive and two other officials be banned from doing business with the government. His superiors declined to act on the suggestion but, by that stage, the lack of official action hardly mattered. At the end of a long, much-admired and, until recently, distinguished career, David Lewis had become the worst kind of liability. If his company hoped to restore its good name, it was going to take a good deal more than the newly drafted code of ethics distributed to employees and the press in a display of qualified contrition. Lewis' withdrawal from the firing line was both the first step in that direction and the company's only voluntary concession to its critics.

It was also, as Wall Street noted, a move that was long overdue. Distracted by the investigations, General Dynamics appeared about to let a golden opportunity slip through its fingers. For more than two years, research and development contracts for President Reagan's Strategic Defense Initiative had constituted some of the largest items in the Pentagon budget. Boeing had cornered $131 million worth of work in just twelve months while Stanley C. Pace had snapped up another $58 million for his former employer, TRW. General Dynamics, by comparison, held a mere $11 million and appeared likely to be squeezed from the table before the feast proper had begun. Pace moved rapidly to correct the situation. In October, the new Valley Systems division was launched, acquiring some of Pomona's missiles and a specific brief to concentrate its energies on developing the theories, test procedures, and equipment that would support the military's march on space.

The late decision to pursue a larger share of the Star Wars' billions drew little attention outside of the aerospace press, so many and so distracting were the other examples of corporate pruning and grafting. A few months earlier, Saint Louis had announced the abrupt and final closure of the Quincy shipyard. Despite the yard's long association with General Dynamics and, before it, with the Electric Boat of Henry Carse and Isaac Rice, there was not a trace of sentimentality in the decision. As the official statements described it, the closure was an unfortunate but unavoidable consequence of the general decline in the maritime industries. With no prospect of further Navy work to follow the last of the TAKX ships and the near-total collapse of the liquid natural gas market on which the yard's executives had once pinned their long-term hopes, Quincy was doomed. The unions complained bitterly, insisting that Veliotis and his successors had ignored repeated suggestions that the yard be readied to turn its hand to heavy engineering projects other than the production of ships and barges. It was a poignant cry, particularly with some 4,000 former workers out looking for new jobs and company officials talking of turning the sprawling site into a real estate development, but there was little basis for the degree of conviction with which the union argument was advanced. Even if it had produced bridge girders or boilers, Quincy could never have been more than marginally profitable. The money, and the managerial expertise, could be better applied in some other, more promising field.

The direction of the company's latest ambitions became clear in mid-September with an acquisition made very much in the spirit of John Jay Hopkins. In return for roughly $670 million, General Dynamics took charge of a 51 percent stake in the Cessna Corporation, perhaps the world's best-known producer of light aircraft. As usual, there were questions about the price. Cessna had spent the last few years floundering through a series of savage losses and while 1985 was shaping up as a comparatively good year, the first nine months had produced profits of only $2.7 million. This was encouraging news for the Wichita work force but no more than small change by General Dynamics' standards. Earlier in March, for example, the company had agreed to withdraw claims against the Navy for all of $23 million, an amount that represented disputed billings for "little" items like country club fees, haircuts and, of course, the infamous dog-boarding expenses. If the move had been prompted by Saint Louis' desire to expand its civilian activities, as some observers asserted, Cessna's stable of light planes and small business jets appeared an unlikely place to find yet another cash cow.

The other, less obvious attractions were, however, very much in keeping with the defense industry's traditional faith in political geography and obsolescence. As a corporate citizen of Kansas, Cessna was represented in Washington by Senators Robert Dole, the Republican majority leader, and Nancy Kassebaum, the chairman of the aviation subcommittee and a member of the Select Committee on Ethics. As *Forbes* commented at the time: "What better way to get them in your corner than by being a major employer in the state they represent? Not every corporate asset shows on the balance sheet." The wisdom of this observation was soon confirmed by Senator Dole's efforts to withhold further funding for rival Fairchild Republic's T-46 trainer. The alternative, never mentioned in his blunt letter to Air Force Secretary Russell A. Rourke, was Cessna's own T-37 jet trainer, the second reason why General Dynamics considered the company a worthwhile proposition. "I, for one, will not sit idle and watch the Pentagon waste additional taxpayers' money on the T-46," Dole wrote. The fact that one more contract to General Dynamics would make the company that much harder to discipline—Senators Goldwater and Grassley's complaint—was likewise never mentioned.

Much more than the civilian programs, it was the T-37 that promised the brightest future. Although it had been designed almost thirty years earlier, the plane remained an entirely satisfactory basic trainer despite predictable Air Force misgivings that it was not quite as sophisticated as it should have been. Since the Fairchild Republic T-46, which was an attempt to produce a "more capable" successor, was much delayed and considerably over budget, it was reasonable to believe that the T-37 could be made to look good by comparison with the addition of a few major changes and a good many more small, updated refinements. Cessna had explored the possibility but its weakened financial position left no room for speculative research and development. General Dynamics had no such problems, and the rewards certainly warranted the risks. Cessna's military work was generating no more than a meager 1 percent of its annual business; an updated trainer, to be followed later by an entirely new aircraft, had the potential to increase the military's contribution to the division's well-being many times over.

What made Quincy's closure and the Cessna purchase so remarkable were the scandals that continued to rage about the company's more prominent divisions. Within weeks of Lewis and MacDonald's appearance in March before the Dingell subcommittee, General Dynamics was suspended from bidding for further Pentagon contracts and fined almost $700,000, ten times the value of the gifts it had bestowed on Admiral Rickover who, in deference to his age and achievements, was let off with a mild rebuke. Just as before, there were stern pronouncements about the need to root out contractor fraud and the usual chorus of self-congratulatory voices praising the strength of the administration's commitment to reform. Once again, however, things were not entirely as they seemed.

The day after the suspension was announced, the Navy sought permission to sell Pomona's Phalanx rapid-fire radar gun system to Britain's Royal Navy in a deal that would earn General Dynamics an immediate $60 million. A few months later, in August when the suspension was lifted, some $900 million worth of delayed orders rolled down the Pentagon pipeline to Groton and the missile operations on the West Coast. This was punishment a masochist might have ordered, a little humiliation and discomfort followed by the customary reward.

Still, the embarrassments kept coming. In October, at about the same time Land Systems' George A. Sawyer was made the subject of a grand jury investigation, the Dingell subcommittee began to probe General Dynamics' relationship with a certain South Korean businessman, Eung-Yul Yoon. When questions first had surfaced earlier in the year about the $2.4 million which General Dynamics' records indicated Yoon had received since 1976, Lewis explained that the sum represented payments for "consultancy work and translation services." Further records, unearthed over the months that followed, led several members of the committee to conclude that while the Korean might be a gifted linguist, many of his transactions were conducted in the universal language of cold, hard cash. In Athens, Takis Veliotis told visiting investigators that Yoon was the Seoul bagman for several American aerospace firms and that he had played a major role in persuading his own country to adopt the F-16 jet fighter. The reflexive denials from Saint Louis and the four other defense contractors involved were called into question by their own internal correspondence.

In addition to General Dynamics, Yoon's largest client, he also received a total of some $4 million from United Technologies, Martin Marietta, General Electric and LTV. All denied that the former general in the South Korean Air Force was their conduit to the decision makers in the Seoul regime, yet none could describe the exact nature of the services he rendered. In one of his cables, Yoon instructed United Technologies to deposit a large sum in a Dallas bank account, ending the message with an enigmatic demand that his employer delete all mention of the arrangement from future correspondence. In General Dynamics' case, the documents indicated that he had received at least one payment via a curiously circuitous route, $250,000 deposited with Credit du Nord in Paris. Initial reports from South Korea quoted local investigators as saying they could find no record of these sums in any of the Yoon company's accounts.

With the investigation still in its early days, it was too soon to say if the Yoon affair would see a replay of the Lockheed and Northrop hearings of a decade earlier. It was certain, however, that there would be at least two rounds of further disclosures before the questions were put to rest. While Dingell's people pressed Credit du Nord for records of Yoon's deposits and withdrawals, South Korean authorities began their own inquiry to determine if there had been any breaches of their country's foreign exchange and financial-reporting laws. In Washington, the State Department was asked to explain why Yoon had been recommended to General Dynamics and the other companies by the U.S. embassy in Seoul.

It was ironic that the Yoon hearings gained as few inches of column space as they did. While the suggestions of payoffs were enough to have congressional legal advisers rereading the Foreign Corrupt Practices Act of 1977 the newspapers were drawn to a fresh specter looming on the West Coast. On December 2, 1985 a Los Angeles grand jury charged one former and three current General Dynamics executives with conspiracy to defraud the Army of $7.5 million in bogus billings on a research program concluded five years earlier. More than the rhetoric of the Los Angeles federal attorney who described the indictments as an effort to "crack down on defense contractors who are pillaging the U.S. Treasury," the story was made even more remarkable by the names and positions of those involved. The list began with James M. Beggs, a former Pomona chief who had left in 1981 to administer NASA for the Reagan administration. He was followed by Ralph E. Hawes, Jr., a current vice president and the man named to take charge of the newly created Valley Systems division. Two senior program administrators, James C. Hansen and David L. McPherson, rounded out the quartet, the first representatives of the company's executive elite to be summoned into court despite almost eighteen months of hearings and continuous denunciations.

As usual, General Dynamics insisted that the matter revolved about a question of bookkeeping rather than any intention to defraud. The issue concerned a "highly sophisticated regulatory and accounting matter which should be resolved in a civil forum, not in a criminal case," the company argued. "We are confident that when our side is heard, we will prevail."

As portrayed to the grand jury, the alleged improprieties, while complex in execution, were relatively straightforward. In 1979, the Army handed $40 million to General Dynamics and a similar amount to Ford Aerospace with instructions to develop two competing versions of a radar-guided, rapid-fire gun system to be known as Divad (Division Air Defense) or, less formally, the Sgt. York in honor of the sharpshooting hero of World War I. General Dynamics entered the race a clear favorite. Its design was based in large part on the proven Phalanx gun system and promised, at least on paper, greater firepower and range as well as the capacity to store a substantially larger stockpile of ammunition. These virtues were to prove largely meaningless when Ford went on to win the competition for reasons that observers concluded were directly related to the poor health of aerospace division's automotive parent. Like the Sgt. York's long and expensive path to its ultimate cancellation just two weeks before Beggs and the others were charged, Ford's eventual triumph lay in the future. In the meantime, while the rivals strove for a prize then estimated to be worth almost $2 billion, General Dynamics sowed the seeds of its later, and latest, embarrassment.

U.S. attorney Robert C. Bonner was to assert that Pomona spent every penny of its $40 million without bringing the prototype up to the promised levels of performance. Perhaps worried that the Army and Congress might indeed prefer to see the contract go to Ford, the division's chiefs chose to plow a further $7.5 million into the program before subjecting it to the critical scrutiny of the source selection committee. What they neglected to do, at least according to testimony before the grand jury, was to inform the Army of the overruns or, more important, accept the financial consequences of failure by financing the additional computer programs and mechanical refinements with corporate funds. Instead the work was allegedly disguised with a variety of misleading descriptions and sent off to Washington for payment against two other outstanding accounts.

Bells began to ring in 1983 when a routine Pentagon audit turned up a number of discrepancies. Of the $7.5 million worth of bills submitted, only $3.2 million had been approved. Now these too were under investigation along with several other "cross-charged" claims submitted by at least two other aerospace firms. Beggs, who had long since left for NASA, seemed an unlikely target right up until the grand jury announced its findings. He was generally liked at the space agency and credited by some with putting the shuttle program back on the rails. "You're either lying or joking, and I don't think it is very funny," one of his aides told the *Washington Post* when its reporter rang to relay news of the charges. A day or so later, Beggs announced that he would step down until he had cleared his name in court. "I intend to vigorously defend the case," he said in his one official statement. "I am confident when all the evidence is aired I will be exonerated."

In Washington, there were howls of anger as various officials strove to outdo each other's condemnations of the much sullied arms maker. And of course, inevitably, there was another suspension, this latest no more severe than the first. Having announced that General Dynamics would get no more business "pending completion of the legal proceedings initiated against it," the Navy was soon issuing a string of qualifications. As first described, the suspension would have stopped Electric Boat bidding for the latest batch of 688-class attack submarine contracts while simultaneously eliminating Fort Worth from an Air Force quest to find a builder for a new, extraordinarily sophisticated airplane being considered as the eventual replacement for the F-15. Neither of these grim prospects was allowed to eventuate.

The next day, the Navy postponed an upcoming deadline for bids on the 688-class boats, much to the jaded amusement of critics and cynics who had expected no less. If no contracts were to be let, they asked, what was the point in stopping General Dynamics from bidding for them? A Navy spokesman tried to explain: "We feel we aren't protecting General Dynamics. We're keeping the competition option open, which we feel will give us the best buy for the American taxpayer." As obtuse as it was to begin with, the logic of this argument was made less comprehensible by the recent comments of Navy secretary John Lehman, who had argued that the public would be better served if Newport News was placed on a more competitive footing with its larger rival. An uncontested award of a sole-source contract for several new boats might have done something to further these aims, as an inducement to see the yard's facilities expanded to handle the giant Tridents, long an Electric Boat monopoly about which Lehman also claimed to hold serious misgivings. Navy officials soon discounted the possibility though. Eliminating Electric Boat at this late stage, they argued, would confront Newport News with the opportunity, and the temptation, to pad its bid with some inflated, last-minute costs and expenses. Those knowledgeable about the bidding process disagreed. Since the final submissions were due within a matter of weeks, they insisted it would have been too late for any postdated profiteering. It came as no surprise when, a few days later, the Air Force followed the Navy's lenient lead.

If any further confirmation of the suspension's lack of severity was needed, it could be found on Wall Street, where the stock price fluttered briefly and slightly before reverting to its buoyant norm. This was just as it always had been. No matter how unsettling the news, the market's faith in General Dynamics' monolithic immunity remained unshaken. In November, for example, the Defense Legal Services Agency concluded its six-month investigation of Lester Crown with the recommendation that he be stripped of his security clearance for concealing his involvement in the 1972 attempt to bribe Illinois state legislators. Crown's denial that he had played an active role in the scheme was rejected by the DLSA's investigators, who reviewed the evidence and concluded he lied during an earlier security review in 1983. With any other company, the prospect that the heir apparent to the largest single block of stock would be banished from all future board meetings might have been enough to spark at least a flutter of uncertainty. Once again, there was barely a ripple. If the past was any guide, the recommendation might well be overturned on appeal. And even if it was allowed to stand, General Dynamics' balance sheet would remain a picture of prosperity well into the new decade, so great was its funded backlog.

As if determined to show that things always turn out for the best, providence brightened the cloud hanging over Lester Crown by concluding another investigation on a cheerful note. In mid-December, two months after commencing its investigation of George A. Sawyer and the charge that he had improperly concealed his job negotiations with General Dynamics while serving as assistant Navy secretary, the Alexandria, Virginia, grand jury declined to press charges. Sawyer, who had been defended by his former boss, John Lehman, was delighted at the outcome. "This restores my faith in the American legal system," he was reported to have said on learning of the verdict.

For anyone open to the rhythms of history, the scandals surrounding not just General Dynamics but the entire American defense industry ring with disquieting reminders of an earlier age. Exactly one hundred and one years to the day before James Beggs and the others were charged with fraud in the Divad case, Britain's House of Lords authorized the first in a series of vast and unprecedented increases in the size of the Royal Navy's operating budget. Two years later, with its own shipyards now hopelessly overloaded, the Admiralty turned to private manufacturers for help in distributing the overabundance of funds. From that point on, unnoticed through a succession of petty wars and distant, dusty victories, Britain's path was pretty much downhill.

At first the arrangement had suited everyone. Although Britain's leading manufacturers had grown rich at the forefront of the early Industrial Revolution, most of them had been finding the going much tougher by the end of the century. Foreign competitors, cheaper, and boasting more efficient methods, were stealing the export markets while a resulting depression at home threw hundreds of thousands out of work. Since many of the unemployed lost their jobs not long after gaining the right to vote, increased arms budgets seldom encountered strong opposition in Parliament.

The thirty years that passed until the eruption of World War I saw what had once been an alliance of convenience harden into a union incapable of easy dissolution. By 1906 the livelihood of perhaps one family in six had come to rest in part or in whole upon the services' continued ability to consume ever larger budgets. No prime minister or cabinet member could consider cutting back, though unsuccessful attempts were made several times. There was always a new dreadnought or some improved armor plate to squeeze a few more pounds from the treasury, each successive innovation promising to provide that elusive margin of superiority about which the game revolved. Meanwhile, the signs of what would today be called a senescent economy were growing more obvious with every passing year.

Innovation was limited by the military caste's narrow horizons and aversion to change. In factory towns like Sheffield and Coventry any incentive to gamble, to take chances, was banished by easy access to the public purse. Vickers rejected a proposal to build a line of small, cheap family cars when it became clear that the profits would not match those generated by a single battleship. In the United States, Henry Ford took the same idea and revolutionized both the automobile and the national economy. As for the weapons, they reflected the builders' and purchasers' obsessions with tradition rather than the quest for imperial supremacy which they were generally believed to represent. While each generation of battleships was larger, bigger gunned and better protected than the one before, none proved to be of the slightest value in discouraging the coming war or bringing it to a rapid conclusion. Too costly to risk losing and too cumbersome to be of any great value, the floating fortresses spent most of the conflict tied up in harbor or, on those rare occasions when they ventured out to sea, being sunk by submarines and mines against which their might and majesty proved entirely impotent.

The flawed complexity of Victorian battleships is perhaps the most obvious parallel with today's high-tech and high-priced weapons systems. Disturbing as it may be, however, it is certainly not the only similarity. Far more chilling is the apparently paradoxical relationship between mounting defense expenditures and the shrinking number of large contractors. When Britain launched its unrestrained naval buildup, there were at least half a dozen firms to share more or less equally in Whitehall's bounty. Within a few years after the conclusion of World War I, in a direct reversal of the civilian marketplace's usual standards, there was only one of any note, an increasingly troubled conglomerate formed by the forced union of the two largest and last survivors, Vickers and Armstrong. One by one, the smaller firms had been swallowed in their entirety, those that remained absorbing not just the contracts and political connections, but also their inherent weaknesses. Anyone wondering why the Quincy shipyard was shut down at a time of record defense budgets might look to the British example and take note. One might also consider the fact that while America boasted at least five yards capable of building submarines in 1945, there are now but two; similarly, the number of airframe firms has undergone a marked, though less drastic, reduction, while only General Dynamics continues to build tanks.

It is popular in some circles to picture the Pentagon and its suppliers as partners in the systematic fleecing of the national wealth. Like most broad political arguments, the critique boils down to a series of all too easy generalizations and shallow perceptions. Just as with its contractors, the military is composed for the most part of decent, dutiful public servants who go about their business in accordance with the prevailing norms. More than the scandals and the periodic outbreaks of "procurement horror stories," it is the nature and consequences of the system itself that pose the greatest threat to the society it may someday be called upon to defend.

While the innovations of American researchers are appropriated and commercially exploited by foreign rivals, buses and trolleys made by U.S. defense contractors are withdrawn in broken-down disgrace from the streets of New York and Boston. Improving the efficiency of production, the old ideal on which America's economic rise was based, has been replaced by the pursuit of short-term profits. Why risk the vagaries and competition of the civilian market when a military contract is likely to be both more lucrative and less risky? Locked into an easy and seductive relationship with the Pentagon, companies like General Dynamics have proven no less impervious to rebuke than immune to reform, the shortcomings mitigated by the ironic assumption that their continued survival is one of the most important components of the nation's security.

It is these hallowed misconceptions rather than the nickel and dime malfeasance of rigged bids and doctored books that represent the greatest and least-recognized danger. In this sense, General Dynamics' long march to its apparently invulnerable perch atop the procurement system remains a lesson about the folly of government policies which elevate expedience above sound, long-term goals. Above all else, it stands as an example of the present sad state of affairs and a beacon illuminating a deeply disturbing vision of the future.

Notes

1. Fenian Fantasies

PAGE 3. **Anthony made a tactful decision:** *Devoy's Post Bag* (Dublin: C. J. Fallon, 1953), vols. 1 and 2. Devoy was one of the principal Irish-American leaders, and his collected correspondence provides much information about the rebels' activities in the U.S. Also, W. J. Laubenstein, "The Emerald Whaler" (New York, 1960), for further reading on the *Catalpa* expedition.

PAGE 4. **native-born Australian criminals:** Since no further mention is made of the Australian prisoners, we can only assume they found freedom in the United States.

PAGE 4. **life on the *Catalpa:*** Captain Anthony forgave and forgot, later using his Irish contacts to secure a position as Inspector General of Customs at New Bedford. He also wrote a book about the rescue mission, *The* Catalpa *Expedition* (New York, 1897).

PAGE 5. **chief British agent:** Britain's campaign against the Irish nationalist movement in the U.S. is described by Richard Deacon in *A History of the British Secret Service* (London, 1969). Beach was so successful in his deceptions he was once invited to present the Fenians' case to President Andrew Johnson at the White House. "My sympathies are entirely with you," was the way Beach reported Johnson's comments to Whitehall.

PAGE 5. **Irish government in exile:** *Devoy's Post Bag,* vols. 1 and 2.

PAGE 5. **Breslin was to learn:** Anyone researching the life of John Holland and the birth of Electric Boat must start with the inventor's most energetic biographer, R. K. Morris. Morris' father, Charles, was a close friend of the inventor and an occasional employer. See *John P. Holland 1841–1914: Inventor of the Modern Submarine* (Naval Institute Press, 1966). Morris has written extensively on Holland, and his work was my starting point in researching much of the material for this chapter.

PAGE 5. **O'Donovan Rossa:** See Rossa entry in *Encyclopedia Americana* (1951 edition), p. 706. Rossa was a favorite of newspaper cartoonists, particularly the illustrators for *Puck,* where his caricature appeared regularly.

PAGE 6. **best way of fighting England:** *New York Gaelic American,* July 1927.

PAGE 6. **Drebbel's boat:** For a general history of submarine development, Drew Middleton, *Submarine* (New York: Playboy Press, 1976).

PAGE 7. **rejected all the earlier nostrums:** Morris, *John K. Holland,* chap. 2, "The Irish Years." Also, Frank Cable, *The Birth and Development of the American Submarine* (New York, 1924). Cable was an associate of Holland's and a major contributor to submarine development in his own right. Many of Holland's papers, plans, and letters are available for inspection at the Rogers Annex of the Paterson Museum, Paterson, N.J.

PAGE 8. **Sympathies were with my own country:** Cable, *Birth and Development of the American Submarine,* p. 38.

PAGE 9. **at Coney Island:** Ibid., chap. 4.

PAGE 9. **trio played boats:** LeCarron reported the submarine experiments to Whitehall, so it is safe to assume the attempts at secrecy were no more successful than usual. See Deacon, *History of the British Secret Service,* pp. 135–38, 140–42.

PAGE 9. **We can do it:** *Devoy's Post Bag,* vol. 1, p. 230.

PAGE 10. **without even the aid:** *Paterson Daily Press,* May 12, 1878.

PAGE 11. **a much larger sum:** Estimates of the total cost vary. LeCarron told London it had cost $47,500, whereas Rossa later put the figure at twice that sum. *Galway Archeological and Historical Society Journal,* May 1967.

PAGE 11. **do the greatest damage:** *Devoy's Post Bag,* vol. 2, p. 515.

PAGE 12. **soared several hundred feet:** Cable differs with Holland in his estimation of the projectile's range, and his opinion may well be the more accurate. Holland had a tendency to embroider his early achievements with postdated flourishes.

PAGE 12. **submarine was a "Fenian Ram":** The reporter was Blakely Hall of the *New York Sun. Galway Journal,* op. cit., p. 32.

PAGE 13. **no longer on speaking terms:** *Galway Journal,* p. 35.

PAGE 13. **It went to the bottom:** Knowles' research led him to conclude that the smaller boat went down in the East River off Whitestone Point. If so, it remains there today, waiting beneath 110 feet of water for some enterprising soul to resurrect it.

PAGE 13. **to rot on their hands:** Morris, p. 47.

PAGE 14. **equally ill-starred *Holland I:*** The *Ram* did make a small contribution to the nationalist cause. For ten years after the 1916 Easter uprising in Dublin, it was carted around the U.S. to raise money for the rebels and their families. By the 1930s it was located in a Paterson park, where it remained until the early seventies. Paterson Museum official Tom Peters explained that it was then moved inside when students at a local high school persisted in painting it yellow as an annual end-of-year prank in honor of the Beatles song.

PAGE 14. **Best captured in a photograph:** The most commonly reproduced picture of Holland, it can be seen in *Dynamic America,* an official corporate history of General Dynamics published jointly by GD and Doubleday in 1960.

PAGE 14. **beyond his ken:** Cable. See the postscript chapter supplied by Holland's staunch friend, Admiral W. W. Kimball.

PAGE 14. **tome of performance specifications:** Holland once asked: "What will the Navy next require; that my boat should climb a tree?" Morris, p. 89.

PAGE 16. **on designing dredging equipment:** The company was owned by Charles Morris, father of the Holland biographer.

PAGE 16. **rabbit and the dove died:** Cable, p. 101.

PAGE 17. **result was a stalemate:** Morris, p. 69.

PAGE 18. **never be made to work:** Kimball chap. in Cable, p. 325.

PAGE 19. **right amount of blarney:** *New York Times,* May 17, 1897. Also, New York *Herald,* May 18, 1897.

PAGE 19. **destruction of Manhattan:** Under Frost's direction, Holland never passed up an opportunity to promote his submarine in the press. Stories beneath his by-line appeared in *Cassier* magazine's "Marine Number," 1897. Also, *North American Review,* December 1900.

PAGE 19. **recent sinking of the *Maine*:** *New York Sun,* May 27, 1898. Submarines were very much in the news, the Hearst papers even attributing the *Maine's* sinking "to an enemy's infernal machine."

PAGE 20. **commune with the fishes:** Copies of tickets and other promotional materials in *Dynamic America;* Roosevelt's letter praising Holland's boat is on p. 69.

PAGE 20. **Holland Company's cause:** Morris, p. 103.

PAGE 20. **unbreakable submarine monopoly:** Lake hearings, House of Representatives Special Committee, Mar. 9, 1908.

PAGE 21. **The diatribe continued:** Kimball chap. in Cable.

PAGE 21. **blockade the place:** House Committee on Naval Affairs. Budgetary Item 6966, Apr. 23, 1900.

2. The Navy Blockade

PAGE 22. **exploited to the hilt:** Holland's obit, *New York Gaelic American,* Aug. 22, 1914.
PAGE 22. **totally oblivious:** The cartoon by McAllister may well have been the inspiration for *Mad* magazine's long-running gag.
PAGE 23. **take total control:** *Encyclopedia Americana,* vol. 23 (1951 edition).
PAGE 23. **conspiring to defeat:** Jack London, *The Iron Heel.*
PAGE 24. **Holland's permanent replacement:** Cable, chap. 9.
PAGE 25. **buy Rice out of business:** *Dynamic America,* p. 25.
PAGE 25. **very high figures indeed:** Ibid.
PAGE 25. **new corporate entity:** The merger took place on Feb. 7, 1899.
PAGE 26. **Mr. Frost will not find:** Morris, chap. 8. Also, the collected letters of the J. P. Holland Torpedo Boat Company, 1899, at the Paterson Museum.
PAGE 27. **Dear Mr. Rice:** Morris, chap. 8.
PAGE 27. **bore the label:** *Newark Evening News,* May 3, 1906.
PAGE 27. **supervised the construction:** Cable, chap. 15.
PAGE 27. **rumors persisted:** *New York Gaelic American,* Aug. 22, 1914.
PAGE 28. **£26 million a year:** B. R. Mitchel, *British Historical Statistics* (Cambridge University Press, 1971).
PAGE 28. **a damned un-English weapon:** J. D. Scott, *Vickers: A History* (London: Weidenfeld and Nicolson, 1962), p. 106.
PAGE 29. **unacknowledged mainstay:** William Ashworth, *An Economic History of England* (London, 1960), p. 236–37.
PAGE 29. **carved up the world:** Scott, p. 64. Also testimony before the Nye panel, U.S. Senate, 1934 (Sept. 7).
PAGE 30. **international trade in weapons:** See chap. 4, "Dirty Laundry."
PAGE 31. **unwitting and unwilling participant:** Cable, pp. 216–27.
PAGE 31. **mutual and mortal enemies:** Cable, chap. 19. Also, *Dynamic America,* pp. 81–85.
PAGE 31. **unshakeable submarine monopoly:** See House Select Committee (convened under Resolution 288), hearings commencing Mar. 9, 1908.
PAGE 31. **yards of subcontractors:** This was the start of EB's long association with the Fore River Shipyard, which was finally purchased from Bethlehem in 1963.
PAGE 31. **dummies been retired:** Scott, p. 66.
PAGE 32. **ninety dollars per week:** 1908 hearings, vol. 1.
PAGE 33. **disadvantage not of our choosing:** Morris, p. 117.
PAGE 34. **discomfort to their crews:** Ibid.
PAGE 34. **almost six miles:** E. A. Grey, *The Devil's Device* (London, 1975), for history of torpedo developments.
PAGE 34. **succumbed to the chest ailments:** *New York Gaelic American,* op. cit.
PAGE 34. **sinking all three:** The ships went down on September 22, prompting recriminatory debates in Parliament and the press. The sinkings also marked the start of the heavy-handed censorship that was to remain a part of British life until the war's end.

3. Feast and Famine

PAGE 35. **chess and music buffs:** Rice was a prominent chess patron and a gifted player, although never recognized as a master. His will included a bequest to fund a number of annual competitions. See David McKay and Anne Sunnucks, *The Encyclopaedia of Chess* (New York: St. Martin's Press, 1970), p. 404.
PAGE 36. **donated a million dollars:** *New York Times,* Nov. 22, 1915; Dec. 12, 1915. Also *New York Herald,* Nov. 3, 1915.
PAGE 36. **Sloan's sketches:** Reproduced in *Chess Life,* April 1983.

PAGE 37. **license in Seattle:** Gaddis Smith, *Britain's Clandestine Submarines* (New Haven, Conn.: Yale University Press, 1964), p. 20–21.

PAGE 37. **Electro Dynamic faced:** Details of Electric Boat's financial performance drawn from *Moody's Analysis of Investments,* 1914–1920.

PAGE 38. **loophole in the law:** Smith is a prime source for this chapter.

PAGE 39. **Rice promised to promote:** J. D. Scott, *Vickers* (London: Weidenfeld and Nicholson, 1962), pp. 66, 133–38.

PAGE 39. **Schwab had crossed:** Smith, pp. 25–28.

PAGE 40. **it went to the bottom:** Jellicoe, *The Grand Fleet* (New York, 1919), p. 150.

PAGE 41. **under an assumed name:** Schwab was booked under the name Alexander MacDonald, a pseudonym he used often. As usual, the deception did little good, for his current alias, his London address, and the purpose of his trip were disclosed in the *New York Times,* Nov. 4, 1914.

PAGE 41. **German intelligence remained skeptical:** There was reason for the German uncertainty. On Nov. 6, before rumors of the ship's loss became widespread, the British Secret Service arrested six German spies, effectively dismantling the enemy's intelligence network in the course of an afternoon. See Richard Deacon, *A History of the British Secret Service* (London, 1969), p. 244.

PAGE 42. **fivefold increase:** Winston Churchill, *The World Crisis,* vol. 1, (London, 1923–1929), p. 496.

PAGE 42. **life and career:** Stewart Holbrook, *The Age of the Moguls* (New York, 1953).

PAGE 43. **The deal was sealed:** Smith, chap. 3. Also, *Wall Street Journal,* Nov. 11, 1914.

PAGE 43. **decoded at the British Embassy:** Douglas Brownrigg, *Indiscretions of a Naval Censor* (New York, 1920), pp. 227–31.

PAGE 43. **a gigantic deal:** *Fear God and Dread Nought: The Correspondence of Admiral Lord Fisher of Kilverstone* (London, 1952–59), vol. 3, p. 66.

PAGE 44. **The neutrality laws:** *A Diplomatic History of the United States* (New York, 1955), pp. 410–13. Also, Smith, pp. 31–33, 37, 45, 49–50.

PAGE 45. **Hayden could have expected:** Walter Millis, *Road to War* (New York, 1935). Bryan figures prominently throughout the book.

PAGE 45. **a favorable verdict:** Ibid., pp. 26–68; 87–88; 118; 242.

PAGE 45. **word of the decision:** Smith, chap. 3.

PAGE 46. ***Times* gleaned:** *New York Times,* Nov. 10, 1914. *Wall Street Journal,* Nov. 11, 1914.

PAGE 46. **fraught with uncertainty:** *New York Times,* Nov. 20.

PAGE 47. **Bryan the winner:** Smith, pp. 40–51.

PAGE 48. **This closes:** *New York Times,* Dec. 8, 1914.

PAGE 49. **British were apologetic:** Smith, pp. 56–63.

PAGE 49. **Capable of performing:** *Wall Street Journal,* Dec. 14, 1914.

PAGE 50. **regarded as a ruse:** *Wall Street Journal,* Jan. 12, 1915. Also, *Montreal Star,* Dec. 24, 1914.

PAGE 50. **On the West Coast:** *San Francisco Chronicle,* Dec. 24–Jan 20, 1914. Also, Smith, p. 86.

PAGE 50. **Paul Cravath:** James B. Stewart, *The Partners* (New York, 1983), pp. 55–56.

PAGE 51. **his opinion in writing:** Smith, p. 88.

PAGE 51. **making no purchases:** Ibid., p. 89.

PAGE 52. **short term commercial credits:** By the time the U.S. entered the war, American business had issued some $2 billion in loans and credits to the Allies and only $27 million to Germany. See Richard Heffner, ed., *A Documentary History of the United States* (New York, 1952).

PAGE 52. **persons of large means:** *New York Times,* July 25, 1915.

PAGE 53. **surging past $275:** *New York Times,* July 24, 1914.

PAGE 53. **submarines figured prominently:** *New York Times,* July 16, 1914.

PAGE 54. **started the stampede:** "Stampede" is an understatement. In one year the price went from $20 to $420.

PAGE 54. **a majority of stock:** *New York Times,* July 14, 15, 1915; *Wall Street Journal,* July 15, 1915.
PAGE 54. **$20 million worth:** *New York Times,* July 25, 1915.
PAGE 54. **gross earnings totaled:** *Moody's Analysis.*
PAGE 55. **scant $1 million:** Ibid.
PAGE 56. **known all over Europe:** Nye hearings, 1934.
PAGE 56. **most of the ships:** *Dynamic America;* also, *Moody's Analysis.*
PAGE 56. **It was such a small amount:** Nye hearings, 1934.

4. Dirty Laundry

PAGE 58. **a foe of human liberty:** *Saturday Evening Post,* Sept. 1, 1934.
PAGE 58. **tonic for the national spirit:** CBS news transcripts. Nye made several radio addresses over the period. The transcripts are available at the New York Public Library.
PAGE 58. **an isolationist of the deepest hue:** Cordell Hull, *Memoirs* (London, 1948), vol. 1, p. 398.
PAGE 59. **insane racket:** John E. Wiltz, *In Search of Peace: The Senate Munitions Inquiry 1934–36* (Louisiana State University Press, 1963).
PAGE 59. **lifted its profits:** *New York Times,* Sept. 4, 1934.
PAGE 59. **Department of Peace:** A favorite demand of disarmament lobbyists, calls for a Peace Department still arise today.
PAGE 60. **weapons Who's Who:** *New York Times,* Sept. 2, 1934.
PAGE 61. **annoyance in the Oval Office:** *New York Times,* Sept. 3, 1934. Also, Hull, *Memoirs.*
PAGE 61. **redeem the nation's lost honor:** *New York Times,* Sept. 6, 1934.
PAGE 62. **depraved minds:** Wiltz, p. 29.
PAGE 62. **Shearer:** The Shearer scandal was covered in H. C. Engelbrecht and F. C. Hannighen, *The Merchants of Death.* The book was so popular it became a Book-of-the-Month Club selection.
PAGE 62. **his patrons:** Shearer also approached EB, but his assistance was declined. Nye hearings, U.S. Senate, 1934, vol. 1, p. 435.
PAGE 63. **at least one member:** Ibid.
PAGE 63. **brash, good-natured:** Quoted in Engelbrecht, p. 206.
PAGE 63. **no mean feat:** Engelbrecht, pp. 207–9.
PAGE 63. **vast amounts:** Shearer's activities at Geneva were investigated by the Senate's Navy Affairs Committee in 1931, "Alleged Activities at the Geneva Conference." Wythe Williams was a witness along with several other reporters and Navy personnel.
PAGE 64. **fool the simple Irish:** Ibid.
PAGE 64. **ebb and flow of debate:** Ibid.
PAGE 64. **one of his weekend excursions:** Ibid.
PAGE 65. **wining and dining:** Ibid.
PAGE 65. **Upton Sinclair:** *New York Times,* Sept. 3, 1934.
PAGE 66. **a theatrical flourish:** By the time the hearings started, committee investigators had amassed several hundred thousand documents. Nye used them as a backdrop for newspaper photos. *New York Times,* Sept. 3, 1934.
PAGE 66. **merchant of death:** Russell Warren Howe, *Weapons* (New York: Doubleday, 1980). Howe's account of Zaharoff's life is as good as any.
PAGE 67. **We do not pay him:** Senate Munitions Inquiry, Nye hearings, 1934, vol. 1, p. 21. The panelists were fascinated by Zaharoff and raised his name constantly throughout the EB witnesses' three days at the witness table.
PAGE 68. **Zaharoff's contacts:** Nye hearings, vol. 1, pp. 65–74.

PAGE 68. **The licensing deal:** Nye hearings, p. 14. For a less critical perspective, see J. D. Scott's official history of Vickers, pp. 244–46.

PAGE 69. **an essay competition:** Nye hearings—Zaharoff correspondence collected as exhibits, p. 67.

PAGE 70. **The Spanish authorities:** Ibid.

PAGE 70. **The senators wondered:** *New York Times,* Sept. 7, 1934.

PAGE 70. **Techel's willingness:** Nye hearings, vol. 1.

PAGE 71. **Third Reich's new U-boats:** William Manchester, *Arms of Krupp* (Boston: Little, Brown, 1964), pp. 336–37, for further background on the clandestine German submarine operation. Prototypes of future U-boats were built, tested, and perfected by German crews based in Holland before being delivered to Japan, Spain, Italy, Finland, and Holland. The Dutch government's tacit protection was won through the generous distribution of stock certificates and inflated dividends that continued through the worst years of the Depression. Manchester, *Arms of Krupp.*

PAGE 71. **Electric Boat battled:** The Germans attempted to defeat the patent action by pointing out that their design included an additional compensatory ballast tank, the so-called *Tank F.* With the aid of the French Navy, Koster proved that tank F was a mere refinement rather than the major improvement claimed by the defendants. Nye hearings.

PAGE 71. **Doesn't it pretty nearly:** *New York Times,* Sept. 7, 1934.

PAGE 73. **It makes one wonder:** Nye hearings, vol. 2. The comments were inspired by another war supplier, the Driggs Ordnance Co.

PAGE 74. **the real foundation:** *New York Times,* Sept. 6, 1934. For a more sympathetic view, see J. D. Scott's *Vickers,* pp. 241–45.

PAGE 74. **It is too bad:** Nye hearings.

PAGE 74. **Carse had forgotten:** Scott, p. 244.

PAGE 74. **Parliament was soon obliged:** Royal Commission into the Production and Sale of Munitions, 1935–1936. Also, Sampson, pp. 82–89.

PAGE 75. **there would not now be:** Nye hearings, vol. 1.

PAGE 76. **Washington Hotel:** Ibid

PAGE 76. **The weeks that passed:** ıbid.

PAGE 77. **millions of yen:** Ibid.

PAGE 77. **within the month:** Ibid.

PAGE 77. **a friendly note:** Ibid., and Exhibit No. 45.

PAGE 78. **The only man:** An official from the Office of Navy Intelligence journeyed up from Washington to interview EB's executives about the Japanese rumors. Carse denied any knowledge of the deal. Nye hearings, vol. 1.

PAGE 78. **These things are always:** Critics of recent U.S. arms sales in the Middle East would still agree with Nye. See Andrew J. Pierre, *The Global Impact of Arms Sales* (Princeton University Press, 1982).

PAGE 78. **Successfully managed:** *New York Times,* Sept. 7, 1934.

PAGE 78. **Apparently Joyner:** Ibid.

PAGE 79. **Speaking personally:** Ibid.

PAGE 79. **I sincerely promised:** Ibid.

PAGE 80. **Even Basil Zaharoff:** Nye hearings, Exhibit 195.

PAGE 81. **As Electric Boat:** *New York Times,* Sept. 7, 1934.

PAGE 81. **reports that demonstrated:** Nye hearings, Exhibit 194. Also, Spear testimony.

PAGE 81. **a rather strange development:** *New York Times,* Sept. 7, 1934.

PAGE 82. **hierarchy's support:** It did little good. The contracts went to the Navy's own shipyards. Nye hearings, vol. 1.

PAGE 83. **whatever guilt accrued:** Sampson, chap. 4, for an examination of the ways in which British and U.S. governments faced the moral and ethical dilemmas involved in supervising their native arms industries.

5. War and Redemption

PAGE 87. **What made the turnabout:** *Moody's Analysis of Investments,* 1927–1941, for details of Electric Boat's performance.

PAGE 88. **The naval budget:** Clark R. Mollenhoff, *The Pentagon* (New York: Putnam, 1967), pp. 67–68. Also, Richard N. Current, Frank Freidel, and Harry Williams, *A History of the United States* (New York: Knopf, 1959), pp. 527–50, for details of Roosevelt's battle to win increased defense expenditures.

PAGE 88. **no later than 1944:** The Vinson-Trammel Act (1934) purported to authorize a gradual buildup of the national arsenal. However, it was largely meaningless because Congress consistently refused to allocate sufficient funds.

PAGE 88. **fleet-class boats:** Electric Boat built 57 S-series boats before the expiration of the final contract in 1923.

PAGE 89. **abandon submarine warfare:** The S-5 sank in 1920; the S-36 and S-39 in 1923. The S-51 went down about eighteen months later, and in 1927 the S-4 sank after a collision with a Coast Guard cutter. Details of the sinkings and the often heroic rescue attempts are in Edwin P. Hoyt, *Submarines at War* (New York: Stein and Day, 1983), chaps. 3, 4.

PAGE 89. **equal measure of politics:** The fleet-class boats were consistently opposed by those in charge of the surface fleet. The S-series boats were powered by engines known as H.O.R.s, an acronym of their inventors' initials. The new General Motors engines in the fleet boats retired one obvious and disparaging nickname, a label Hoyt called "a libel on the ladies of the evening." Hoyt, pp. 87–89.

PAGE 89. **unrestricted underwater warfare:** Hoyt, pp. 55–56. Also, Drew Middleton's *Submarine,* p. 64.

PAGE 89. **Tucked into a space:** Theodore Roscoe, *Pig Boats* (New York: Bantam, 1958), p. 17.

PAGE 90. **Twice as many:** *New York Times,* Aug. 16, 1940.

PAGE 91. **most productive cogs:** *Dynamic America,* pp. 283–95.

PAGE 91. **prefabricated hull sections:** Ibid., p. 285.

PAGE 92. **Every time we set off:** *New York Times,* May 17, 1942.

PAGE 92. **Mantiowoc:** *Dynamic America,* p. 287. The EB supervisor at the site was Eric Ewertz, a veteran employee who had worked on the *Holland I* almost forty-five years before.

PAGE 93. **go out and dig it up:** *New York Times,* May 17, 1942.

PAGE 93. **conversing in a foreign tongue:** *Dynamic America,* p. 287.

PAGE 94. **John Jay Hopkins: Hopkins' life and career:** *Current Biography,* 1952.

PAGE 94. **The new boy:** One story maintains that Hopkins became involved with EB after meeting Henry Carse at an organ recital in a church on Fifth Avenue in New York.

PAGE 94. **Groton had built:** *Dynamic America,* p. 287.

PAGE 94. **Japanese entered the war:** See Roscoe's *Pigboats* for a comprehensive account of the underwater war in the Pacific.

PAGE 95. **oddly matched rivals:** Bruce Catton, *Warlords of Washington* (New York: Harcourt Brace, 1948). Catton was a reporter in Washington. Also, see I. F. Stone, *Business as Usual,* for a view of the war effort devoid of propaganda.

PAGE 96. **What price patriotism:** *Report of the Temporary National Committee Investigating Concentration of Economic Power in the United States,* 75th Congress, 1939.

PAGE 96. **come out of the war:** *Emporia Gazette,* 1943. The quotes were reproduced in Albert E. Kahn, *High Treason,* (Lear, 1950).

PAGE 96. **Deal with the government:** Address to the National Association of Manufacturers, *New York Times,* Sept. 13, 1942.

PAGE 96. **a front for the rest:** Stone, *Business as Usual.*

PAGE 96. **reporter Sidney Shallett:** Shallett's story appeared in the *New York Times,* May 17, 1942. Cost of improvements and extensions is the author's estimate drawn from newspaper reports through World War II.

PAGE 98. **German U-boats:** Between April 1940 and March 1941, U-boats sank 2,314,000 tons of shipping in the Atlantic. Later, between August 1, 1942, and May 21, 1943, the figure leapt to 3,760,722 tons.

PAGE 98. **much more effective:** The 6 million tons is probably an overestimate. Japanese merchant losses totaled only 4,859,634 tons, and account for a further 397,412. Of the total a small percentage credited to U.S. boats was actually sunk by Dutch and British boats. In the Atlantic, U.S. boats failed to sink a single German. See Roscoe.

PAGE 98. **failure of imagination:** See Clay Blair, *Silent Victory* (Philadelphia: J. B. Lippincott, 1975).

PAGE 99. **without scoring a single hit:** Max Hastings and Simon Jenkins, *The Battle for the Falklands* (New York: Norton, 1983), p. 160.

PAGE 99. **submariner's greatest virtue:** Captains who lost their boats during peacetime maneuvers were often stripped of their commands, particularly if it was the result of too much aggression. See Hoyt, pp. 89–94. Also, Roscoe, pp. 56–57.

PAGE 99. **machine-gun survivors:** "Anyone who had not experienced the feel of life on the deck of a surfaced submarine with all the deck guns firing and all machine guns going has not lived," Morton enthused after one machine-gunning of Japanese survivors. See Hoyt, pp. 190–93, 200–201.

PAGE 100. **impenetrable secrecy:** Hoyt, pp. 68–71.

PAGE 101. **clean bill of health:** Ibid.

PAGE 102. **incident was witnessed:** Roscoe, p. 192.

PAGE 102. **to bear down:** Some believe Morton and his boat were destroyed by one of their own torpedoes. See Hoyt, p. 269.

PAGE 102. **so many mistakes:** The Mark XIV always ran at least 10 feet deeper than set and often went straight to the bottom. Roscoe, pp. 148–51.

PAGE 103. **makeshift bracket:** Hoyt, pp. 209–13.

PAGE 103. **stream of complaints:** Germany also began the war with similar torpedoes but scrapped the design within months.

PAGE 103. **all-electric torpedoes:** This was the new Mark XVIII.

PAGE 103. **own research programs:** Work had begun on an electric torpedo in the late 1920s but was stopped because of the Mark XIV's purported infallibility.

PAGE 103. **Failure to provide:** Inspector General to Commander in Chief, U.S. Fleet, April 1943. Roscoe, p. 245.

PAGE 104. **own private Navy:** Hoyt describes the feuding at length in *Submarines at War.*

PAGE 104. **Soviet Union's coastal defenses:** "John Lehman: Fool of Ships," *New Republic,* June 3, 1985.

PAGE 105. **ninety-four official tests:** Fred Kaplan, "The Little Plane That Could Fly if the Air Force Would Let It," *Boston Globe,* Mar. 14, 1982.

PAGE 105. **the White House:** Roosevelt had been enthusiastic supporter of the PT concept since his days as Navy Secretary during World War I.

PAGE 105. **acknowledged leader:** Elco boats, aged though they are, still command premium prices on the secondhand boat market. A price of $50,000 for a 35-foot, 30-year-old boat is not unusual.

PAGE 106. **more promising design:** *At Close Quarters—PT Boats at War* (Naval Institute Press), pp. 45–48.

PAGE 107. **said design is known:** Ibid. It was not Hart's only misjudgment. Several years earlier he had campaigned to have the fleet-type boats scrapped, insisting that a submarine's only duty was coastal defense.

PAGE 108. **more than $600,000:** Ibid.

PAGE 108. **By the end of the war:** For further reading on PT boats, see Bryan Cooper, *Battle of the Torpedo Boats* (New York: Stein and Day, 1970).

6. The Promised Land

PAGE 109. **accelerated demobilization:** Partly to win the armed forces' support for the reductions, Truman agreed to lower the enlisted man-officer ratio from 10 to 1 down to 6 to 1. See Fred J. Cook, *The Warfare State* (New York: Macmillan, 1962), p. 99. Today the ratio is about the same.

PAGE 110. **permanent peacetime draft:** The Army launched a massive campaign to "sell the draft program to the public with the hope that the public would sell it to Congress." Assistant Secretary of War Howard C. Petersen testifying before the House Subcommittee of the Committee on Expenditures in the Executive Department, June 1947.

PAGE 110. **renovated chicken coop:** "The House That Levitt Built," *Esquire,* December 1983.

PAGE 110. **backlogged orders:** See Catton, *Warlords of Washington,* pp. 293–96.

PAGE 111. **unqualified U.S. commitment:** Marshall's speech was summarized in *Fortune,* December 1945, p. 56.

PAGE 112. **bombers' days were numbered:** "The Atomic Frame of Reference or Else," *Aviation News,* September 1945, p. 107.

PAGE 112. **submarine of tomorrow:** *Fortune,* December 1945.

PAGE 113. **reduced Pentagon budget:** The battle for the largest share of the reduced defense budget culminated in the celebrated "Revolt of the Admirals" in 1949, when the Navy assailed the Air Force's B-36 bomber as "a billion-dollar blunder." See Gregg Herken, *The Winning Weapon* (New York: Knopf, 1981). Herken explains that this "guaranteed the doctrinal and budgetary ascendency of air power."

PAGE 114. **prefabricated girders:** "So They Called It General Dynamics," *Fortune,* April 1953.

PAGE 115. **a vanishing hope:** Electric Boat was not alone in trying its hand at civilian production—Grumman began building canoes while Fairchild tried mopeds and Northrop, in an ironic departure from the arms game, toyed with the idea of setting up a division to produce artificial limbs.

PAGE 115. **$23 million:** Transcripts of Canadian House of Commons debates Mar. 20, 1947, p. 1540. Questions to Minister for Reconstruction C. D. Howe.

PAGE 116. **assembled in the United States:** Ibid.

PAGE 116. **surplus DC-3s:** Ibid.

PAGE 116. **multitude of problems:** Michael Hardy, *World Civil Aircraft Since 1945* (New York: Scribner's, 1979), p. 35.

PAGE 117. **back to their original owner:** Ibid.

PAGE 117. **Merlin engine:** House of Commons, Apr. 29, 1949, pp. 2763–65.

PAGE 117. **routine maintenance:** Ibid.

PAGE 118. **disrupted journeys:** Ibid.

PAGE 118. **take out a lease:** *Business Week,* Mar. 22, 1952.

PAGE 118. **so much as given away:** "Canadair's Rags to Riches Story," *Canadian Financial Post,* May 5, 1980. The charge that Canadair had been "given away" figured as a prominent campaign issue in the 1949 election.

PAGE 118. **real capitalist country:** *Business Week,* Mar. 22, 1952.

PAGE 119. **manner, or the terms:** House of Commons, Mar. 20, 1947, pp. 1541–44.

PAGE 119. **realize the Canadian capital:** House of Commons debate on foreign exchange conservation, speech by Mr. Probe, Feb. 4, 1947.

PAGE 119. **every form of steel:** Ibid.

PAGE 119. **know-how:** House of Commons, question time, Feb. 4, 1947, question to C. D. Howe from A. M. Nicholson.

PAGE 120. **Canadian nationalism:** Hopkins summed up his feelings about Canada in a speech before the Edmonton Chamber of Commerce on Jan. 28, 1953: "We in Canada and the U.S. have a common language and literature, common institutions, common ideals and common purposes which have been handed down to us by our common ancestors. We need each other."

PAGE 120. **a world leader:** Much of the information in this chapter is drawn from author interviews with Canadair's public relations officer Ron Pickler and former executives Dick Faucher and Ralph Stopps.

PAGE 121. **any self-governing people:** *Time* cover story, Jan. 19, 1948. Also, "The Wildest Blue Yonder Yet," *Fortune,* February 1948, and "Bigger Air Force Is Prime Need for U.S. Survival," *Aviation Week,* Jan. 14, 1948.

PAGE 122. **$122.45:** *Time,* Jan. 19, 1948.

PAGE 122. **would have to be appeased:** Ibid.

PAGE 123. **repurchase the Sabres:** For accounts of Canadair's Sabres' later careers on the international arms market, see Tom Gervasi, *Arsenal of Democracy II,* pp. 51, 120–21.

PAGE 124. **given to Canadair:** *Fortune,* April 1953.

PAGE 125. **Canadair's primary importance:** Interview, June 1984.

PAGE 126. **about number 174:** Interview, June 1984.

PAGE 126. **another whale:** "How a Great Corporation Got Out of Control—Part I," *Fortune,* January 1962.

PAGE 126. **Atlas rocket:** The first Americans to venture into space did so atop the much smaller Redstone rockets. The Atlas carried four times as much fuel and was responsible for the first orbital flight. See Tom Wolfe, *The Right Stuff* (New York: Bantam), p. 248.

PAGE 127. **grandiose ambitions:** "Convair Control Gets Shuffled Again," *Business Week,* May 2, 1953.

PAGE 127. **Hopkins' aspirations:** *Fortune,* April 1953. Also, *New York Times,* Apr. 25, 1952.

PAGE 128. **campus laundry service:** *Current Biography,* 1941, p. 629.

PAGE 129. **time of his life:** *Saturday Evening Post,* July 10, 1937. Also, *New Yorker,* Aug. 26, 1933.

PAGE 129. **Depression phenomenon:** *Fortune,* September 1935.

PAGE 130. **publicity we couldn't buy:** Cochrane's recollections: Oral History Section, Columbia University.

PAGE 130. **resourcefulness, ingenuity:** I. F. Stone, *Business as Usual.* Also, *PM* magazine, Sept. 11, 1944.

PAGE 131. **financial and military oligarchy:** Bruce Catton, *Warlords of Washington* (New York: Harcourt Brace, 1948), pp. 28–40.

PAGE 131. **fifty-six firms:** "The leading 100 corporations in America received approximately 70 percent of all the war contracts." Statistics contained in *Report of the Smaller War Plants Corporation to the Special Committee to Study Problems of American Small Business,* U.S. Senate, 1946.

PAGE 131. **The giants themselves:** Catton, p. 33.

PAGE 131. **contributed little:** The tires were good only for retreads. Catton, pp. 158–67.

PAGE 132. **return entirely to fat:** "Hap Arnold's Legacy," *Air Force Magazine,* February 1950.

PAGE 132. **bombing of Germany:** After the war the Air Force sent an appraisal team to Germany and Japan to inspect bomb damage and assess its contribution to the final victory. Much to their surprise, the survey members found that Allied bombing had done little to slow Axis war production and may have slowed the Allies' advance through Europe by strewing rubble in the paths of advancing tanks, trains, and troops. The peak of the bombing offensive against Germany coincided with the peak of German war production. See Herken, *The Winning Weapon,* p. 209. Also, *U.S. Strategic Bombing Survey* (Washington D.C., 1946).

PAGE 133. **kitchen stoves:** For Convair's early history: *50: Convair Aerospace Division of General Dynamics 1923–1973.*

PAGE 133. **take charge of Convair:** Ibid. Also, *Business Week,* May 2, 1953.

PAGE 135. **Air Force order:** Ibid. Also, *Aviation Week,* Jan. 19 and Jan. 26, 1948.

PAGE 135. **main construction hangar:** *50,* pp. 48–49.

PAGE 135. **Willow Run factory:** Ibid., pp. 46–47.

PAGE 135. **San Diego concentrated:** Ibid., pp. 56–57, 60–61.

PAGE 136. **bizarre round:** Donald Barlett and James B. Steele, *Empire: The Life, Legend and Madness of Howard Hughes* (New York: Norton, 1979), pp. 162–66.

PAGE 136. **Rickover's decision:** Rickover was a man who was to figure prominently in General Dynamics' triumphs and tribulations over the next thirty-five years, and his gift of not one but two submarine contracts was the result of a decision made in true Rickover style. According to Navy legend, he went to the Portsmouth Naval Yard to discuss the construction of the *Seawolf* with Rear Admiral R. E. McShane. On being told that Portsmouth was too busy to give the project the priority Rickover demanded, his visitor reached across the desk and called O. Pomeroy Robinson to ask if Electric Boat could handle a second ship, this one with a liquid-sodium–cooled reactor. Robinson gave an immediate and favorable response, and Electric Boat acquired its N-submarine monopoly. See Norman Polmar and Thomas B. Allen, *Rickover* (New York: Simon & Schuster, 1982), pp. 146–48.

PAGE 137. **arsenal of atomic conflict:** "Convair Does Its Own Planning," *Business Week,* Mar. 15, 1952.

PAGE 137. **before it was scrapped:** *50,* p. 70.

PAGE 137. **won an award:** Ibid. The Pogo is now on display at the Air Force Museum along with several other GD products, including the last surviving B-36.

PAGE 138. **Age of the Atom:** For further reading on the outbreak of atomic intoxication that swept through the military, business, and the press, see Gerard H. Clarfield and William M. Wiecek, *Nuclear America* (Harper & Row, 1984). Also, Stephen Hilgartner, Richard C. Bell, and Rory O'Connor, *Nukespeak,*—chaps. 2 and 4 in particular.

PAGE 138. **Mr. Atom:** *Time,* Apr. 6, 1953.

PAGE 138. **donate "simple" reactors:** Hopkins' address to International Chamber of Commerce meeting, Tokyo, 1954.

PAGE 140. **setting fire to the countryside:** During the last weeks of 1960 and through January 1961, the trade magazine *Aviation Week* ran a 20,000-word, four-part series in which various "experts" opined that Convair's atom-bomber would definitely be airborne no later than the summer of 1965. The articles were as ill-timed as they were misconceived. Within three months the newly installed Kennedy Administration took the first steps toward canceling the project entirely. See *Aviation Week,* Dec. 19 and Dec. 26, 1960, and Jan. 2, Jan. 9, 1961. Also, *Astronautics,* November 1962.

PAGE 140. **stereoscopic periscopes:** *Aviation Week,* Jan. 2, Jan. 9, 1961.

PAGE 141. **safety-related issues:** Incredible though it may seem, the problems of radioactive contamination were simply glossed over. See *Aviation Week,* Dec. 26, 1960.

PAGE 141. **congressionally sponsored merger:** For the N-plane's supporting bureaucracy, see R. W. Bussard, *Fundamentals of Nuclear Flight* (New York: McGraw-Hill, 1965).

PAGE 141. **lacked the power:** *Aviation Week,* Dec. 26, 1960. Also, papers on the *Tory II* and other nuclear aviation programs at the American Institute of Aeronautics and Astrophysics, New York.

PAGE 141. **pattern of action:** Philip Stern, *The Oppenheimer Case* (Harper & Row, 1969).
PAGE 142. **rumbled about the skies:** *50,* p. 59.
PAGE 142. **obliged to withdraw:** *Aviation Week,* Dec. 19, 1960.
PAGE 143. **three Russian authors:** *Applications of Atomic Energy in Aviation,* published by Military Press of the Ministry of Defense of the U.S.S.R., Moscow, 1957. Copy at the AIAA library, New York.
PAGE 143. **a sickening shock:** Herbert York, *Race to Oblivion: A Participant's View of the Arms Race* (New York: Simon and Schuster, 1971), p. 146.
PAGE 143. **speeches won the N-bomber:** Clarfield, pp. 147, 172–74.
PAGE 143. **kind of deal:** *Fortune,* May 1953, p. 107. Also, *Business Week,* Apr. 4, 1953.
PAGE 144. **$8.7 million:** *Business Week,* May 2, 1953, pp. 94–98.
PAGE 144. **Convair has its problems:** *Fortune,* May 1953, p. 107.

7. The Flying Edsel

PAGE 145. **none of the big airlines:** Interview, March 1984. Also *New York Times,* Apr. 7, 1984.
PAGE 146. **according to gossip:** *Life,* June 6, 1949.
PAGE 147. **Eleven months later:** Michael Hardy, *World Civil Aircraft Since 1945* (New York: Scribner's, 1979).
PAGE 148. **holding it back:** "Jet Airliners—Parts I and II," *Fortune,* April and May, 1953. Also, *American Aviation,* Feb. 28, 1955, and Nov. 21, 1955.
PAGE 148. **Howard Hughes:** Much of the information in this chapter is drawn from the transcripts of one of the longest and most costly civil trials in U.S. history, *TWA* v. *Hughes* (U.S. Federal Court, Manhattan, Docket 61 Civ 2324, June 30, 1961). More accessible background to the Hughes fiasco can be found in Donald L. Barlett and James B. Steele, *Empire: The Life, Legend and Madness of Howard Hughes* (New York: Norton, 1979).
PAGE 149. **hundreds of tiny points:** Rummel testimony, *TWA* v. *Hughes* transcript, pp. 1895–99. Huge passenger aircraft were something of a fixation with General Dynamics during the 1950s. Just before the talks with Hughes began, Fort Worth lobbied hard within the company to produce a two-deck passenger version of the B-36, complete with bedrooms and dining lounges.
PAGE 150. **we may even be denied:** Report to TWA directors by President R. S. Damon, Exhibit 309, *TWA* v. *Hughes.*
PAGE 150. **a self-imposed limitation:** *Fortune,* May 1953, p. 131.
PAGE 150. **Was Jay Hopkins:** Richard Austin Smith, "How a Great Corporation Got Out of Control," 2-part series appearing in *Fortune,* January and February 1963.
PAGE 151. **lose the initiative:** Ibid.
PAGE 151. **back to Howard Hughes:** Rummel testimony, p. 1863.
PAGE 152. **Hughes was capable:** Testimony of A. V. Leslie, *TWA* v. *Hughes,* pp. 6967–7200.
PAGE 152. **actress Janet Leigh:** Barlett and Steele, chap. 8.
PAGE 154. **contribute just 10 percent:** *Fortune,* January 1962.
PAGE 155. **"wasp" waists:** Clay Blair, Jr., "The Man Who Put the Squeeze on Airplane Design," *Air Force Magazine,* January 1956.
PAGE 155. **Canadair applied the lesson:** Jacqueline Cochran's recollections: Oral History Section, Columbia University.
PAGE 155. **future General Motors:** *Fortune,* January 1962, p. 180.
PAGE 156. **joined in the war effort:** *Current Biography,* 1952, p. 345.
PAGE 157. **If anything will call a halt:** Ibid., p. 346.
PAGE 157. **defense spending actually stiffened:** Pace praised defense spending in a speech before the American Bankers Association, September 1957. It results,

he said, "in the accession to our society of economic, scientific and cultural benefits of enduring nonmilitary value."

PAGE 157. **Wars will be fought:** "Building a Business for War or Peace," *Business Week,* Oct. 18, 1952.

PAGE 157. **We had to get someone:** *Fortune,* January 1963.

PAGE 157. **impeccable credentials:** *Current Biography,* 1950, p. 435.

PAGE 158. **Well, I just bought:** *Fortune,* February 1963.

PAGE 158. **present him with a medal:** *New York Times,* Mar. 3, 1957.

PAGE 159. **rejected for the job:** *Fortune,* January 1963.

PAGE 160. **died soon after:** *New York Times,* May 4, 1957.

PAGE 160. **TWA's Golden Arrows:** Hardy, *World Civil Aircraft,* p. 44. The "880" tag was appropriate since it was also a descendent of the prop-driven 220s and 440s.

PAGE 160. **more secure assembly site:** Rummel testimony, p. 1936.

PAGE 161. **wide enough for five:** *Fortune,* February 1963.

PAGE 161. **orders balancing larger cancellations:** Ibid.

PAGE 162. **No engineer would have made:** Ibid.

PAGE 162. **modification of the 880:** Ibid.

PAGE 163. **selling half a dozen:** Barlett and Steele, p. 223.

PAGE 164. **strangest hijacking:** Deposition of J. William Bew, TWA's liaison at Convair on the date of the seizure. *TWA* v. *Hughes,* pp. 3065–75. Also Rummel, pp. 2949–87.

PAGE 164. **covers to protect them:** No one at San Diego doubted that Hughes' men carried guns. In testimony, TWA witnesses almost always referred to them as "armed" guards.

PAGE 164. **his financial problems:** Desk diary of E. O. Cocke, *TWA* v. *Hughes,* Exhibit 328.

PAGE 165. **surrendering his voting stock:** TWA seized Hughes' records as evidence in the case. Hughes would dictate the memos, often by phone to aides in adjoining rooms, to relay them to and from their destinations. Hughes memos dictated on July 5 and 6, 1960. There are five pages of typed memos, all single-spaced and almost always incoherent. Copies of memos in author's possession.

PAGE 165. **less than three weeks:** Bew deposition, *TWA* v. *Hughes.*

PAGE 165. **towed it away:** Rummel testimony, *TWA* v. *Hughes,* pp. 2797–2827.

PAGE 165. **Hughes did arrange:** "Hughes Signs TWA Financing," *Aviation Week,* Dec. 12, 1960. Also "CAB Plays Major Role in TWA Financing," *Aviation Week,* Jan. 9, 1961, pp. 38–39.

PAGE 166. **$425 million worth:** *Fortune,* January and February, 1963.

PAGE 166. **tendency to "flap":** Hardy, p. 45. Also *Interavia* magazine, February 1961, pp. 176–77.

PAGE 166. **sent back to General Electric:** *Aviation Week,* Dec. 12, 1960, p. 29.

PAGE 167. **Convair resisted:** *Fortune,* February 1963.

PAGE 167. **a very small dark horse:** Arthur Sylvester, "Of Horses Dark and Otherwise," published in *Candidates 1960* (New York: Basic Books, 1959).

PAGE 168. **187 retired officers:** Subcommittee of Special Investigations of the House Committee on Armed Services. "Employment of Retired Commissioned Officers by Defense Department Contractors," 1960, p. 9.

PAGE 168. **off-the-record party:** Fred Cook, *The Warfare State* (New York: Macmillan, 1962), p. 187. Also, *New York Times,* July 9, 1959.

PAGE 169. **These properties:** Ovid Demaris, *Captive City: Chicago in Chains* (New York: Lyle Stuart, 1969), p. 214.

PAGE 170. **$33 a thousand:** Ibid. pp. 219–20.

PAGE 170. **$96 million:** *Fortune,* February 1963.

PAGE 170n. **Watch for dramatic upheaval:** *Aviation Week,* Apr. 17, 1961.

Chapter 8. The Birth of an Aardvark

PAGE 172. **Air Force II:** *Fort Worth Star-Telegram,* Dec. 12, 1962. Also, *Report of the Permanent Subcommittee on Investigations of the Committee on Government Operations: TFX Investigation 1963,* the so-called McClellan committee hearings, are a prime source for this and later chapters. Henceforth called "McClellan hearings."

PAGE 173. **LBJ Has Saved:** McClellan hearings, p. 2658.

PAGE 174. **political considerations:** Memorandum to Frank Pace from Gilpatric, McClellan hearings, pp. 2646–47.

PAGE 174. **No. 1 position:** Ibid., p. 2647.

PAGE 174. **all-time low:** An analysis of Fort Worth's employment roles from World War II until the early seventies is contained in Jacques Gansler, *The Defense Industry* (Cambridge, Mass.: MIT Press, 1980), p. 52.

PAGE 175. **cost of electrical components:** GAO Report on B-58 Electrical Systems, 1964.

PAGE 175. **got to have friends:** *Fort Worth Star-Telegram,* Dec. 12, 1962.

PAGE 175. **Plane That Changed:** This was soon to become the subject for another two-part series in *Fortune,* "The $7 Billion Contract That Changed the Rules," Richard Austin Smith, *Fortune,* March and April, 1963.

PAGE 175. **spectacular failures:** Before going to the Defense Department, McNamara was at Ford for the Edsel debacle. Later he moved to the World Bank in time for the international debt crisis of the early 1980s.

PAGE 176. **open to challenge:** *Current Biography,* 1961, pp. 291–93.

PAGE 176. **$40 billion enterprise:** *New York Times,* Jan. 22, 1961.

PAGE 177. **commonality:** William W. Kaufman, *The McNamara Strategy* (New York: Harper & Row, 1964), pp. 183–85.

PAGE 177. **In a further break:** Design background and service history of the F-4 Phantom: See Ray Bonds, ed., *The U.S. War Machine* (New York: Crown, 1983), pp. 194–95.

PAGE 178. **tested and found wanting:** Pierre Sprey, *The Impact of Avionics on the Effectiveness of Tactical Air,* a study prepared for the Assistant Secretary of Defense, 1968. Sprey estimates the effectiveness—i.e., the lethality—of both missiles at around 10 percent of all missiles fired. See also James Fallows, *National Defense* (New York: Random House, 1981), pp. 5, 99.

PAGE 179n. **little son of a bitch:** Fred Kaplan, *Wizards of Armageddon* (New York: Simon and Schuster, 1983), pp. 255–56.

PAGE 180. **Missileer proved:** The Missileer experiment is described in Bill Gunston's *F-111* (New York: Scribner's, 1978), p. 13. Although an ardent admirer of the F-111, Gunston's book is second to none in its command and collation of technical data about the plane. His book was a prime source for the technical information in this and later chapters.

PAGE 181. **comparison with the F-105:** Ibid., pp. 10–12.

PAGE 182. **shepherded through Congress:** See *Fortune,* March and April 1963, for the political background to the F-111 scandal. Also, "The McNamara Mobarchy," *Saturday Evening Post,* Mar. 9, 1963.

PAGE 182. **degree of flexibility:** Gerard H. Clarfield and William M. Wiecek, *Nuclear America* (New York: Harper & Row, 1984), p. 233.

PAGE 184. **Our experience:** Speech by Robert A. Frosch to Electronics and Aerospace Systems Convention, Defense Department news release, Oct. 17, 1967.

PAGE 185. **Air conditioned and pressurized:** "F-111 Systems, Sub-systems Fact Sheet," General Dynamics news release, 1967.

PAGE 185. **Of the F-111s lost:** Of the first F-111s sent to Vietnam in 1968, three crashed within the first month. The last crash, the one with survivors, was attributed to faulty welds in the tailplane control assembly. See Gunston, "The Proof of the Pudding" chapter.

PAGE 186. **first concession:** See Everest's and LeMay's testimony before McClellan's committee. Also, LeMay quoted on "compromises" in Trevor Armbrister, "Is This Plane a Billion Dollar Blunder?," *Saturday Evening Post,* May 17, 1968.

PAGE 186. **voluntary concession:** Gunston, chapter "Exit the Navy."

PAGE 187. **could not be considered:** Ibid., p. 16.

PAGE 188. **1,700 aircraft:** *Fortune,* March 1963, pp. 186–88.

PAGE 188. **Ten contenders:** *Fortune,* April 1963.

PAGE 189. **We thought we had done:** Ibid.

PAGE 190. **three hours and thirty-five minutes:** Ibid.

PAGE 190. **embarrassing problem:** *Fortune,* March 1963.

PAGE 190. **Boeing had concluded:** Gunston, pp. 16–19.

PAGE 191. **sanctity of the panel's opinion:** "Memorandum for the Record," written by Albert Blackburn. McClellan hearings, pp. 1203–7.

PAGE 191. **was not one plane:** Blackburn memo, McClellan hearings, ibid.

PAGE 191. **a big black Packard:** *Fortune,* April 1963.

PAGE 191. **Boeing and General Dynamics:** Ibid.

PAGE 193. **Increasingly suspicious:** Clark R. Mollenhoff, *The Pentagon,* chaps. 28, 29.

PAGE 194. **McNamara shunted:** Pirie was retired while Anderson was not reappointed to the position of Chief of Naval Operations. See Hanson W. Baldwin, "F-111: An All Purpose Plane That Is Provoking a Political Debate," *New York Times,* June 7, 1966.

PAGE 195. **a common mean:** *Fortune,* April 1963.

PAGE 195. **commonality problem:** Blackburn was also impressed by General Dynamics' final sales presentation, if not by the actual design. "The GD presentation was inspired," he told McClellan. "Histrionically it could not have been better paced nor more interestingly presented." Boeing's, he said, was "dull by comparison." McClellan hearings, p. 1206.

PAGE 196. **clear and substantial advantage:** Gunston, p. 19.

PAGE 196. **Boeing's designers:** McClellan hearings, p. 1206.

PAGE 197. **Did we get it?:** *Fortune,* April 1963.

9. Skewered on the Hill

PAGE 198. **persistent whispers:** "GD TFX Contract in the Bag," *Fort Worth Press,* Oct. 25, 1962. This was a full month before the decision.

PAGE 198. **The TFX program:** As a general reference source, see *The TFX Decision* (Boston: Little, Brown, 1968).

PAGE 199. **unrealistic cost estimates:** McNamara before McClellan Committee, pp. 430–32.

PAGE 199. **84 percent commonality:** *Fortune,* April 1963.

PAGE 199. **these were parts:** See Mollenhoff, p. 305.

PAGE 200. **threaten the nation's security:** Ibid., p. 316.

PAGE 201. **stock worth $160,000:** Ibid., p. 317.

PAGE 201. **charges of unethical conduct:** *Congressional Record,* July 28, 1963, p. 19768; July 24, 1963, p. 12585.

PAGE 202. **Texas is going to have:** *Fort Worth Star-Telegram,* Dec. 2, 1962.

PAGE 202. **McNamara's business:** Interview, February 1985.

PAGE 203. **Boeing's people:** Interview, confidential source.

PAGE 204. **Good grief:** McClellan hearings, pp. 435–37. Also, "Air Force Hits TFX Probe Tactics," *Washington Star,* Mar. 21, 1963.

PAGE 204. **You don't approve:** Ibid., p. 435.

PAGE 205. **close personal relationship:** "Statement of Mr. Gilpatric's Prior Associations with General Dynamics," p. 419.

PAGE 205. **recruiting agent:** Gilpatric's recruiting efforts on General Dynamics' behalf presented in a series of letters to Pace and others, pp. 2660–2669. Also, attendance at board meetings 1958–1960, p. 2569.

PAGE 206. **$45,000 a year:** Ibid., pp. 2574–80.

PAGE 206. **leave of absence:** Mollenhoff, p. 318. Also, McClellan hearings, p. 2641.

PAGE 207. **today's new account list:** Ibid., pp. 319–20.

PAGE 207. **best fighter system:** I. F. Stone, *New York Review of Books,* Jan. 2, 1969.

PAGE 208. **a ballpark figure:** McClellan hearings. Blackburn testimony and memo, pp. 1188–1225.

PAGE 209. **much cheaper aircraft:** There was precedent for Boeing's proposal, since it had built much of the 707 and 727 airframe with tools also used on the military KC-135. McClellan hearings, p. 1199.

PAGE 209. **When the announcement was made:** Blackburn, McClellan hearings, p. 1206. For General Dynamics' rebuttal of criticisms, see *Aviation Week,* May 20, 1963.

PAGE 210. **subdued landing:** Gunston, pp. 23–26.

PAGE 211. **stable course:** Ibid., p. 24.

PAGE 211. **not one aircraft but many:** Ibid., p. 53. Also *Aviation Week,* Dec. 18, 1972.

PAGE 211. **If foreigners were prepared:** At various times McNamara's spokesmen predicted sales to Britain, Germany, New Zealand and Canada. See Gunston, pp. 82–87.

PAGE 212. **keep British workmen:** Anthony Sampson, *Arms Bazaar* (London: Hodder and Stoughton, 1977), pp. 104–5, 155–56. Also, Mary Kaldor, *Disintegrating West* (New York: Hill and Wang, 1978), pp. 134–39, for a general discussion of trans-Atlantic arms sales agreements.

PAGE 212. **more wonderful airplane:** This was the MRCA Tornado, a pan-European effort to build something similar to the F-111 in "multi-mission capability." It has been called "the milk giving, wool growing, egg-laying sow"—Kaldor, *Baroque Arsenal* (New York: Hill and Wang, 1981), p. 189.

PAGE 212. **Prime Minister Menzies:** McClellan hearings, p. 2529.

PAGE 213. **government's willingness:** For further reading on the 1963 election, Indonesia, and Australia's growing dependence on the U.S., see Russel Ward, *The History of Australia: The Twentieth Century* (Harper & Row, 1977), pp. 340–51.

PAGE 214. **Flying Opera House:** *Melbourne Age,* Oct. 13, 1967.

PAGE 214. **reducing commonality:** Gunston, "Exit the Navy" chapter. Table of the F-111B's deficiencies on p. 31. Also, "McNamara's 10 Billion Dollar Gamble," *U.S. News and World Report,* Aug. 15, 1966.

PAGE 215. **When all of the elements:** Speech by Robert A. Frosch to Electronics and Aerospace Systems Convention. Also, "Why the Flak Around the F-111," *New York Times* Sunday Magazine, Apr. 2, 1967.

PAGE 216. **the type was grounded:** *Jane's All the World's Aircraft, 1968–69* (New York: McGraw-Hill).

PAGE 216. **disappeared without trace:** *Time,* Dec. 4, 1972, p. 16–17.

PAGE 216. **First, there were all:** Interview with former TAC officer.

PAGE 217. **running out of fuel:** Gunston, p. 42.

PAGE 217. **Would it be a very serious matter:** Senate Appropriations Hearings on 1969 Department of Defense budget, p. 103.

PAGE 217. **the Russians unveiled:** Andrew Cockburn, *The Threat* (New York: Random House, 1983), p. 148.

10. The Combative Colonel

PAGE 221. **William Proxmire:** "The F-111 Mystery," *Time,* Dec. 4, 1972.

PAGE 221. **The only thing worse:** "Who Wants General Dynamics? Henry Crown, That's Who," *Fortune,* June 1970.

PAGE 222. **a full complement:** Gunston, p. 77

PAGE 222. **the high priest:** Barbara Tuchman's observation, quoted in Ronald Brownstein and Nina Easton, *Reagan's Ruling Class* (Presidential Accountability Group, 1982); Nitze portrait, pp. 518–24.

PAGE 222. **one million dollars:** Tom Gervasi, *Arsenal of Democracy,* 2d ed. (New York: Grove, 1981), p. 91.

PAGE 223. **structural flaws:** The Selb trial was covered by reporter Derryn Hinch for the Sydney Morning Herald group. His notes and unedited stories are the source for much of the information on the Selb trial.

PAGE 223. **F-111 in a Nixon administration:** Quoted in I. F. Stone, "The War Machine Under Nixon," *New York Review of Books,* June 5, 1969.

PAGE 223. **Against military advice:** Quoted in I. F. Stone, "Nixon and the Arms Race," *New York Review of Books,* Jan. 2, 1969.

PAGE 224. **cut from 250:** See Ernest A. Fitzgerald, *High Priests of Waste* for a participant's view of the C-5 scandal. For a nutshell account of the affair, see Gervasi, *Arsenal of Democracy,* p. 162.

PAGE 225. **$240 million:** *Fortune,* June 1970.

PAGE 225. **repurchased the division:** Details of the sale in Canadair Annual Report, 1976. "Note to Consolidated Financial Statements for the years ended December 31, 1976 and 1975." For Canadair's subsequent adventures, see "Canadair's Rags to Riches Story," *Financial Post,* June 31, 1980.

PAGE 225. **all-time low:** Earnings per share were $3.68 in 1968; $5.46 in 1967, and $5.59 in 1966.

PAGE 225. **cheap alternative:** McNamara launched the bomber version of the F-111 in December 1965, by announcing the purchase of a further 263 planes. Three years later, after the predictable cost overruns and development problems, Nixon's Defense Secretary Melvin Laird cut the number to just 76. See Gunston, p. 88.

PAGE 225. **to cover excess costs:** *Aviation Week,* May 5, 1969.

PAGE 226. **lump-sum settlement:** *New York Review of Books,* June 5, 1969.

PAGE 226. **the highway engineers:** Interview, March 1985. Also, *Chicago Tribune,* Oct. 23 and Dec. 5, 1974.

PAGE 226. **two-bit cigar chompers:** Mike Royko, "Pentagon Hustle Smells of Chicago," *Chicago Tribune,* Mar. 6, 1985.

PAGE 227. **above normal market levels:** Demaris, p. 229.

PAGE 228. **Henry Crown was named:** Ibid., p. 230.

PAGE 228. **he accepted defeat:** *Fortune,* June 1970.

PAGE 229. **same merger agreement:** The details of the merger were covered by the McClellan hearings, Gilpatric testimony, pp. 2544–55. Also *Fortune,* June 1970.

PAGE 229. **The chief motive:** *Fortune,* June 1970.

PAGE 231. **Tell Roger:** *Fortune,* June 1970.

PAGE 232. **526,400 shares:** "Out Again, In Again," *Forbes,* Mar. 15, 1970.

PAGE 233. **He should redirect:** *Business Week,* July 11, 1970.

PAGE 233. **Frequent major technological changes:** General Dynamics *Annual Report,* 1982.

PAGE 233. **David S. Lewis:** "General Dynamics Finds a New Boss," *Business Week,* Oct. 24, 1970. Also, *Time* and *Newsweek* issues dated Nov. 2, 1970.

PAGE 233. **white-haired:** For further Lewis background, see "David S. Lewis," *Sky* magazine, April 1981.

11. The F-16: Too Good to Survive

PAGE 234. **It was the most awful:** Interview, confidential source.

PAGE 234. **We had layers:** Ibid.

PAGE 235. **overblown managerial system:** The F-111 and the multilayered bureaucracy that it produced figured prominently in an enlightening study. According to the Commission on the Organization of the Government for the Conduct of Foreign Policy (Washington, 1975), vol. 4, App. K, "Even relatively minor reformulations would require the concurrence of layer upon layer of contractor and government authorities." See also Robert Coulam, *Illusion of Choice* (Princeton University Press, 1977).

PAGE 235. **eliminate the FB-111:** Model numbers and designations are confusing. The FB-111 was the long-range bomber version; the F-111 B was the aborted Navy model. For detailed descriptions of all models, see Gunston, pp. 110–12.

PAGE 236. **Now we are finally:** *Business Week,* Oct. 24, 1970.

PAGE 236. **inability to pay:** *Aviation Week* and *Aerospace Technology,* Oct. 26, 1970.

PAGE 236. **We have chosen:** *New York Times,* Feb. 7, 1970.

PAGE 238. **The number of weapons systems:** Quoted in James Fallows, *National Defense* (New York: Random House), p. 43.

PAGE 239. **"vectored" those:** Pierre Sprey's "Impact of Avionics on Tactical Effectiveness" study prepared for Assistant Secretary of Defense, 1968. Sprey reports that U.S. pilots equipped with beyond-visual-range missiles often fired on fellow U.S.A.F. planes, and once on an Australian destroyer that was mistaken for a helicopter.

PAGE 239. **It is curious:** "The Nature of a Fighter Aircraft," *Interavia* magazine, February 1973.

PAGE 239. **one of the western world's:** *New York Times,* Oct. 14, 1984.

PAGE 240. **twenty-eight minor computers:** Tom Gervasi, *Arsenal of Democracy,* 2d ed. (New York: Grove Press, 1981), pp. 116–19.

PAGE 240. **The lightest, cheapest:** *Interavia,* February 1973.

PAGE 241. **It is a curious result:** Ibid.

PAGE 241. **formal tactical doctrine:** This is the "Double Attack Doctrine." Simple in theory yet immensely complicated for a layman to comprehend, Riccioni's theory aims to get greater offensive utility from an attacking pilot's wingman. The Navy has a somewhat similar combat doctrine known as the Loose Deuce. Interviews, U.S.A.F. pilots

PAGE 242. **A/ Outgamed:** Fallows, *National Defense,* p. 102.

PAGE 242. **This is all right:** Sprey interview.

PAGE 243. **It wasn't as if:** Interview, confidential source.

PAGE 243. **even sanctified:** Arnold Kanter, *Defense Politics* (University of Chicago Press, 1979), p. 4.

PAGE 243. **We were just lucky:** Sprey interview.

PAGE 245. **Skunk Works:** For a pithy description of the advantages to a closed-shop development team, see Edward Luttwak, *The Pentagon and the Art of War* (New York: Simon and Schuster, 1984), p. 180.

PAGE 245. **It was easier:** Interview, confidential source.

PAGE 245. **As Oestricher explained:** Oestricher interview, *Sale of the Century,* documentary by Andrew Cockburn for WGBH, Boston.

PAGE 247. **Northrop believed:** See Anthony Sampson, *Arms Bazaar,* chap. 15

PAGE 247. **parceled out many millions:** Ibid., chap. 16.

PAGE 248. **It was unquestionably:** Sprey interview.

PAGE 248. **against Dassault's Mirage:** *Sale of the Century* transcript.

PAGE 248. **Lower-cost alternatives:** Fred Kaplan, "The Little Plane That Could Fly if the Air Force Would Let It," *Boston Globe,* Mar. 14, 1982.

PAGE 249. **Air Force would agree:** Ibid.

PAGE 250. **F-16 could drop:** Ibid. Also, "U.S. Will Give F-16 Nuclear Capacity in Move to Reassure Allies," *New York Times,* Nov. 1, 1976.

PAGE 250. **bat turns:** "Return of the Gunfighter," *Air Combat,* September 1980.

PAGE 252. **We basically used:** "Superfighter," *High Technology,* April 1984.

PAGE 252. *too* **maneuverable:** "Jets' Acceleration Reported to Cause Pilots to Black Out," *New York Times,* Mar. 25, 1985. Also, "G-Force Gremlins," *Business Week,* p. 109.

PAGE 254. **The cakes:** *Sale of the Century* transcript.

PAGE 254. **F-16 is virtually limited:** Ibid.

PAGE 255. **offered a large bribe:** Sampson, chap. 15.

PAGE 256. **$6.09 million:** The actual sum was $6,091,000 million. Also, Russell Warren Howe, *Weapons* (New York: Doubleday, 1980).

PAGE 256. **establish themselves:** "F-16's into Service," *Flight International,* Jan. 5, 1980.

PAGE 257. **Could you imagine:** Sampson, chap. 16.

PAGE 257. **Early in June:** Cockburn, *Sale of the Century.*

PAGE 258. **nonstandard sales techniques:** Ibid.

PAGE 258. **Three months before:** Ibid.

PAGE 259. **stock price began:** Early in 1985, General Dynamics shares had risen to $84. Scandal and the subsequent purchase of Cessna lopped $20 off the figure in the course of the next five months.

PAGE 259. **almost four thousand:** "Major U.S. and European Suppliers for the F-16 Program," General Dynamics press release, June 1980.

PAGE 259. **Son, we have:** McClellan hearings, p. 2659.

PAGE 260. **cassette player installed:** Cockburn, *Sale of the Century.*

PAGE 260. **multirole warplane:** GAO report: "The Multi-National F-16 Aircraft Program: Its Progress and Concerns," June 1979. Also, "Co-production Programs and Licensing Arrangements in Foreign Countries," December 1975.

PAGE 260. **Our NATO allies:** Department of Defense report, quoted in Mary Kaldor, *Disintegrating West* (New York: Hill and Wang, 1978), p. 135.

PAGE 260. **We in the United Kingdom:** Quoted in Howe, p. 585.

12. Two Weapons

PAGE 262. **$336.1 million:** "Striking It Rich on Defense," *Business Week* cover story, May 3, 1982.

PAGE 263. **the twin plants:** These are the two main plants. There is also a third facility on the outskirts of Cleveland and a fourth in Pennsylvania, an example of the way in which a single defense operation can cover many congressional districts.

PAGE 263. **only 230 tanks:** Hearings of the Subcommittee on Tactical Warfare of the Committee on the Armed Services, Mar. 5, 1982.

PAGE 263. **target of opportunity:** *Business Week,* May 3, 1982.

PAGE 264. **go the Russians:** Andrew Cockburn, *The Threat* (New York: Simon and Schuster, 1981), pp. 125–29.

PAGE 264. **over my dead body:** Quoted in Tom Gervasi, *Arsenal of Democracy,* 2d ed. (New York: Grove Press, 1981), p. 53.

PAGE 265. **The real reason:** Defense Secretary Donald Rumsfeld, interviewed on ABC's *20/20,* June 20, 1980.

PAGE 265. **hailed the acceleration:** Testimony of Army and Department of Defense witnesses, Tac-War Subcommittee hearings, Mar. 5, 1982.

PAGE 266. **high-speed dashes:** Tank engagements were analyzed by Pierre Sprey. His criticisms and observations were answered by the Army in a resulting set of questions submitted to the Tactical War Subcommittee by the Military Reform Caucus. See Mar. 5, 1982, transcripts, pp. 2391–2433.

PAGE 266. **Testing has not been done:** Tac-War Subcommittee hearings, Mar. 5, p. 2408. While the Army has never tested the tank's dash ability, other writers maintain that it is satisfactory. For a more cheerful view of the M-1, see Jeff Cooper, "Dragon of Liberty," *Armed Conflict,* vol. 1, no. 1, 1985. Even Cooper, an avowed enthusiast for the M-1, concedes that fuel economy is a problem. For further discussion of the M-1's established deficiencies and attributes, see Edward Luttwak, *The Pentagon and the Art of War* (New York: Simon and Schuster, 1984), pp. 223–24.

PAGE 266. **vulnerable pillboxes:** Tac-War Subcommittee hearings, p. 2388.

PAGE 267. **supplying frontline forces:** Army officials insist two M-1s will fit in a C-5. What they fail to mention is that support equipment and other gear normally consigned with the M-60 will not. The Army's solution to the difficulties involved in moving the M-1 by rail is to build a fleet of new king-size flatcars. See Sheila Tobias, Peter Goudinoff, Stefan Leader, and Shela Leader, *What Kind of Guns Are They Buying for Your Butter?* (New York: 1984).

PAGE 267. **super-dozer:** "New Tank Needs an ACE to Dig a Hole," *Washington Post,* Feb. 9, 1982.

PAGE 267. **Washington agreed:** Mary Kaldor, *Baroque Arsenal* (New York: Hill and Wang, 1982), p. 199. Also, Gervasi, *Arsenal of Democracy,* p. 230. Also, Tac-War hearings, p. 2378.

PAGE 268. **one of the most powerful:** See any copy of *Soviet Military Power,* a Pentagon publication that appears yearly to heap praise on the Russian armed forces. It is inevitably released prior to annual defense authorization bills.

PAGE 268. **it breaks down regularly:** Cockburn, *The Threat,* pp. 122–27.

PAGE 268. **the T-72s, which burned:** Ibid., pp. 273–74.

PAGE 268. **pilloried in the press:** Opening remarks, Tac-War hearings.

PAGE 268. **dirigible engine:** Cockburn, *The Threat,* p. 126.

PAGE 269. **conditioning further appropriations:** GAO report released in December 1981 and reported in most U.S. papers the next day. See "Another GAO Warning," *St. Louis Post-Dispatch,* Dec. 31, 1981.

PAGE 269. **power train components:** Tac-War hearings, pp. 2329–31.

PAGE 269. **Fulda Gap:** Cooper, op. cit.

PAGE 270. **between two and three times:** Jacques Gansler, *The Defense Industry* (Cambridge, Mass.: MIT Press, 1980), p. 15. A former Assistant Secretary of Defense, Gansler maintains the price rise is justified since the M-1 "performs much better than the M-60." The Army disputes the threefold increase, insisting that the figure is inaccurate since it includes the cost of tooling, R&D, new manufacturing facilities, etc. Since the M-60 has long since amortized these expenses, its production overhead should be much lower. The Army, however, treats the M-60 as if it too was new and thus loads the figure with startup costs that are impossible to justify. Even on this basis, $1 billion still buys 769 M-60s but only 555 M-1s.

PAGE 270. **Quincy maintained:** General Dynamics *Annual Report,* 1982, p. 3.

PAGE 271. **shuttle supplies:** Ibid., p. 17.

PAGE 271. **waiting for trouble:** For the importance of delivering troops to a beachhead and keeping them supplied in the field, see Luttwak, pp. 240–44.

PAGE 271. **bureaucratic mind:** For a general introduction to the TAKX decision, see *Forbes,* June 20, 1983.

PAGE 271. **This is an example:** Candidate position paper, Sept. 22, 1980. "A Program for the Development of Effective Maritime Strategy," available through the Republican National Committee.

PAGE 272. **Twelve years later:** Figures and statistics pertaining to Sealift Command are drawn from a complaint for injunctory relief filed with the U.S. District Court for the District of Columbia by the Joint Maritime Council (83-2003). I am indebted to attorney Joe Wager, who successfully presented MSC's case to have the *Nordaway* contracts overturned, for his research. His papers, court filings, and correspondence with the Justice Department on other matters pertaining to Quincy were immensely valuable.

PAGE 272. **competitive procurement:** Ibid.

PAGE 272. **at high standards:** Request for Proposal, May 28, 1982, no. N0003382R3004.

PAGE 273. **President's preference:** "Report on the Evaluation of Military Sealift Command Bidding Procedures for Manning Ships," prepared by Logistics Support Programs Directorate under the signature of Joseph H. Sherrick, Inspector General, Oct. 20, 1983. Sherrick suspended the award, noting charitably that while there was no direct evidence of a Navy conspiracy to repel bidders, "information that was important to offerers for preparation of their proposals and that was readily available within MSC was not provided on an equal basis to all offerers."

PAGE 273. **beyond the cost:** Quoted in a Wager reply to a request by Senator Exon for further information about Sealift Command's plans and procedures. Obtained by the author through Hill contacts. The Joint Taxation Committee heard GAO evidence that estimated a 12 percent increase in costs. *Forbes,* June 20, 1983.

PAGE 274. **Wall Streeters:** Ibid.

PAGE 274. **devoted to capital improvements:** Ibid.

PAGE 275. **recompense General Dynamics:** Decision of the Comptroller General, Jan. 28, 1983. "Navy Industrial Fund: Obligations in Connection with Long-Term Vessel Charters." This report gave the Navy the right to go ahead with the scheme. However, the report also pointed out: "We do not address the more fundamental question of whether the Navy Industrial Fund is a proper source for funding such long-term lease agreements. . . . Similarly, the wisdom of long-term leasing as opposed to purchase of TAKX vessels."

13. One Last Scandal

PAGE 277. **worth $13 million:** The history of the 688 claims dispute was laid out in splendid detail by Richard F. Kaufman, researcher for Senator Proxmire's Joint Economic Committee. See "Navy Shipbuilding at General Dynamics: The SSn 688 Class Submarine Program, Flights I and II," a briefing paper supplied to the Subcommittee on International Trade, Finance and Security Economics, Apr. 2, 1985.

PAGE 277. **pure theater:** This was a favorite Rickover expression, also used in reference to Newport News. See Rickover testimony before Joint Economic Committee, June 3, 1976. Veliotis' performance with the steel plates took place before the Sea-Power Subcommittee of the House Armed Services Committee. In this case he was responding to Rickover's general criticisms and more specific charges leveled before the same committee by Admiral Earl Fowler.

PAGE 278. **approved Navy shipbuilders:** Lehman quoted in "Beware of Greeks Bearing Gifts," *Time,* Feb. 20, 1984.

PAGE 278. **a valued man:** Lewis quoted in "Defense Company Giant Now a Man on the Run," *Philadelphia Inquirer,* June 25, 1984.

PAGE 278. **his first success:** "Striking It Rich on Defense," *Business Week,* May 3, 1982.

PAGE 279. **it was ridiculous:** *Time,* Feb. 20, 1984; *Business Week,* June 25, 1984.

PAGE 279. **only five days:** *Business Week,* June 25, 1984.

PAGE 280. **We built every ship:** "The One That Got Away," *Forbes,* Jan. 16, 1984.

PAGE 280. **Overheads soared:** "Who Wants General Dynamics? Henry Crown, That's Who," *Fortune,* June 1970.

PAGE 281. **General Dynamics' involvement:** Don Mickak, "Elliot Richardson and the Shipyard Fraud Case," *Hartford Advocate,* Feb. 29, 1984. Also, *Fortune,* March 1977.

PAGE 282. **to be cleaned up:** *Wall Street Journal,* Oct. 21 and Dec. 10, 1976.
PAGE 282. **Burmah's other ships:** *Washington Post,* June 17, 1977.
PAGE 283. **not the product:** *Financial Times,* Apr. 12, 1976.
PAGE 283. **his last day:** *Hartford Advocate,* Feb. 29, 1984.
PAGE 284. **the SEC declined:** *Business Week,* Sept. 6, 1976.
PAGE 284. **he was rewarded:** Veliotis was hired to run Quincy in January 1973; he moved to Groton in October 1977; he was named vice-president in charge of General Dynamics' Marine Division in May 1980. In November 1981, shortly before his departure for Greece, he was placed in charge of both Marine and International Operations.
PAGE 284. **tankers' annual maintenance costs:** Comments by Burmah spokesman Peter Granger. *Hartford Advocate,* Feb. 29, 1984.
PAGE 284. **ten times Ogden's worth:** *Forbes,* Jan. 16, 1984.
PAGE 284. **$85 million a year:** "Sherlock Bernstein," *Forbes,* Jan. 30, 1984.
PAGE 285. **dog and pony show:** Testimony of G. G. Davis and others, U.S. Federal Court (Manhattan), *U.S.* v. *G. G. Davis* (henceforth referred to as Frigitemp trial).
PAGE 286. **decided not to pursue:** Ibid.
PAGE 286. **We looked askance:** Ibid., testimony of Blackwell and McGowan.
PAGE 286. **finder's fee:** Ibid., Davis and Bose testimony.
PAGE 286. **Frigitemp conduit:** Ibid., Bose testimony.
PAGE 287. **paid for renovations:** *Philadelphia Inquirer,* June 25, 1984.
PAGE 287. **They moved from:** Blackburn interview, *NBC Nightly News,* June 26, 1984.
PAGE 288. **the trees were still growing:** *NBC Nightly News,* June 27, 1984.
PAGE 288. **Pascagoula executive:** A bagman who distributed cash to cooperative shipyard executives later testified in bankruptcy court that he had delivered money to Edward F. Loganero, later vice-president of Litton's Ingalls Division. See *Wall Street Journal,* Oct. 12, 1984. Also, *NBC Nightly News,* June 27, 1984.
PAGE 288. **Lee and Silver:** "Defunct Firm's Chief Tells of Murder Plot," *Wall Street Journal,* May 7, 1984.
PAGE 289. **arrange a murder:** Ibid. Also, Silver testimony at Davis trial.
PAGE 289. **O'Sullivan might disappear:** Ibid.
PAGE 289. **reassigning the contracts:** The preliminaries of the contract transfer were laid out in a civil action brought against General Dynamics, Veliotis, Davis, their wives, and other (unnamed) individuals by Lawson Bernstein, Frigitemp's trustee in bankruptcy. Civil action 83-294 before the U.S. District Court in Delaware (1983). Also, *Philadelphia Inquirer,* June 25, 1984.
PAGE 290. **a giant jigsaw:** Frigitemp trial. Testimony of former Frigitemp and IDT draftsman, D. R. Dammiano.
PAGE 290. **It wouldn't look good:** Frigitemp trial, Davis testimony.
PAGE 291. **Veliotis mounted:** *Business Week,* June 25, 1984.

14. Messing Around with Boats

PAGE 294. **None was found:** Ashton testimony before Representative John Dingell's Subcommittee on Oversight and Investigation of the Committee on Energy and Commerce, March 25, 1985.
PAGE 295. **The tradesmen:** Norman Polmar and Thomas Allen, *Rickover* (New York: Simon & Schuster, 1982), p. 373.
PAGE 295. **a new generation:** Ibid., pp. 405–18. Rickover was typically colorful in his denunciation of the surface fleet. "The Navy's idea [of a balanced fleet] is like a chicken-horseburger sandwich: One horse, one chicken. The Navy has many surface ship officers who see promotion and command opportunities denied them if we get more nuclear submarines."

PAGE 295. **Rickover so often presented:** Ibid. See chapter titled "A Fascinating Experience" for Zumwalt's recollections and those of many other veterans of the Rickover interview technique.

PAGE 298. **Few additional pennies:** Ibid., p. 406.

PAGE 298. **$69.1 million:** "Report on US Nuclear-Powered Attack Submarine Program," House Armed Services Committee, 1979.

PAGE 298. **Electric Boat Division has not developed:** Haines memo, Oct. 6, 1977. Obtained by Proxmire investigators and released at hearing before the Subcommittee on International Trade and Security Economics of the Joint Economic Committee, Apr. 2, 1985.

PAGE 299. **an aggressive effort:** Letter from Navy Supervisor of Shipbuilding to Electric Boat General Manager J. D. Pierce, summarized in briefing paper supplied to Proxmire subcommittee by Proxmire's chief researcher, Richard Kaufman (Kaufman report), p. 33–37.

PAGE 300. **for every two weeks:** Ibid.

PAGE 300. **had to be farmed off:** Report to Electric Boat executive N. D. Victor, Aug. 15, 1973. Also Ashton testimony before Dingell.

PAGE 300. **We are still better:** Report to N. D. Victor, Aug. 15, 1973.

PAGE 300. **little less than six:** "Schedule Chronology," Kaufman, pp. 53–56, 61.

PAGE 301. **$908 million: Trident background:** Morton Mintz, "Depth Charge: Cost Overruns on New Trident Sub Leave a Muddied Wake," *Washington Post,* Oct. 4, 1981.

PAGE 301. **39.8 million man-hours:** Kaufman, pp. 53–56.

PAGE 301. **about our ability:** Report to N. D. Victor.

PAGE 302. **construction experience:** Summarized, Kaufman report, p. 36.

PAGE 302. **David Lewis himself:** Kaufman, p. 20.

PAGE 302. **We'll never be able:** Summarized in Kaufman, p. 36.

PAGE 302. **We must recognize:** Memo from Hyman to Pierce and Curtis, Jan. 14, 1974.

PAGE 304. **revising the figures:** Mintz, "Depth Charge." Also, Polmar and Allen, *Rickover,* pp. 564–581.

PAGE 304. **internal peace treaties:** Polmar and Allen, *Rickover,* pp. 565–69.

PAGE 305. **more than eighteen months:** Kaufman, pp. 53–56.

PAGE 305. **forcing a number:** Arthur Andersen memo to head office from Electric Boat representative, July 1, 1976.

PAGE 306. **clean up the mess:** Ibid.

PAGE 306. **It is almost laughable:** Lewis memo to MacDonald, Jan. 27, 1977.

PAGE 307. **process accelerated:** Lewis interview, *Philadelphia Inquirer,* June 25, 1984.

PAGE 308. **seldom the case:** Kaufman analysis of settlement, pp. 12–19.

PAGE 308. **Catch us:** Quote of Representative Gerry Sikorski during hearings before Subcommittee on Oversight and Investigation of the Committee on Energy and Commerce, Feb. 28, 1985 (Dingell hearings).

15. "Catch Us if You Can"

PAGE 309. **troglodyte Republicans:** "The $3 Billion Cap," *New Republic,* Apr. 22, 1985, p. 4.

PAGE 309. **new generation of welfare queens:** Grassley speech to Conservative Union Conference, Washington, March 2, 1985.

PAGE 310. **necessary to stop work:** Minutes of General Dynamics board meeting, Aug. 4, 1977, and Executive Committee meeting, Aug. 31, 1977.

PAGE 310. **ad hoc agreement:** Newport News actually stopped work on a cruiser in August 1975 but was soon back at work when the Navy obtained a court order. See Norman Polmar and Thomas Allen, *Rickover* (New York: Simon and Schuster, 1982), p. 505.

PAGE 310. **achieve organizational objectives:** Quoted by Kaufman in "Summary of Documents Relating to Navy Shipbuilding at the Electric Boat Division of G.D." Proxmire-Grassley joint committee hearing, Apr. 2, 1985.

PAGE 311. **You know who:** MacDonald memo, quoted ibid., p. 22.

PAGE 312. **1971 and 1973 contracts:** Hidalgo testimony before Proxmire-Grassley subcommittee, Apr. 2, 1985 (henceforth referred to as Proxmire hearing).

PAGE 312. **a further two months:** Minutes of General Dynamics board of directors, Apr. 4, 1978.

PAGE 313. **any assistance:** Memo to David Lewis from M. Golden and G. E. MacDonald.

PAGE 313. **for factual accuracy:** Kaufman report, p. 25.

PAGE 314. **Lewis was still complaining:** Proxmire hearing, Apr. 2, 1985.

PAGE 315. **Future Cost Growth:** Kaufman report, witnesses' testimony, Apr. 2, 1985.

PAGE 315. **Electric Boat had earned:** Hidalgo testimony, ibid.

PAGE 315. **estimated the savings:** Kaufman report.

PAGE 316. **enough to dissuade:** Hidalgo testimony, Apr. 2, 1985.

PAGE 316. **bothered him not:** Ibid.

PAGE 317. **four weeks' work:** Ibid.

PAGE 317. **George A. Sawyer:** "Dynamics Hiring Questioned," *New York Times,* Mar. 29, 1985. Also "General Dynamics Official Assailed Over Job Talks with Navy," *Washington Post,* Mar. 29, 1985. Also, testimony before Representative John Dingell's Subcommittee on Oversight and Investigations of the Committee on Energy and Commerce, Mar. 25, 1985.

PAGE 318. **company has been plagued:** Lewis was referring to antiwar and antinuclear protesters who gather at Groton's front gate every time a submarine is launched. See "Submarine Building City Protesting the Protestors," *New York Times,* May 21, 1984. He may also have been thinking of the attempt to kidnap Henry Crown by a group of Puerto Rican separatists in April 1980. The plot was foiled when Crown's neighbors became suspicious of the six track-suited Hispanics loitering in the neighborhood and called the police. See *Chicago Sun-Times,* Dec. 7, 8, and 14, 1980.

PAGE 319. **only so faithless:** *New York Times,* Mar. 31, 1985.

PAGE 321. **steps to subpoena:** "Attorney General Faces Contempt of Congress Motion," *Chicago Tribune,* Nov. 1, 1984.

PAGE 321. **favorite weaponsmith:** For a full list of all the investigations under way, see *Business Week,* Mar. 25, 1985.

PAGE 321. **kind the admiral favored:** MacDonald testimony before Dingell panel.

PAGE 322. **not an oversight:** Testimony before Dingell panel, Feb. 28, 1985. Also, *Business Week,* Mar. 25, 1985.

PAGE 322. **They looked at everything:** *St. Louis Post-Dispatch,* Sept. 23, 1984.

PAGE 323. **It was a big investigation:** Ibid.

PAGE 323. **had dug through:** *Business Week,* June 25, 1984.

PAGE 323. **militated against:** Kaufman briefing, pp. 72–76. Also, Federal Contracts Report 7-30-84, "Proxmire Raps Justice Department's Investigation of General Dynamics Claims."

PAGE 324. **Had the precedents:** Kaufman briefing.

PAGE 324. **If this silly bastard:** One of several Veliotis tapes played for Dingell hearing, Mar. 25, 1985.

PAGE 325. **failed to replace:** Ashton testimony, Dingell hearings, Mar. 25, 1985.

PAGE 325. **progress in 1981:** General Dynamics *Annual Report,* 1980.

PAGE 326. **most pugnacious display:** Lehman speech to National Press Club, Aug. 19, 1981. Reported in *Washington Post,* Aug. 20, 1981.

PAGE 326. **$444 million:** Dingell hearings, both Feb. 28 and Mar. 25, 1985.

PAGE 327. **Even the Internal Revenue Service:** "Corporate Taxpayers and Corporate Freeloaders," report compiled and released by Citizens for Tax Justice Group, Washington, 1985.

PAGE 327. **Lewis and LeFevre:** Frequent references to the Lewis-Meese meeting were made throughout the Dingell hearings. White House replies to written questions submitted by Senator Proxmire and other congressmen prior to Meese's confirmation as attorney general were the chief sources for this section. Also, Meese confirmation hearings and statements.

PAGE 329. **Lehman got mad:** Veliotis tapes, Dingell hearings, Mar. 25, 1985.

PAGE 329. **What small contractor:** Sikorski statement, Dingell hearings, Mar. 25, 1985.

PAGE 330. **You've got to trust me:** Veliotis tapes, taped Aug. 25, 1981. Dingell hearings, Mar. 25, 1985.

PAGE 330. **round-trip airline ticket:** Lewis testimony, Dingell hearings, Mar. 25, 1985.

PAGE 332. **one last meeting:** Ibid.

PAGE 332. **employment negotiations:** Dingell hearings, Mar. 25, 1985. Also, "GD Chairman Assailed Over Job Talks with Navy Official," *Washington Post,* Mar. 26, 1985.

PAGE 333. **override Rickover:** Polmar and Allen, *Rickover,* p. 654.

PAGE 334. **The *Journal-Bulletin* asked:** The *Journal-Bulletin* story was picked up across the country. See *New York Times,* Oct. 27, 1984.

PAGE 335. **FBI had made:** "Chronology of Electric Boat Investigation," a private briefing paper prepared for members of Grassley's subcommittee on Judicial Oversight and obtained by the author.

PAGE 335. **impossible to establish:** Ibid.

PAGE 336. **I have a job:** Author interview with Grassley, April 1985.

PAGE 337. **haphazard attitude:** Briefing papers compiled by Grassley's staff.

PAGE 338. **that "scintilla":** Ibid. Also, Proxmire hearing, Apr. 4, 1985.

Index